NEUROMETHODS

Series Editor
Wolfgang Walz
University of Saskatchewan
Saskatoon, SK, Canada

For further volumes:
http://www.springer.com/series/7657

Dopamine Receptor Technologies

Edited by

Mario Tiberi

Ottawa Hospital Research Institute and University of Ottawa, Ottawa, ON, Canada

Humana Press

Editor
Mario Tiberi
Ottawa Hospital Research Institute
and University of Ottawa
Ottawa, ON, Canada

ISSN 0893-2336 ISSN 1940-6045 (electronic)
ISBN 978-1-4939-2195-9 ISBN 978-1-4939-2196-6 (eBook)
DOI 10.1007/978-1-4939-2196-6
Springer New York Heidelberg Dordrecht London

Library of Congress Control Number: 2014957165

Cover illustration: The image depicts the high resolution of the structure of the dopamine D3 receptor (Protein Data Bank # 3PBL) with its helices and loops (purple) and dopamine (shown as cyan, blue and red spheres) on top of the receptor (Covert Art: Boyang Zhang).

Printed on acid-free paper

Humana Press is a brand of Springer
Springer is part of Springer Science+Business Media (www.springer.com)

Series Preface

Experimental life sciences have two basic foundations: concepts and tools. The *Neuromethods* series focuses on the tools and techniques unique to the investigation of the nervous system and excitable cells. It will not, however, shortchange the concept side of things as care has been taken to integrate these tools within the context of the concepts and questions under investigation. In this way, the series is unique in that it not only collects protocols but also includes theoretical background information and critiques which led to the methods and their development. Thus it gives the reader a better understanding of the origin of the techniques and their potential future development. The *Neuromethods* publishing program strikes a balance between recent and exciting developments like those concerning new animal models of disease, imaging, in vivo methods, and more established techniques, including, for example, immunocytochemistry and electrophysiological technologies. New trainees in neurosciences still need a sound footing in these older methods in order to apply a critical approach to their results.

Under the guidance of its founders, Alan Boulton and Glen Baker, the *Neuromethods* series has been a success since its first volume published through Humana Press in 1985. The series continues to flourish through many changes over the years. It is now published under the umbrella of Springer Protocols. While methods involving brain research have changed a lot since the series started, the publishing environment and technology have changed even more radically. Neuromethods has the distinct layout and style of the Springer Protocols program, designed specifically for readability and ease of reference in a laboratory setting.

The careful application of methods is potentially the most important step in the process of scientific inquiry. In the past, new methodologies led the way in developing new disciplines in the biological and medical sciences. For example, Physiology emerged out of Anatomy in the nineteenth century by harnessing new methods based on the newly discovered phenomenon of electricity. Nowadays, the relationships between disciplines and methods are more complex. Methods are now widely shared between disciplines and research areas. New developments in electronic publishing make it possible for scientists that encounter new methods to quickly find sources of information electronically. The design of individual volumes and chapters in this series takes this new access technology into account. single protocol makes it possible to download single protocols separately. In addition, Springer makes its print-on-demand technology available globally. A print copy can therefore be acquired quickly and for a competitive price anywhere in the world.

New York, NY, USA *Wolfgang Walz*

Preface

A Historical Perspective of the Study of Dopamine and Its Receptors

Dopamine was first synthesized in 1910 at the Wellcome laboratories in London, England, by two chemists, George Barger and Arthur James Ewins. In the same year, Sir Henry Dale, British physiologist and pharmacologist, reported that dopamine had very weak sympathomimetic activity suggesting at that time an insignificant physiological role for this amine. Up to 1952, dopamine was referred by its original name, 3,4-dihydroxyphenylethylamine. In fact on that year, it was Sir Henry Dale, then a Physiology and Medicine Nobelist, who first coined the word dopamine from the contraction of 3,4-*d*ihydr*o*xy*p*henylethyl*amine*. Pioneering work spanning from the early 1910s to the late 1930s unraveled the discovery of l-DOPA (also known as levodopa), a naturally occurring isomer of the amino acid 3,4-dihydroxyphenylalanine and the enzyme DOPA decarboxylase, which allowed paving the way for the still in force principles of the catecholamine biosynthetic pathway. Up to the late 1950s, dopamine's role was solely confined to that of a precursor generated from l-DOPA decarboxylation to serve as a critical intermediary in the synthesis of the sympathomimetic active catecholamines, norepinephrine and epinephrine. Seminal biochemical, histoimmunochemical, and radioligand studies done notably from the late 1950s to the late 1970s in the laboratories of Arvid Carlsson (2000 Physiology and Medicine Nobelist), Oleh Hornykiewicz, Paul Greengard (2000 Physiology and Medicine Nobelist), Maynard H. Makman, Pier Franco Spano, Leslie Iversen, Solomon Snyder, Philip Seeman, and others gave credence to a role of dopamine beyond that of just being a mere intermediary in the biosynthesis of norepinephrine but also on its own a true biogenic amine neurotransmitter involved in a plethora of physiological effects in brain and peripheral tissues. Importantly, work done in the late 1950s and the early 1960s by Oleh Hornykiewicz and Arvid Carlsson also hinted to an important role of a deregulated dopamine activity in the etiology of Parkinson's disease and schizophrenia. Furthermore, these studies strengthened the view that dopamine played a critical role in the signal transduction via two receptor subtypes. Initially, the view was that one subtype, D_1, was linked to the stimulation of adenylyl cyclase and cAMP production while the other subtype, D2, exhibited high affinity for antipsychotics but was not linked to adenylyl cyclase. Molecular cloning studies in the late 1980s and the early 1990s uncovered the existence of a gene family coding for dopamine receptor proteins larger than anticipated.

Nowadays, it is well established that dopamine actions are chiefly mediated through the binding and activation of six cell surface seven-transmembrane proteins that belong to the large family of G protein-coupled receptors or GPCRs. The dopaminergic GPCRs are catalogued in two major classes: D_1-class (D_1 and D_5) and D_2-class ($D_{2short/long}$, D_3, and D_4) receptors. These receptors can regulate locomotion, cognition, reward, natriuresis, vascular tone, gastrointestinal motility, heart function, and respiratory activity. Besides Parkinson's disease and schizophrenia, impaired activity of dopamine and its receptors is also implicated in the etiology or phenotypic expression of several other hallmark human brain illnesses and conditions such as Huntington's disease, attention deficit and hyperactivity disorder, and substance abuse. Whether the disorders arise from death of neurons in substantia nigra pars compacta (Parkinson's disease) or striatum (Huntington's disease), or impairment in dopamine reuptake (cocaine addiction), a large body of studies suggest that these conditions likely culminate in compromised signal transduction through dopamine receptors.

Consistent with this view, drugs targeting dopamine receptors are currently used in clinical settings to manage and treat symptoms of diseases associated with impaired dopamine activity. However, these drugs are not magic bullets as they also cause undesirable side effects or are unable to alleviate all disease symptoms linked to dopamine dysfunction. To address these issues, important facets of dopamine receptor biology remain to be further addressed at the functional and mechanistic levels: drug selectivity, high-resolution structure predictions, subtype-specific signaling properties, posttranslational modifications, and receptor gene expression. Potentially, a better understanding of dopamine receptor functionality will help in developing new pharmacological tools to improve our knowledge of in vivo roles played by each receptor subtype and synthesis of prospective lead compounds for drug discovery.

The primary objective of this *Neuromethods* book is to lay an interface between updated classical methods and new emerging technologies heretofore not available to new or seasoned researchers, who are keen to further our understanding of dopamine receptor biology. The book is divided into five sections dedicated to experimental approaches investigating different facets of dopaminergic signal transduction: epigenetic and posttranscriptional analysis, computational and biochemical techniques, visualization and imaging methods, molecular and cell biological tools, and behavioral assessment. The book provides insightful step-by-step protocols and methodological reviews that readers will find useful. Furthermore, this book will be a complement to existing literature experimental protocols to study dopamine function.

In closing, I would like to express my sincere gratitude to all scientists who took time to contribute a chapter in spite of their busy schedule. I want also to thank Boyang Zhang, Michael Beazely, Jean-Claude Béique, Diane Lagace, and Kursad Turksen for their help and valuable advice during the editing of this book. Finally, I want to thank Wolfgang Walz, the series editor, and Patrick Marton from Springer New York for their encouragements and support during the making of this book.

Ottawa, ON, Canada ***Mario Tiberi***

Contents

Contributors

PAUL R. ALBERT • *Ottawa Hospital Research Institute (Neuroscience Program), University of Ottawa, Ottawa, ON, Canada*

PER E. ANDREN • *Biomolecular Imaging and Proteomics, Department of Pharmaceutical Biosciences, National Center for Mass Spectrometry Imaging, Uppsala University, Uppsala, Sweden*

SRDJAN D. ANTIC • *Department of Neuroscience Stem Cell Institute, Institute for Systems Genomics, UConn Health, Farmington, CT, USA*

JA-HYUN BAIK • *Molecular Neurobiology Laboratory, Department of Life Sciences, Korea University, Seoul, South Korea*

DIPANNITA BASU • *Department of Psychiatry and Behavioural Neurosciences, McMaster University, Hamilton, ON, Canada*

JEROME BAUFRETON • *CNRS, UMR 5293, Institut des Maladies Neurodégénératives, Université de Bordeaux, Bordeaux, France*

MICHAEL A. BEAZELY • *Faculty of Science, School of Pharmacy, University of Waterloo, Kitchener, ON, Canada*

GLENN S. BELINSKY • *Department of Neuroscience Stem Cell Institute, Institute for Systems Genomics, UConn Health, Farmington, CT, USA*

ALESSANDRA BONITO-OLIVA • *Department of Neuroscience, Karolinska Institutet, Stockholm, Sweden*

MI-HYUN CHOI • *Molecular Neurobiology Laboratory, Department of Life Sciences, Korea University, Seoul, South Korea*

LANI S. CHUN • *Molecular Neuropharmacology Section, National Institute of Neurologic Disorders and Stroke, National Institutes of Health, Bethesda, MD, USA; Cell, Molecular, Developmental Biology & Biophysics Program, Johns Hopkins University, Baltimore, MD, USA*

SAMANTHA R. COTE • *Department of Pharmacology and Physiology, Rutgers-New Jersey Medical School, Newark, NJ, USA*

KATHLEEN VAN CRAENENBROECK • *Laboratory of GPCR Expression and Signal Transduction (LEGEST), Department of Physiology, Ghent University, Ghent, Belgium*

MIREILLE DAIGLE • *Ottawa Hospital Research Institute (Neuroscience Program), University of Ottawa, Ottawa, ON, Canada*

RITESH DAYA • *Department of Psychiatry and Behavioural Neurosciences, McMaster University, Hamilton, ON, Canada*

MARTA DZIEDZICKA-WASYLEWSKA • *Institute of Pharmacology, Polish Academy of Sciences, Kraków, Poland*

AGATA FARON-GÓRECKA • *Institute of Pharmacology, Polish Academy of Sciences, Kraków, Poland*

LAURA M. FIORI • *Ottawa Hospital Research Institute (Neuroscience Program), University of Ottawa, Ottawa, ON, Canada*

GILBERTO FISONE • *Department of Neuroscience, Karolinska Institutet, Stockholm, Sweden*

T. CHASE FRANCIS • *Department of Anatomy and Neurobiology, University of Maryland School of Medicine, Baltimore, MD, USA*

R. Benjamin Free • *Molecular Neuropharmacology Section, National Institute of Neurologic Disorders and Stroke, National Institutes of Health, Bethesda, MD, USA*
Lionel Froux • *CNRS, UMR 5293, Institut des Maladies Neurodégénératives, Université de Bordeaux, Bordeaux, France*
Maurice Garret • *UMR 5287, INCIA, Université de Bordeaux, Bordeaux, France; CNRS, UMR 5287, INCIA, Bordeaux, France*
Bruno Giros • *Department of Psychiatry, Douglas Mental Health University Institute, McGill University, Montréal, QC, Canada; INSERM, UMRS 1130, Paris, France; CNRS, UMR 8246, Paris, France; Neuroscience Paris-Seine, Sorbonne University UPMC, Paris, France*
Richard J.A. Goodwin • *Drug Safety and Metabolism, (DMPK), Cambridge, UK*
Victor Gorgievski • *Department of Psychiatry, Douglas Mental Health University Institute, McGill University, Montréal, QC, Canada; INSERM, UMRS 1130, Paris, France; CNRS, UMR 8246, Paris, France; Neuroscience Paris-Seine, Sorbonne University UPMC, Paris, France*
Nicholas Grimberg • *Departments of Medical Science and Biology, Western University, London, ON, Canada*
Piotr Gruca • *Institute of Pharmacology, Polish Academy of Sciences, Kraków, Poland*
Kyeong-Man Kim • *Department of Pharmacology, College of Pharmacy, Chonnam National University, Gwang-Ju, South Korea*
Katarzyna Kmiotek • *Institute of Pharmacology, Polish Academy of Sciences, Kraków, Poland*
Magdalena Kolasa • *Institute of Pharmacology, Polish Academy of Sciences, Kraków, Poland*
Azita Kouchmeshky • *Faculty of Science, School of Pharmacy, University of Waterloo, Kitchener, ON, Canada*
Jeff S. Kruk • *Faculty of Science, School of Pharmacy, University of Waterloo, Kitchener, ON, Canada*
Maciej Kuśmider • *Institute of Pharmacology, Polish Academy of Sciences, Kraków, Poland*
Eldo V. Kuzhikandathil • *Department of Pharmacology and Physiology, Rutgers-New Jersey Medical School, Newark, NJ, USA*
Caroline Lefebvre • *Ottawa Hospital Research Institute (Neuroscience Program), Departments of Medicine, Cellular and Molecular Medicine, and Psychiatry, University of Ottawa, Ottawa, ON, Canada*
Fang Liu • *Department of Neuroscience, Centre for Addiction and Mental Health, Toronto, ON, Canada; Department of Psychiatry, University of Toronto, Toronto, ON, Canada*
Mary Kay Lobo • *Department of Anatomy and Neurobiology, University of Maryland School of Medicine, Baltimore, MD, USA*
Abraham Martín • *Molecular Imaging Unit, CIC biomaGUNE, San Sebastian, Guipuzcoa, Spain*
Michele L. McGovern • *Department of Neuroscience Stem Cell Institute, Institute for Systems Genomics, UConn Health, Farmington, CT, USA*
Mayako Michino • *Department of Physiology and Biophysics and Institute for Computational Biomedicine, Weill Medical College of Cornell University, New York, NY, USA*
Chengchun Min • *Department of Pharmacology, College of Pharmacy, Chonnam National University, Gwang-Ju, South Korea*
Ram K. Mishra • *Department of Psychiatry and Behavioural Neurosciences, McMaster University, Hamilton, ON, Canada*

KATERINA D. OIKONOMOU • *Department of Neuroscience Stem Cell Institute, Institute for Systems Genomics, UConn Health, Farmington, CT, USA*

MARIUSZ PAPP • *Institute of Pharmacology|, Polish Academy of Sciences, Kraków, Poland*

MARINA REZKELLA • *School of Kinesiology and Health Science, York University, Toronto, ON, Canada*

ALICIA RIVERA • *Facultad de Ciencias, Instituto de Investigación Biomédica, Universidad de Málaga, Málaga, Spain*

PIETER RONDOU • *Laboratory of GPCR Expression and Signal Transduction (LEGEST), Department of Physiology, Ghent University, Ghent, Belgium*

KEYVAN SEDAGHAT • *Ottawa Hospital Research Institute (Neuroscience Program), Departments of Medicine, Cellular, and Molecular Medicine, and Psychiatry, University of Ottawa, Ottawa, ON, Canada*

MOHAMMADREZA SHARIATGORJI • *Biomolecular Imaging and Proteomics, Department of Pharmaceutical Biosciences, National Center for Mass Spectrometry Imaging, Uppsala University, Uppsala, Sweden*

LEI SHI • *Department of Physiology and Biophysics and Institute for Computational Biomedicine, Weill Medical College of Cornell University, New York, NY, USA*

DAVID R. SIBLEY • *Molecular Neuropharmacology Section, National Institute of Neurologic Disorders and Stroke, National Institutes of Health, Bethesda, MD, USA*

MANDAKINI B. SINGH • *Department of Neuroscience Stem Cell Institute, Institute for Systems Genomics, UConn Health, Farmington, CT, USA*

KAMILA SKIETERSKA • *Laboratory of GPCR Expression and Signal Transduction (LEGEST), Department of Physiology, Ghent University, Ghent, Belgium*

JOANNA SOLICH • *Institute of Pharmacology, Polish Academy of Sciences, Kraków, Poland*

CHRISTAL D.R. SOOKRAM • *Department of Psychiatry and Behavioural Neurosciences, McMaster University, Hamilton, ON, Canada*

GIADA SPIGOLON • *Department of Neuroscience, Karolinska Institutet, Stockholm, Sweden*

PING SU • *Department of Neuroscience, Centre for Addiction and Mental Health, Toronto, ON, Canada*

DIANA SUAREZ-BOOMGAARD • *Facultad de Ciencias, Instituto de Investigación Biomédica, Universidad de Málaga, Málaga, Spain*

KINGA SZAFRAN-PILCH • *Institute of Pharmacology, Polish Academy of Sciences, Kraków, Poland*

BOGUSLAW SZCZUPAK • *Molecular Imaging Unit, CIC biomaGUNE, San Sebastian, Guipuzcoa, Spain*

ANNE TAUPIGNON • *CNRS, UMR 5293, Institut des Maladies Neurodégénératives, Université de Bordeaux, Bordeaux, France*

MARIO TIBERI • *Ottawa Hospital Research Institute (Neuroscience Program), Departments of Medicine, Cellular and Molecular Medicine, and Psychiatry, University of Ottawa, Ottawa, ON, Canada*

ELENI T. TZAVARA • *INSERM, UMRS 1130, Paris, France; CNRS, UMR 8246, Paris, France; Neuroscience Paris-Seine, Sorbonne University UPMC, Paris, France*

ALBERT H.C. WONG • *Department of Neuroscience, Centre for Addiction and Mental Health, Toronto, ON, Canada; Department of Psychiatry, University of Toronto, Toronto, ON, Canada*

XIAODI YANG • *Ottawa Hospital Research Institute (Neuroscience Program), Departments of Medicine, Cellular and Molecular Medicine, and Psychiatry, University of Ottawa, Ottawa, ON, Canada*

Sehyoun Yoon • *Molecular Neurobiology Laboratory, Department of Life Sciences, Korea University, Seoul, South Korea*
Boyang Zhang • *Ottawa Hospital Research Institute (Neuroscience Program), Departments of Medicine, Cellular and Molecular Medicine, and Psychiatry, University of Ottawa, Ottawa, ON, Canada*
Mei Zheng • *Department of Pharmacology, College of Pharmacy, Chonnam National University, Gwang-Ju, South Korea*
Dariusz Žurawek • *Institute of Pharmacology, Polish Academy of Sciences, Kraków, Poland*

Part I

Genetic, Epigenetic, and Post-transcriptional Analysis of Dopamine Receptors

Chapter 1

Genetic and Epigenetic Methods for Analysis of the Dopamine D2 Receptor Gene

Laura M. Fiori, Mireille Daigle, and Paul R. Albert

Abstract

Dys-regulation of the dopamine system is strongly implicated in schizophrenia and addiction, and the dopamine D2 receptor (DRD2) is a critical regulator of dopaminergic activity and mediates dopamine actions on motivation and reward. We have identified a single nucleotide polymorphism in the DRD2 gene (rs2734836) that reduces the binding of the transcriptional repressor Freud-1/CC2D1A and in turn alters expression of the dopamine D2 receptor (Rogaeva et al., J Biol Chem 282:20897–20905, 2007). Increasing evidence shows that promoter methylation and demethylation during the early postnatal critical period can establish lifelong change in hippocampal glucocorticoid receptor expression and stress reactivity (Meaney and Szyf, Trends Neurosci 28:456–463, 2005). The early postnatal period is also that time during which expression of the DRD2 matures and reaches adult levels. Hence, we hypothesize that differences in DNA methylation of the DRD2 promoter may correlate with altered expression of the receptor in adults, and predispose to schizophrenia. In this chapter, we review methods for both genotype analysis of key polymorphisms of the DRD2 gene (rs2734836 and the Taq1A variant rs1800497), as well as methodology for analysis of the DNA methylation. These methods can be easily adapted to any gene of interest using the methods for oligonucleotide primer selection and PCR conditions suggested.

Key words SNP, Polymorphisms, PCR, Bisulfite, DNA methylation, Epigenetic, Genetic, Methods, Dopamine D2 receptor, CpG island, CG-rich, Promoter, Repressor, Transcription factor

1 Introduction

The dopamine system is implicated in a variety of mental illnesses, including schizophrenia and addiction to psychotropic drugs like heroin, cocaine, and methamphetamine [1, 2]. All clinically used antipsychotics antagonize dopamine-D2 receptors (DRD2), and non-D2 antipsychotics have not been developed [3]. Furthermore, there is a direct correlation between the clinical effective dose of antipsychotic compounds and their affinity as antagonists for D2 receptors [1], implicating hyperactivation of the dopamine system in schizophrenia. Dopaminergic hyperactivity could result from either over-expression of post-synaptic DRD2s or from hyperactivity of dopamine neurons. Ligand binding studies of postmortem

Mario Tiberi (ed.), *Dopamine Receptor Technologies*, Neuromethods, vol. 96,
DOI 10.1007/978-1-4939-2196-6_1, © Springer Science+Business Media New York 2015

brain tissue indicate that DRD2-like binding levels are up to sixfold higher in the nucleus accumbens of schizophrenic patients [1, 4]. PET imaging studies of DRD2 occupancy in the nucleus accumbens following dopamine depletion indicates that schizophrenics have twofold higher DRD2 occupancy than normal subjects, indicating a higher level of dopaminergic neurotransmission [5]. Similarly, schizophrenics show increases in amphetamine-induced dopamine release [6–9]. Thus hyperactivity of the dopamine system that is implicated in schizophrenia could be due to higher post-synaptic D2 receptor levels and/or increased activity of dopaminergic neurons.

We hypothesized that increased DRD2 receptor expression could be due to de-repression of the DRD2 gene. In order to understand factors that might alter the regulation of the DRD2 gene, we focused on the discovery of homologue of the 5-HT1A receptor dual repressor element (DRE) in intron 2 of the DRD2 gene [10]. We showed that this site binds the repressor Freud-1/CC2D1A that we identified as a major repressor of 5-HT1A autoreceptors [11]. We showed that the level of expression of Freud-1 in DRD2 dopaminergic and non-dopaminergic cells was inversely related to levels of DRD2 RNA and D2 binding levels. Close to one of the DRE sites we identified a polymorphism (rs2734836, position 113291239) in the DRD2 gene that reduces the binding of the transcriptional repressor Freud-1/CC2D1A and in turn increases transcriptional activity [10]. We hypothesize that this functional DRD2 polymorphism may be associated with schizophrenia or drug addiction. One study has shown association of the adjacent DRD2 SNP rs2734835 (position 113291343) and its haplotype block with alcoholism [12]. In addition, the rs27334836 polymorphism lies within a haplotype block that affects alternate splicing to generate short or long forms of the DRD2 [13].

Early life stress including reduced maternal care or childhood physical or sexual abuse can result in promoter methylation of the glucocorticoid receptor gene in the brain and can establish lifelong change in hippocampal GR expression and stress reactivity [14, 15]. Increasingly, the interaction between stress-driven DNA methylation and risk alleles is being found to account in part for the proposed gene × environment interaction in susceptibility to severe mental illness [16]. Schizophrenia is thought to involve developmental stress that occur in the early postnatal period and could lead to lifelong alterations in brain DNA methylation. Although it has yet to be specifically addressed, differences in DNA methylation of the DRD2 promoter or intronic sequences may strengthen the association of the DRD2 risk allele with addiction or schizophrenia. Specific alterations in DNA methylation of the DRD2 gene may correlate with altered expression of the receptor in adults, and predispose to schizophrenia. Here we review methods for genotype analysis of the rs2734836 and TaqIA (rs1800497) polymorphisms of the DRD2 gene, as well as for analysis of the DNA methylation. The methods presented below can be readily adapted to address genotype or DNA methylation changes for any gene of interest.

2 Materials and Methods

2.1 Analysis of CpG Methylation

2.1.1 DNA Extraction

The first step in the analysis of genomic DNA methylation patterns is the extraction of large quantities of high quality DNA from the tissue of interest. As the optimal extraction method depends upon the type of tissue to be analyzed, the protocol to be used for this step must be determined by the reader. Numerous commercially available kits exist for this task, or readers may wish to use their own protocols and reagents. Regardless of the method selected, it is imperative that the tissue itself is pure (i.e., the sample only contains the type of tissue that is of interest to the researcher), as methylation patterns can vary greatly between different types of samples. Furthermore, as the bisulfite conversion process described in the next section is very harsh on the DNA, it is best to start with as high quality DNA as possible.

2.1.2 Bisulfite Conversion

Bisulfite treatment of DNA is the most commonly used method for allowing the detection of methylated cytosine residues. This process specifically converts all unmethylated cytosine residues within the DNA to uracil, while leaving methylated cytosines unaltered. Several kits are commercially available (for instance, QIAGEN's EpiTect Bisulfite kit) which allow for both the bisulfite treatment and purification of converted DNA. This process may also be performed without the use of kits, using the protocol described below. A critical aspect of bisulfite conversion is to obtain >95 % conversion (*see* Sect. 3.1).

1. Digest 2.0 μg of genomic DNA with HindIII for 18 h per manufacturer's instructions.
2. Incubate digested DNA at 99 °C for 5 min to produce single-stranded DNA.
3. Incubate for 30 min at 39 °C in 0.31 M NaOH.
4. Add bisulfite mixture (4 M sodium bisulfite, 6 μM hydroquinone, 0.3 M guanidine HCl, 0.24 M NaOH) and incubate at 55 °C for 16 h.
   (a) Note: the bisulfite mixture is light-sensitive.
5. Purify DNA to remove salts. This can be done using a commercial kit (such as the Promega Wizard DNA Cleanup Kit) or any other preferred method. Ensure a final volume of 100 μL in ddH_2O.

6. Desulfonate DNA by adding 5 μL of 6.3 M NaOH (0.3 M) and incubating for 15 min at 37 °C.
7. Add 90 μL of 5 M NH_4OAc, 20 μg of linear acrylamide, and 400 μL of 100 % EtOH.
8. Incubate overnight at −20 °C.
9. Centrifuge sample for 10 min at 16,000 ×*g* at 4 °C.
10. Remove supernatant and wash pellet with 450 μL of ice-cold 70 % EtOH.
11. Centrifuge sample for 10 min at 16,000 ×*g* at 4 °C.
12. Remove supernatant and dry pellet.
13. Resuspend in 50 μL of ddH_2O.

2.1.3 Amplification of Target Sequences

Target Selection for PCR Amplification

The ability to amplify, and later clone, sequences of interest from bisulfite-treated DNA represents one of the most challenging aspects of methylation analysis. Both the harsh conditions required for the bisulfite conversion, and the resulting highly AT-rich DNA, necessitate the selection of relatively small target regions of the DNA. Although larger genomic regions can be successfully cloned, we suggest aiming for no more than 400 bases within each region. If the analysis of methylation patterns in a larger region is desired, it may be optimal to divide the region into smaller sections [17].

Primer Design

A comprehensive description of strategies for the design of primers directed towards bisulfite-converted DNA is beyond the scope of this chapter. The reader is referred to Clark et al. [18] or Fraga and Esteller [17] for a detailed explanation on primer design for DNA methylation studies (also *see* Sect. 3.2). Ultimately, whichever methods the reader uses for designing primers, there are several key points that must be considered. Most importantly, primer sequences must be designed against the target DNA template in which all cytosines have been converted to uracils. Additionally, the decreased CG content and complexity of the DNA sequence will require longer primers to ensure specificity (*see* Sect. 3.3). Finally, although this may not always be possible depending on the DNA sequence, it is preferable to avoid choosing primers that anneal to potential methylation sites (i.e., CG dinucleotides), as this will create mismatches in the event of methylation, and could skew the results by preferentially amplifying non-methylated strands of DNA. Although it is preferable to perform only one round of PCR, in cases where the yield is poor, a nested PCR strategy may also be employed. In this case, the suggestions listed above should be considered for both pairs of primers. For addition considerations in primer design *see* Sect. 3.2.

Polymerase Chain Reaction

The conditions for this reaction will depend greatly upon the length of the DNA template, as well as the amplification strategy (single or nested) and primers which have been selected.

Readers will need to optimize their conditions with regard to the amount of input DNA, PCR reagents, and buffers, as well as the cycling parameters for the PCR. When selecting PCR reagents, the use of a high-fidelity polymerase is essential, as products will ultimately be sequenced. Special consideration of PCR conditions needs to be given when amplifying GC-rich sequences that are typical of gene promoters (*see* Sect. 3.3). Furthermore, the reader must ensure that the products generated in this reaction will be compatible with the cloning system to be used in the next section, particularly with regard to the requirement for A-overhangs. Following the reaction, products should be purified prior to cloning. We strongly recommend the use of a gel extraction method for this step.

2.1.4 Cloning and Sequencing

The ultimate goals of cloning and sequencing PCR products are: to allow for the identification of the presence of methylated cytosines on a single molecule of genomic DNA and to determine the percentage of methylation at a particular site within an individual sample. Accordingly, a cloning system should be selected based on its capacity for quick ligation reactions and screening for successful insertion of PCR products. Additionally, the suitability of a vector for sequencing reactions (i.e., sites for common sequencing primers) should also be considered. In our hands, we have had success with the use of the pGEM-T Easy system (Promega), which requires A-overhands on PCR products, and allows for blue-white screening of colonies. After the ligation and cloning of PCR products, white colonies can be selected and plasmids extracted. As with other steps in this process, the plasmid extraction step can be performed using a commercial kit (such as Qiagen's QIAprep Miniprep kit), or any other method preferred by the reader, provided that products are of sufficient purity and concentration for sequencing reactions. For the sequencing reactions, given the low AT content of the target sequence, optimization of DNA and primer concentrations, as well as sequencing conditions themselves, may be required.

2.1.5 Analysis

The methods used for presenting and statistically analyzing the results from the sequencing reactions depend completely on the aims and hypotheses of the reader, which may represent exploratory analyses of methylation patterns in a particular sample, or more quantitative analyses of methylation percentages at specific sites or overall within the target region.

Regardless of the experimental goals, several additional points must be considered in order to improve the reproducibility and strength of the conclusions arising from these experiments. First, although the strong conditions of the bisulfite reactions are generally sufficient to ensure complete conversion of unmethylated cytosines to uracils, this is not guaranteed. Evidence suggesting incomplete conversion would be the presence of methylated cytosines at non-CG dinucleotides (which is rare in animals), as well as clones in which all cytosines appear to have been methylated. Although not required, readers may wish to include positive and negative control samples for the bisulfite conversion step. These can be prepared by pretreating a vector with SssI methylase (New England Biolabs) to fully methylate all cytosines in CG dinucleotides, as well as propagating the same vector in a non-methylating strain of bacteria (dam−/dcm− *E. coli* are commercially available), to generate a completely unmethylated version of the vector.

Second, we strongly advise performing multiple PCRs for each sample, which can then be combined prior to cloning. This ensures a greater representation of DNA molecules from within each sample, which can be particularly important when assessing methylation rates at specific sites.

Finally, the number of clones which are sequenced from each sample will directly influence the strength of the data. Although power analyses are difficult and likely not appropriate for this type of analysis, researchers must ensure that they are basing their conclusions upon a sufficient quantity of data.

2.2 Analysis of Genetic Polymorphisms

2.2.1 Selection of Polymorphisms

Whereas epigenetic analyses are relatively new, and often have more of an exploratory nature, researchers have been investigating genetic polymorphisms for decades. In this time, large undertakings including the sequencing of the human genome and the HapMap project (http://hapmap.ncbi.nlm.nih.gov/) have allowed researchers to establish the precise location and allelic frequencies of well over a million common genetic variations. Moreover, their frequencies have been assessed in a range of ethnic populations, and their relationship to nearby polymorphisms has also been established. However, numerous other novel or rare variants exist, and the selection of which variants to examine depends largely on the overall experimental objectives of the researcher.

We describe here the method for genotyping of two single nucleotide polymorphisms (SNPs) in the DRD2 locus which have been successfully genotyped in our laboratory. Indeed, the TaqIA variant (rs1800497) has been the most well-studied SNP in the DRD2 locus, and is mapped to a point beyond the DRD2 3′-untranslated region, within exon 8 of the adjacent gene, ANKK1. Although not within the DRD2 gene itself, it is contained within a large haplotype block which encompasses part of this gene, and thus can be used to infer information regarding the status of variations within DRD2. The early discovery and ongoing popularity of this variant lies in its status as a restriction fragment length polymorphism (RFLP), as variation at this site destroys the recognition sequence for the restriction enzyme TaqI. The second SNP we describe, rs2734836, is one which we identified as a novel functional variant which influences the expression of DRD2 by interfering with the binding of the transcriptional repressor

Freud-1/CC2D1A [10]. As opposed to the TaqIA variant, this SNP does not create a variable enzyme restriction site, and thus genotyping must be performed by other methods.

2.2.2 DNA Extraction

Regardless of the method that will be used for genotyping, the first step involves the extraction of genomic DNA. Unlike the analysis of epigenetic effects, the selection of tissue used as a source of DNA for genotyping is relatively unimportant, although as above, the extraction method that is used will depend upon the tissue. However, less DNA is required, and therefore smaller amounts of tissue may be used.

2.2.3 Genotyping by Sequencing

The most direct method for genotyping polymorphisms is by sequencing the DNA itself. For this method, DNA containing the polymorphism of interest is amplified by PCR, products are purified, and samples are sequenced using Sanger sequencing. Unlike the analysis of DNA methylation, it is not advisable to clone products prior to sequencing, as heterozygosity may not be detected. We have used direct sequencing to analyze the rs2734836 polymorphism using this method as follows.

First, we performed PCR using the following primers:

5′ TTCCAGGGCAGCTTAGTAGAGAG

5′ CCCTTCTTTCCTACAAACACTTATT

Using these primers, amplification was accomplished following a step-down PCR amplification program: 92 °C for 5 min, 92 °C for 45 s, 69 °C for 45 s (-0.5 °C/cycle); 72 °C for 90 s (10 cycles); 92 °C for 45 s, 64 °C for 45 s, 72 °C for 90 s (30 cycles); and terminated at 72 °C for 10 min. This generated a 458 bp product, which was purified using a gel extraction method, then sequenced using the primer:

5′-TGGAGAGTAGTTAGGGCTG

Sequencing results were analyzed by eye, and subjects designated as homozygous GG or AA, or as heterozygous carriers of both the G and A alleles.

2.2.4 Genotyping by Enzymatic Digestion

The TaqIA polymorphism may also be genotyped by direct sequencing; however its status as an RFLP allows it to also be detected through simple enzymatic digestion. This method was originally described by Grandy et al. [19], and represents a quick and cost-effective means for genotyping this polymorphism. The first step of this method involves amplification of a 310 bp fragment by PCR, using the following primers:

5′ CCGTCGACGGCTGGCCAAGTTGTCTA

5′ CCGTCGACCCTTCCTGAGTGTCATCA

The PCR reaction can be performed using standard conditions, using annealing and extension conditions appropriate for whichever polymerase is selected by the researcher, and a melting temperature of 50 °C. The PCR products can then be digested overnight using TaqI (Invitrogen) and then analyzed directly on a 2 % agarose gel. Presence of the C allele (also known as A2, the major allele) yields two bands (130 and 180 bp), while PCR products containing the T allele (A1) remain uncut. Subjects who are heterozygous for this polymorphism will show all three bands. It is best to include control samples with known genotypes to verify that complete digestion of the PCR product has occurred. Additionally, researchers may wish to sequence a subset of samples to verify genotypes.

2.2.5 Genotyping by TaqMan® Assay

Although both the enzymatic digestion and direct sequencing methods are effective and relatively quick, a high throughput method may be desired when analyzing large samples. Commercial real time PCR (RT-PCR)-based assays are available, the most common being the TaqMan® Assays produced by Applied Biosystems. A large number of assays have already been developed and optimized to detect many known polymorphisms, including the TaqIA variant and rs2734836. Researchers may also custom design their own assays to detect additional polymorphisms using the same chemistry. The conditions used for the RT-PCR and the downstream analysis will depend upon the assay itself, as well as the RT-PCR instrument being used. As with the enzymatic digestion method, it is advisable to verify genotypes by sequencing a small subset of the samples.

2.2.6 Analysis and Additional Considerations

The steps a researcher takes with the information derived from genotype calling are completely dependent upon their initial experimental objectives, be they to determine the relationship between a variant and gene function or to determine its association with a broader phenotype such as disease. The statistical analyses to be used to address these hypotheses are well beyond the scope of this chapter, and the researcher is strongly advised to fully consider issues of statistical power, multiple testing, and linkage disequilibrium between variants prior to undertaking any genotyping study.

3 Notes

3.1 Bisulfite Conversion

For downstream applications, it is essential that this procedure yields a >95 % rate of conversion of unmethylated cytosines. This conversion can be verified by testing with addition of purified plasmid (non-methylated) DNA to a test sample.

3.2 Primer Optimization

For primer design, several programs for designing bisulfite-conversion-based Methylation PCR primers are available online. For example, specialized software like MethPrimer [20] is useful for designing primers to amplify your DNA sequence of interest and can also predict CpG islands in DNA sequences. This software is freely available at this address:

http://www.urogene.org/methprimer/

For optimal PCR primers, the following conditions need to be met:

- The annealing temperature of both primers must be similar (±3 °C) and always between 55 and 65 °C.
- The PCR product should be between 200 and 400 bp, especially since the bisulfite modification degrades the DNA and results in shorter DNA fragments. In addition, the treatment produces a biased base composition, making sequencing of long DNA fragments difficult.
- Ideally each primer should not contain CpG dinucleotides to avoid methylation-slanted clone amplification.
- In order to avoid amplification of unmodified DNA, primers should contain non-CpG cytosines.
- Each primer should be checked that it does not contain a common SNP which would skew amplification in a genotype dependent manner.

3.3 GC-Rich Sequences

GC-rich sequences are typical of many "housekeeping" promoters studied for DNA methylation changes. GC-rich sequences tend to be repetitive, required higher annealing temperatures, and are difficult to amplify by PCR. To amplify GC-rich sequences, we have tried different solution and polymerase. We got success with use of betaine in PCR reaction and the use of platinum Taq PCRx DNA polymerase (Life Technologies).

For GC-rich PCR products (>60 %), we add betaine to the polymerase buffer at a concentration of 1.6 M (optimize between 1 and 2 M). This concentration has to be optimized for each PCR primer pair. For PCR products with >75 % GC content, we have used the Platinum Taq PCRx DNA polymerase (Life Technology) with success. For GC-rich PCR products, we have also tried adding 10 % DMSO or using so-called GC-buffer systems available from different Taq polymerase suppliers without any success.

Acknowledgments

PRA was supported by funding from the Canadian Institutes of Health Research, Ontario Mental Health Foundation, and Heart and Stroke Foundation Canadian Partnership for Stroke Recovery. LMF was supported by postdoctoral funding from the National Science and Engineering Research Council of Canada.

References

1. Seeman P, Van Tol HH (1994) Dopamine receptor pharmacology. Trends Pharmacol Sci 15:264–270
2. Self DW (2004) Regulation of drug-taking and -seeking behaviors by neuroadaptations in the mesolimbic dopamine system. Neuropharmacology 47(Suppl 1):242–255. doi: 10.1016/j.neuropharm.2004.07.005
3. Miyamoto S, Duncan GE, Marx CE et al (2005) Treatments for schizophrenia: a critical review of pharmacology and mechanisms of action of antipsychotic drugs. Mol Psychiatry 10:79–104
4. Kapur S, McClelland RA, VanderSpek SC et al (2002) Increasing D2 affinity results in the loss of clozapine's atypical antipsychotic action. Neuroreport 13:831–835
5. Abi-Dargham A, Rodenhiser J, Printz D et al (2000) From the cover: increased baseline occupancy of D2 receptors by dopamine in schizophrenia [see comments]. Proc Natl Acad Sci U S A 97:8104–8109
6. Abi-Dargham A, Kegeles LS, Zea-Ponce Y et al (2004) Striatal amphetamine-induced dopamine release in patients with schizotypal personality disorder studied with single photon emission computed tomography and [^{123}I]iodobenzamide. Biol Psychiatry 55:1001–1006. doi: 10.1016/j.biopsych.2004.01.018
7. Frankle WG, Lerma J, Laruelle M (2003) The synaptic hypothesis of schizophrenia. Neuron 39:205–216. doi: 20026302
8. Laruelle M, Abi-Dargham A, Gil R et al (1999) Increased dopamine transmission in schizophrenia: relationship to illness phases. Biol Psychiatry 46:56–72
9. Miyake N, Thompson J, Skinbjerg M et al (2011) Presynaptic dopamine in schizophrenia. CNS Neurosci Ther 17:104–109. doi:10.1111/j.1755-5949.2010.00230.x
10. Rogaeva A, Ou XM, Jafar-Nejad H et al (2007) Differential repression by freud-1/CC2D1A at a polymorphic site in the dopamine-D2 receptor gene. J Biol Chem 282:20897–20905. doi: 10.1074/jbc.M610038200
11. Ou XM, Lemonde S, Jafar-Nejad H et al (2003) Freud-1: a novel calcium-regulated repressor of the 5-HT1A receptor gene. J Neurosci 23:7415–7425
12. Bhaskar LV, Thangaraj K, Non AL et al (2010) Population-based case–control study of DRD2 gene polymorphisms and alcoholism. J Addict Dis 29:475–480. doi:10.1080/10550887.2010.509274
13. Zhang Y, Bertolino A, Fazio L et al (2007) Polymorphisms in human dopamine D2 receptor gene affect gene expression, splicing, and neuronal activity during working memory. Proc Natl Acad Sci U S A 104:20552–20557. doi: 10.1073/pnas.0707106104
14. Meaney MJ, Szyf M (2005) Maternal care as a model for experience-dependent chromatin plasticity? Trends Neurosci 28:456–463. doi: 10.1016/j.tins.2005.07.006
15. McGowan PO, Sasaki A, D'Alessio AC et al (2009) Epigenetic regulation of the glucocorticoid receptor in human brain associates with childhood abuse. Nat Neurosci 12:342–348. doi: 10.1038/nn.2270
16. Klengel T, Mehta D, Anacker C et al (2013) Allele-specific FKBP5 DNA demethylation mediates gene-childhood trauma interactions. Nat Neurosci 16:33–41. doi:10.1038/nn.3275
17. Fraga MF, Esteller M (2007) Use of PCR for DNA methylation analyses. In: Hughes S, Moody A (eds) PCR: methods express, Methods Express, vol Series. Scion Publishing Limited, Bloxham, UK, pp 265–277
18. Clark SJ, Harrison J, Paul CL et al (1994) High sensitivity mapping of methylated cytosines. Nucleic Acids Res 22:2990–2997
19. Grandy DK, Zhang Y, Civelli O (1993) PCR detection of the TaqA RFLP at the DRD2 locus. Hum Mol Genet 2:2197
20. Li LC, Dahiya R (2002) MethPrimer: designing primers for methylation PCRs. Bioinformatics 18:1427–1431

Chapter 2

Characterization of D1 Dopamine Receptor Posttranscriptional Regulation

Eldo V. Kuzhikandathil

Abstract

Posttranscriptional regulation (PTR) of gene expression describes regulatory mechanisms that control the expression of protein from its cognate mRNA. Studies that investigate changes in gene expression, for a variety of reasons, typically focus on measuring levels of mRNA or protein but not both. Even studies that measure both mRNA and protein levels of the gene of interest rarely assess the temporal discordance between the two. Given that PTR provides a mechanism for spatial and temporal regulation of gene expression, it likely plays a major role in physiological and pathophysiological conditions. In this chapter, we describe methods to assess PTR using the D1 dopamine receptor gene as an example. PTR mechanisms can be broadly classified into mechanisms that regulate mRNA turnover and those that control mRNA translation. The mouse catecholaminergic CAD cell line which expresses endogenous D1 dopamine receptor is a tractable model system for deciphering the molecular mechanism of D1 receptor PTR. We describe methods to measure D1 dopamine receptor mRNA stability using actinomycin D and methods using reporter constructs to assess microRNA (miRNA)-mediated regulation of D1 receptor protein translation. Using these methods we demonstrate that the D1 dopamine receptor exhibits PTR in which the expression of D1 receptor protein is regulated by miRNAs. The chapter provides detailed methods for studying potential D1 dopamine receptor PTR during development and in disease states.

Key words Dopamine receptor, Gene expression, Posttranscriptional regulation, mRNA stability, mRNA translation, MicroRNA, 3′ untranslated region, RT-PCR, Western blotting

1 Introduction

PTR of gene expression provides mechanisms by which expression of protein can be regulated spatially and temporally [1]. For example, in the nervous system, mRNA is synthesized in the nucleus of cell bodies and trafficked to axonal or dendritic terminals where local protein translation can be initiated by cues at a later time [2]. This process requires molecular mechanisms that can stabilize and protect the mRNA as well as suppress its translation. Ontogeny studies in rodents have shown that while the expression of D1 receptor mRNA commences around embryonic day 14 and reaches steady state expression level around postnatal day 5, the D1

Mario Tiberi (ed.), *Dopamine Receptor Technologies*, Neuromethods, vol. 96,
DOI 10.1007/978-1-4939-2196-6_2, © Springer Science+Business Media New York 2015

receptor protein levels increase postnatally and reach peak values between postnatal day 7 and 14 [3–5]. This lack of correlation in expression of D1 receptor mRNA and protein is also seen during human brain development [6]. While these studies suggest that D1 receptor expression is regulated at the posttranscriptional level, the molecular mechanisms that mediate the posttranscriptional regulation (PTR) of D1 receptor expression are not well understood. In this chapter, we focus on methods used to determine if the D1 dopamine receptor gene is posttranscriptionally regulated and describe approaches to study two mechanisms of PTR. We will describe the use of actinomycin D to study D1 dopamine receptor mRNA stability and methods used to study the role of miRNA in regulation of D1 dopamine receptor protein translation. To study PTR of any gene, three tools are necessary—(a) a tissue or cell line that endogenously expresses the gene and exhibits PTR of the gene under defined conditions, (b) method to quantitate the level of mRNA, and (c) good quality antibodies that can detect and quantitate the expression of the cognate protein.

In the case of D1 dopamine receptor, we have shown that the mouse CAD catecholaminergic cell line expresses endogenous D1 dopamine receptor mRNA and protein [7]. Furthermore, CAD cells undergo reversible differentiation in serum-free media (Fig. 1) during which there is an increase in D1 receptor mRNA but no concomitant increase in D1 receptor protein levels [7]. Thus the CAD cell line is a useful model to study PTR of D1 dopamine receptor. In this chapter, we will discuss the culture conditions for growing non-differentiated and differentiated CAD cells. The mouse D1 receptor gene has a promoter that is ~6.4 kb, one intron in the noncoding region, as well as 5′ and 3′ untranslated regions (5′UTR and 3′UTR) [7, 8]. We have detected and quantitated D1 receptor mRNA using methods such as RNase protection assay and quantitative real-time PCR [7, 8]. In this chapter we will describe the latter method for quantifying D1 receptor mRNA levels using the Taqman® PCR methods. Antibodies for detecting membrane proteins such as G-protein coupled receptors are notoriously poor, lacking selectivity and exhibiting inconsistency from lot to lot. Following years of testing various D1 receptor antibodies, we have validated a commercially available anti-D1 dopamine receptor rat monoclonal antibody that consistently detects the non-glycosylated and multiple glycosylated isoforms of mouse, rat, and human D1 dopamine receptor protein (Fig. 2).

To measure D1 receptor mRNA stability in non-differentiated and differentiated CAD cells, we quantitate the levels of mature D1 receptor mRNA at different time points after the synthesis of new mRNA is blocked using actinomycin D. To assess regulation of D1 receptor protein expression by miRNA, we use plasmid constructs in which the β-galactosidase reporter gene is fused to D1 receptor 3′UTR with and without mutations in putative miRNA

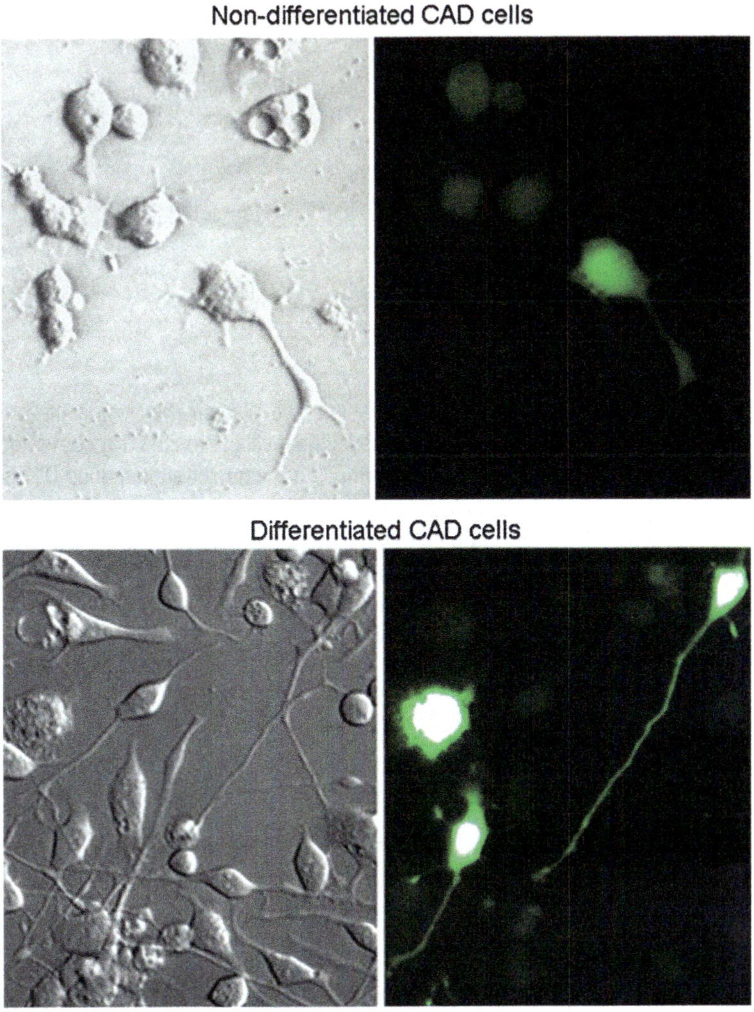

Fig. 1 Representative bright field and fluorescent images of non-differentiated and differentiated CAD cells that were transfected with a plasmid expressing enhanced green fluorescent protein using Lipofectamine 2000 as described in Sect. 3.3. The differentiated cells were cultured in serum-free media for 48 h

binding sites. We describe the spectrophotometric assay used to measure β-galactosidase activity using widely available reagents and equipment.

2 Materials

2.1 Cell Culture Materials

1. Dulbecco's modified Eagle's medium (DMEM)/F12 media (catalog# 12-719Q, Lonza-BioWhittaker, Walkersville, MD, USA).
2. Fetal calf serum (catalog# 26140-079, Invitrogen, Carlsbad, CA, USA).

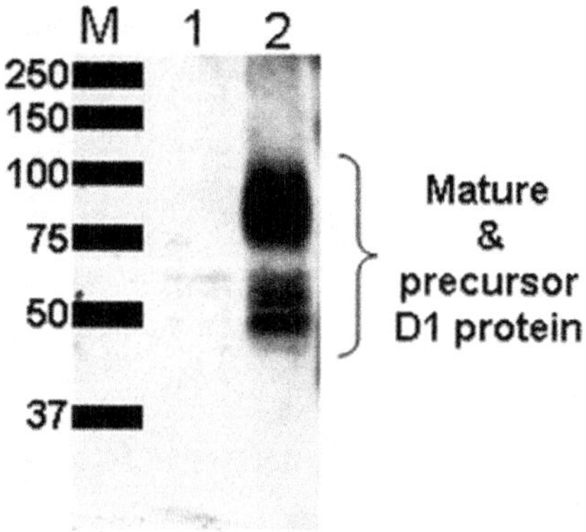

Fig. 2 Representative Western blot using the anti-D1 receptor rat monoclonal antibody shows the level of mouse D1 receptor protein in AtT-20 neuroendocrine cells (which does not express endogenous D1 receptor) transiently transfected with empty plasmid vector (*lane 1*) or a constitutive expression plasmid encoding the mouse D1 dopamine receptor (*lane 2*). Each lane was loaded with 30 μg of total cell protein and the D1 receptor protein detected using the anti-D1 receptor rat monoclonal antibody. Multiple glycosylated forms of mature D1 receptor are detected with this antibody

3. Penicillin/streptomycin (catalog# 15140-148, Invitrogen).
4. 10× Phosphate-buffered saline (catalog# 70011-044, Invitrogen).
5. Sodium selenite (catalog# S9133, Sigma-Aldrich, St. Louis, MO, USA).
6. Transferrin (catalog# T8158, Sigma-Aldrich).
7. Actinomycin D, Streptomyces sp. (catalog# 114666, EMD-Millipore, Billerica, MA, USA).
8. 6- or 12-well Costar tissue culture plates (catalog# 07-200-80 or 07-200-81, Fisher Scientific, Pittsburgh, PA, USA).
9. 100-mm tissue culture plates (catalog# 83. 1802.003) (Sarstedt Inc., Newton, NC, USA).
10. T-25 tissue culture flasks (catalog# 83.1810.001) (Sarstedt Inc., Newton, NC, USA).
11. T-75 tissue culture flasks (catalog# 83. 1813) (Sarstedt Inc., Newton, NC, USA).
12. Cellulose acetate 0.22 μm bottle top filters (catalog# 09-761-50, Corning, Tewksbury, MA, USA).

2.2 Enzymes and Reagents for Cloning D1 3′ UTR

1. Bacterial artificial clone (BAC) containing the entire mouse D1 dopamine receptor gene (catalog# RPCI23.C, clone ID: 47M2, Invitrogen).
2. Primers for amplifying and cloning the 1,277 bp mouse D1 receptor 3′UTR-forward Not I primer 5′-AAGGAAAAA

AGCGGCCGCATATTGGGTTCTCATCTCTGAAGCT ATGAGTTCC-3′, reverse Hind III primer 5′-G GTT TAA AC AAGCTT CACGTG C TTA AGC TCA TTA GCT AGT TTA CCA CTA ACA TTA AAG AGC-3′. The forward primers has a Not I restriction site and the reverse primer has restriction sites for Hind III, Afl II, and Pml I. To generate the mutation in the miR142-3p binding site in the D1 3′UTR, we used a reverse Ssp I/ Kpn I primer 5′-CTTTCCAAAATAT TTTTAGGGGCAGAGCATTGGGTACCTAGTCACTT CTTACC-3′.

3. Restriction enzymes were from New England Biolabs, Ipswich, MA, USA.
4. Advantage® 2 Polymerase Mix (catalog# 639201, Clontech Labs, Mountain View, CA, USA).
5. Deoxynucleotides (dNTP) (catalog# 10297-018, Invitrogen).
6. Enzyme diluent (catalog#1773, BioFire Diagnostics, Salt Lake City, UT, USA).
7. QIAquick Gel Extraction Kit (catalog# 28704, Qiagen, Valencia, CA, USA).
8. T4 DNA ligase (catalog# 15224-017, Invitrogen).
9. MAX Efficiency® DH5α™-T1R competent *E. coli* cells (catalog# 12297-016, Invitrogen).
10. QIAprep Spin Miniprep Kit (catalog# 27104, Qiagen).

2.3 Transfection Reagents

1. Lipofectamine 2000 (catalog# 11668-019, Invitrogen).
2. Opti-MEM (catalog# 31985-070, Invitrogen).
3. p3XFLAG-CMV™-7-BAP transfection efficiency control plasmid (catalog# C7472, Sigma).
4. Carrier plasmid pUC19 (catalog#. 15364-011, Invitrogen).

2.4 RNA Isolation and RT-PCR Reagents

1. RNeasy mini kit (catalog# 74104, Qiagen).
2. TURBO DNA-free™ Kit (catalog# AM1907, Invitrogen).
3. Random primers (catalog#48190-011, Invitrogen).
4. SuperScript® III First-Strand Synthesis System (catalog# 18080-051, Invitrogen).
5. Real-time PCR was performed using the Roche Light Cycler carousel-based system (Indianapolis, IN, USA).
6. Glass capillaries (catalog# 04929292001, Roche).
7. Bovine serum albumin (BSA) (2.5 mg/mL) (catalog#1777, BioFire Diagnostics).
8. TaqMan® gene expression assay for D1 dopamine receptor (Assay ID: Mm01353211_m1, Invitrogen).

9. The TaqMan® gene expression assay for internal control GAPDH (Assay ID: Mm99999915_g1, Invitrogen).
10. TaqMan® Universal PCR Master Mix, No AmpErase® UNG (catalog# 4324018, Invitrogen).

2.5 Protein Isolation and Western Blotting Reagents

1. CelLytic™ M reagent (catalog# C2978, Sigma).
2. Phenylmethylsulfonylfluoride (PMSF; catalog# P7626, Sigma).
3. Protease inhibitor cocktail (catalog# P8340, Sigma).
4. BCA Protein Assay Reagent (catalog# 23227, Thermo-Fisher Scientific).
5. Sample loading buffer (62.5 mM Tris–Cl [pH 6.8], 2%SDS, 10 % glycerol, 10 mM EDTA, 50 mM TCEP bond-breaker (catalog# 77720, Thermo-Fisher), 0.01 % bromophenol blue).
6. SDS-PAGE was performed using the Mini-PROTEAN® Tetra Cell system using homemade gels or commercially available precast gels (Bio-Rad, Hercules, CA, USA).
7. Nitrocellulose membranes (catalog# 88013, Thermo-Fisher Scientific).
8. Nonfat milk dry milk powder (Carnation).
9. Tris-buffered saline (pH 7.4) [20 mM Tris base (catalog# T6066, Sigma), 137 mM sodium chloride (catalog#S9888, Sigma), adjust to pH 7.4 with HCl].
10. Tween®-20 (catalog# P9416, Sigma).
11. Monoclonal anti-D1 dopamine receptor antibody produced in rat (catalog# D2944, Sigma).
12. Monoclonal anti-GAPDH antibody produced in rabbit (catalog# 2118, Cell Signaling Technology, Danvers, MA, USA).
13. Monoclonal ANTI-FLAG® M2 antibody produced in mouse (catalog# F3165, Sigma).
14. Goat anti-rat secondary antibody conjugated to horse radish peroxidase (HRP) (catalog# 31470, Thermo-Fisher Scientific).
15. Goat anti-rabbit secondary antibody conjugated to HRP (catalog# A6154, Sigma).
16. Sheep anti-mouse secondary antibody conjugated to HRP (catalog# A5906, Sigma).
17. Blot stripping buffer—Western ReProbe™ (catalog# 786-119, G-Biosciences, St. Louis, MO, USA).
18. SuperSignal West Dura chemiluminescent substrate (catalog# 34075, Thermo-Fisher Scientific).
19. HyBlot CL™ autoradiography film (catalog# E3018, Denville Scientific, Metuchen, NJ, USA).

2.6 Reagents for β-Galactosidase Assay

1. β-gal lysis buffer (10 mM KCl, 1 mM $MgSO_4$, 2.5 mM EDTA, 0.25 % NP-40 detergent, 50 mM β-mercaptoethanol, and 100 mM sodium phosphate buffer [pH 7.2]).
2. 100 mM sodium phosphate buffer (3.4 mL 1 M Na_2HPO_4 + 1.6 mL 1 M NaH_2PO_4 + 45 mL sterile deionized water).
3. Chlorophenol red-β-D-galactopyranoside (CPRG, catalog# 10884308001, Roche Applied Science, Indianapolis, IN, USA).
4. Coomassie (Bradford) Protein Assay Kit (catalog# 23200, Thermo-Fisher Scientific, Rockford, IL, USA).

3 Methods

3.1 Culture and Actinomycin D Treatment of CAD Cells

Maintain CAD cells in DMEM/F12 media, 8 % fetal calf serum, and 100 U/mL penicillin/streptomycin in T-25 or T-75 tissue culture flasks. Plate and grow the CAD cells used in the experiments in either 6- or 12-well Costar tissue culture plates or 100 mm tissue culture plates. To prepare CAD cells for differentiation, grow CAD cells in serum-containing media for 24–48 h. Subsequently remove the serum-containing media, wash the cells once with phosphate-buffered saline, and treat with serum-free media for 48 h to induce differentiation. Differentiation is induced by treating cells with serum-free medium consisting of DMEM/F12, 20 μg/mL transferrin, 50 ng/mL sodium selenite, and 100 U/mL penicillin/streptomycin for 48 h (*See* **Note 1**).

To determine the stability of D1 receptor mRNA in the non-differentiated and differentiated cells, treat the cells with 1 μg/mL actinomycin D or vehicle (DMSO) control dissolved in either serum-containing (non-differentiated CAD cells) or serum-free (differentiated CAD cells) media (Fig. 3). Harvest cells immediately or after 15 min, 30 min, 1 h, 3 h, 6 h, 12 h, and 24 h. For harvesting, the cells wash with ice-cold 1× PBS twice, scrape, transfer to a prechilled centrifuge tube, and spin in a refrigerated centrifuge at 1,000 × *g* for 5 min. Remove the supernatant and lyse cell pellet in the buffer provided with RNeasy® mini kit. The lysate can be stored at −80 °C or processed immediately and total RNA isolated as described in Sect. 3.4.

3.2 Cloning and Mutagenesis of D1 Receptor 3′ UTR

Amplify the mouse D1R 3′UTR region (the 1,277 bp fragment) using specific primers and a BAC construct containing the entire mouse D1R gene. The primers include Not I and HindIII/AflII/PmlI restriction sites which facilitates the cloning of the amplified D1R 3′UTR into the pcDNAβ-gal reporter plasmid (Fig. 4a). To amplify the 1,277 bp D1 3′UTR from the BAC construct, set up a PCR reaction containing 200 ng of BAC template, 400 nM forward and reverse primers, 1× Advantage2 PCR buffer, 250 μg/mL

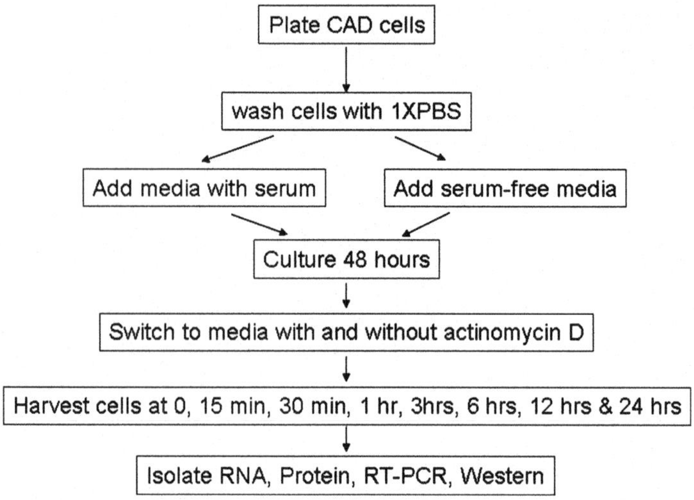

Fig. 3 The experimental design used to measure the stability of D1 receptor mRNA in non-differentiated and differentiated CAD cells. Non-differentiated and differentiated CAD cells are treated with 1 μg/mL actinomycin D and the D1 receptor mRNA and protein levels measured at the indicated time points

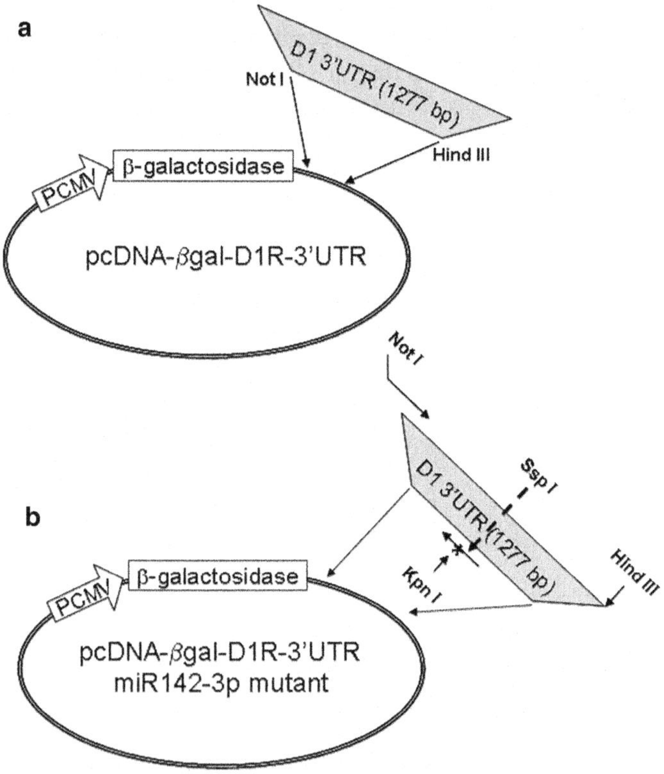

Fig. 4 Schematic representation of the β-galactosidase reporter plasmid under the control of the constitutively active Cytomegalovirus promoter (pCMV) and the wild-type (**a**) or miR142-3p binding site mutated (**b**) D1 receptor 3′UTR. The restriction enzymes used for cloning and the mutagenic primer (*asterisk*, Kpn I) are indicated

BSA, 200 μM dNTP, 0.8 μL enzyme diluent, and 1× Advantage 2 polymerase mix in a final volume of 10 μL. Perform the PCR using a Idaho Technology Light cycler in glass capillary tubes with a 5 min pre-incubation at 94 °C followed by 40 cycles of denaturation (94 °C for 3 s), annealing (57 °C for 3 s), and extension (68 °C for 60 s). At the end of 40 cycles a polishing step (72 °C for 10 min) is included to ensure that all products are full length. Run 16 reactions at one time and pool at the end of the PCR. Run the pooled reactions on a 1 % TBE (Tris-borate-EDTA)-agarose gel and excise the 1,277 bp PCR product from the gel and extract using the QIAquick gel extraction columns. Determine the concentration of PCR product and use 2 μg for restriction digests with Not I and Hind III. In parallel, digest the pcDNA-βgal plasmid vector with Not I and Hind III. Perform the restriction digestions for 2 h at 37 °C and run on a 1 % TBE-agarose gel. Excise the digested products from the gel and extract the fragment with the QIAquick gel extraction columns. Run a portion (1/5th) of the eluted products on a diagnostic 1 %TBE-agarose gel and determine the size and concentration by comparing to molecular weight ladder (*See* **Note 2**). Mix the digested plasmid vector and D1 3′UTR PCR product in a 1:5 ratio with T4 DNA ligase and ligate at 16 °C for 15 h. Use a portion (1/6th) of the ligation reaction to transform the competent *E. coli* cells. Plate the transformed competent cells on Luria broth (LB) plates with 100 μg/mL ampicillin. Next day pick individual colonies, culture in liquid LB media with 50 μg/mL ampicillin, and isolate the plasmid DNA in individual clones using the QIAprep Spin Miniprep Kit. Sequence plasmid DNA isolated from several individual bacterial colonies at commercial sequencing labs using the forward and reverse primers that were used in the PCR amplification step.

Generate the D1R 3′UTR constructs with mutations in the microRNA binding sites using a mutagenic primer with a KpnI restriction site replacing the microRNA seed recognition sequence. To mutate the miR142-3p site in D1 receptor 3′UTR, set up a PCR as described above using the D1 3′UTR as a template and the forward Not I and mutagenic reverse SspI/Kpn I primers (Fig. 4b). The PCR will generate the 5′ end of the D1 3′UTR from the Not I site to the native SspI site. To obtain the 3′end of the D1 3′UTR from the SspI to Hind III site, cut the parent plasmid with the 1,277 bp D1 3′UTR with SspI and Hind III. Ligate the 5′- and 3′- ends of the D1 3′UTR to the parent plasmid cut with Not I and Hind III to generate the D1 3′UTR with the mutated miR142-3p binding site. The ligation, transformation, and screening of plasmid from bacterial clones are performed as above. Sequence all recombinant plasmids and compare to the wild-type D1R 3′UTR sequence, confirming that the sequence matches the sequence in the NCBI database. Isolate all plasmids using the alkaline-lysis plasmid DNA isolation method and purify two sequential CsCl density gradients prior to sequencing and transfection.

3.3 Transfection of CAD Cells

Plate CAD cells and culture for 24 h or more in serum-containing media to about 60 % confluence before transfection. Perform transfections on cells plated in either 6- or 12-well tissue culture plates. For transfection of 6-well plates, for each well, dilute 6 μL of Lipofectamine 2000™ transfection reagent in 250 μL OPT-MEM and mix with 2.0 μg test plasmid, 0.4 μg of BAP-Flag™ transfection control plasmid, and 2.4 μg of pUC19 carrier plasmid in 250 μL OPTI-MEM media and incubate the combined mixture at 25 °C for 30 min. For transfection of 12-well plates, for each well, dilute 2 μL of Lipofectamine 2000™ transfection reagent in 100 μL of OPT-MEM and mix with 1.2 μg of test plasmid, 0.2 μg of BAP-Flag™ transfection control plasmid, and 0.2 μg of pUC19 carrier plasmid in 100 μL of OPTI-MEM media and incubate the combined mixture at 25 °C for 30 min. After a 30-min incubation, overlay the Lipofectamine, DNA, and OPTI-MEM mixture on non-differentiated CAD cells in antibiotic-free serum containing CAD cell culture media for 6 h. After 6 h, replace the media with fresh serum-containing or serum-free media and harvest the cells 48 h later (Fig. 5).

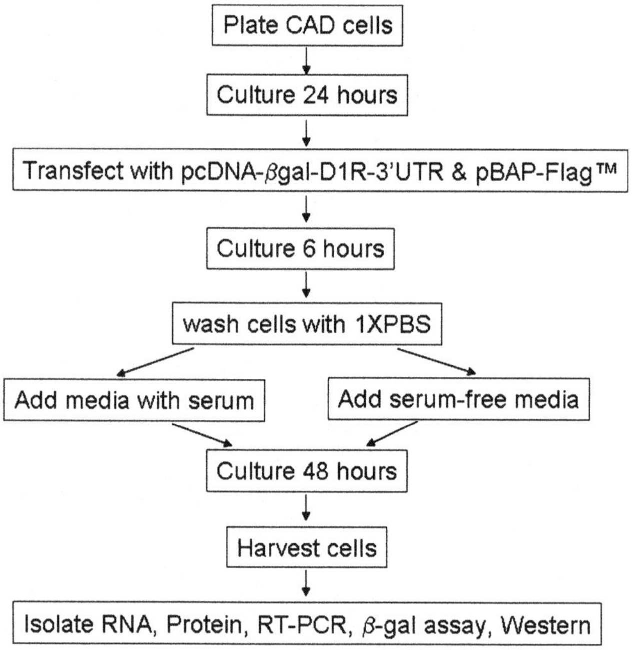

Fig. 5 The experimental design used for transfecting the β-galactosidase reporter plasmid with the wild-type and mutated D1 receptor 3′UTR into non-differentiated and differentiated CAD cells. Non-differentiated and differentiated CAD cells are harvested ~54 h after transfection and the levels of β-gal mRNA, β-gal activity, and BAP-Flag™ transfection control mRNA and protein measured

3.4 Isolation of RNA and RT-PCR

Isolate RNA using the RNeasy® Mini Kit (Qiagen) according to the manufacturers' instructions. Use TURBO DNA-free™ Kit (Invitrogen) to remove DNA contamination from the total RNA sample. Mix 10 μg total RNA, 1× TURBO DNase buffer, and 1 μL TURBO DNase and incubate at 37 °C for 45 min. Inactivate the DNase using the inactivation reagent supplied with the kit (*See* **Note 3**). Confirm the quality and integrity of the DNase-treated RNA by running 2 μg of total RNA on a 1.2 % TBE agarose gel. To set up the reverse transcriptase reaction, use 2 μg of DNase-treated RNA and 300 ng random primers. Incubate this mix at 65 °C for 5 min followed by quick chilling on ice. After the addition of 1× SuperScript III RT buffer (Invitrogen), 0.5 mM deoxynucleotide triphosphates, 10 mM dithiothreitol, 1 U RNaseOut (Ambion), and 200 U SuperScript III RT, incubate the entire mix at 25 °C for 10 min followed by 45 °C for 2 h. Heat-inactivate the reaction at 70 °C for 15 min and then immediately chill on ice (*See* **Note 4**). Perform the real-time PCR using the Roche Light Cycler (Indianapolis, IN, USA) with gene-specific TaqMan® gene expression assays (Applied Biosystems). To measure D1 dopamine receptor cDNA levels use the TaqMan® gene expression assay, Mm0135211. To detect the internal control GAPDH cDNA, use the mM99999915 TaqMan® gene expression assay. For the Light Cycler, use the following amplification parameters: Initial denaturation 95 °C for 5 min, 20 °C/s transition rate, followed by 40 cycles with a denaturation step of 95 °C for 5 s, 2 °C/s transition rate, and a combined annealing and extension step of 60 °C for 60 s, 2 °C/s transition rate with a fluorescence acquisition at the end of the 60 s extension step (*See* **Note 4**). Incorporate two negative and one positive control in each RT-PCR run. Negative control includes a PCR reaction in which water is substituted for the template. For a second negative control, PCR is run with products from a RT reaction in which the SuperScript III RT enzyme is omitted. Use mouse brain cDNA as a positive control.

3.5 Isolation of Protein and Western Blotting

Harvest cells and lyse using the CelLytic™ M reagent supplemented with 1 mM PMSF and 1 % protease inhibitor cocktail (*See* **Note 5** for protein lysate that is to be used for the β-gal assay.). Determine protein amounts in the lysates using the BCA assay. Based on the concentration, mix 20 μg to 50 μg of total cell proteins with sample loading buffer, heat at 37 °C for 10 min, load on to 10 % gels, and separate using SDS-PAGE using the Mini-PROTEAN® Tetra Cell system (*See* **Note 6**). Transfer the proteins on the gel onto nitrocellulose membranes with a Mini-PROTEAN® tank transfer system. Block the nitrocellulose membranes for 2 h at 25 °C in blocking solution (10 % nonfat milk in Tris-buffered saline (pH 7.4) with 0.1 % Tween®-20 (TBS-T)). Wash the membranes three times, 1×10 min and 2×5 min, with TBS-T between blocking and antibody incubations. First detect the D1 receptor using a rat monoclonal anti-D1R antibody (1:2,000 dilution in blocking solution in blocking solution) and goat anti-rat secondary antibody conjugated to horseradish peroxidase (1:20,000 dilution). Following the detection of D1R protein, strip the membrane with Western ReProbe™ for 30 min at 25 °C, wash three times (5 min each) with 0.1 % TBS-T, and block with 5 % nonfat milk in 0.05 % TBS-T for 1 h at 25 °C. After washing the blots with 0.05 % TBS-T, use a rabbit monoclonal antibody (1:5,000 dilution in 5 % blocking solution with 0.05 % TBS-T) to detect GAPDH. Incubate the membranes with GAPDH antibody for 1 h at 25 °C and wash three times (5 min each) with 0.05 % TBS-T. To detect the GAPDH antibody, incubate the membranes for 1 h at 25 °C with goat anti-rabbit secondary antibody conjugated to horseradish peroxidase (1:20,000 dilution in 5 % blocking solution with 0.05 % TBS-T). To detect the Flag™-tagged bacterial alkaline phosphatase protein, which is used as a transfection control, the conditions are identical to those used for the GAPDH antibody except use the M2 anti-Flag™ mouse monoclonal antibody (1:2,000 dilution) and the sheep anti-mouse HRP conjugated secondary antibody (1:20,000 dilution). Detect the horseradish peroxidase conjugated secondary antibodies with the SuperSignal® West Dura extended duration substrate chemiluminescence detection kit using an imaging system or autoradiography film.

3.6 Measurement of β-Galactosidase Reporter Activity

Following transfection of CAD cells (Fig. 5), for the β-galactosidase assay, lyse the CAD cell pellet for 10 min on ice in the β-gal lysis buffer by gently pipetting the mixture five times through a 200 μL micropipette tip. Spin the lysate at $14{,}000 \times g$ for 10 min and determine the total protein concentration in the supernatant using the Bradford protein assay per instructions provided with the kit. Detect the β-galactosidase reporter enzyme activity in the lysate using CPRG as a substrate (*See* **Note 7**). For the β-gal assay, mix 5–10 μg of total protein with 20 μL of CPRG (4 mg/mL) and bring the volume of the assay reaction to 200 μL with the β-gal lysis buffer. Incubate the mixture at 37 °C till there is a visible difference between the blank sample and the test samples. Transfer 100 μL of each sample into a 96 well plate and read the absorbance using a spectrophotometer. The β-galactosidase enzyme cleaves the CPRG substrate to yield a colored product which has maximal absorbance at 575 nm which is detected using a visible wavelength spectrophotometer. Normalize the β-galactosidase reporter enzyme activity to the total protein amount present in the lysate and further normalize for transfection efficiency.

4 Notes

1. *CAD cell culture*: Media is prepared by adding DMEM/F12 with and without fetal calf serum and penicillin/streptomycin and filtering into a sterile glass bottle via 0.22 μm bottle top filter. The prepared media is stored in the dark at 4 °C. Be extremely careful when handling sodium selenite as it is highly toxic. It is also light-sensitive. Dissolve in sterile distilled/deionized water and aliquot 1 mL of the solution (10 μg/mL) into 1.5 mL microcentrifuge tubes and store at −20 °C. Transferrin is dissolved in sterile distilled/deionized water and 1 mL aliquots of the solution (4 mg/mL) are stored at −80 °C. Actinomycin D is dissolved in dimethyl sulfoxide (DMSO) at 1 mg/mL concentration and stored at −20 °C in the dark.

   CAD cells do not adhere to the substrate very well; therefore, to prevent dislodging cells, add solutions to the walls of the culture vessels rather than directly on the cells. Non-differentiated CAD cells divide rapidly; therefore, it is necessary to sub-culture the cells every 2–3 days. Differentiated CAD cells do not divide; therefore, they can be cultured for weeks provided half the media is replaced with fresh serum-free media every 3–4 days.

2. *Cloning*: To detect the nucleic acid bands on the TBE-agarose gels, we use ethidium bromide (10 mg/mL). We typically add 2 μL ethidium bromide to 60 mL of melted gel solution for both RNA and DNA gels. In addition, for DNA gels we add 6 μL ethidium bromide to 200 mL 1× TBE running buffer. Ethidium bromide is a suspected carcinogen and should be handled with care. Used gels and buffers containing ethidium bromide should be disposed by following local regulations. Alternate nontoxic dyes can also be substituted for ethidium where available.

3. *RNA isolation*: Use RNase- and DNase-free water, tubes, and micropipette tips with filters to minimize degradation and contamination. Use RNase-Zap to decontaminate micropipettes and centrifuge rotors and chamber. Always wear clean gloves and change them frequently. RNA pellets are difficult to dissolve when they are overdried. To dissolve RNA pellets, add RNase-free water and heat at 60 °C with intermittent vortexing. Do not pipette up and down to dissolve the pellet as the sticky pellet might get lost in the pipette tip. Store RNA at −80 °C.

4. *RT-PCR*: The cDNA generated in the RT reaction can be stored at −20 °C. We perform the PCR using the Roche Light Cycler and glass capillaries; given the fast transition times with this PCR machine, it will be necessary to alter the amplification parameters for other PCR machines, in particular, the Peltier-based PCR machines. Fortunately, the TaqMan® gene expression assay

probes are optimized for Peltier-based machines and the optimal amplification parameters can be obtained from Invitrogen.

5. *Protein isolation*: Keep all tubes on ice and ensure that they are prechilled. All centrifugation should be at 4 °C. For D1 receptor, BAP-Flag™, and GAPDH the protein isolation can be carried out using the Celytic M™ reagent; however, for the βgal assay it is important to note that the cell lysis has to be performed in the βgal lysis buffer. The cell pellet is resuspended in the lysis buffer and processed as in Sect. 3.6.
6. *Protein sample preparation for WB*: To detect membrane proteins such as the D1 dopamine receptor, the protein lysate should NOT be boiled before loading on the gel as this will aggregate glycosylated membrane proteins. We recommend adding TCEP bond breaker to the sample buffer and protein lysate and heating at 37 °C for 10 min before loading the gel. If necessary, samples can be heated to 50 °C for 5 min.
7. *β-galactosidase assay*: CPRG is dissolved in water and stored as aliquots at −20 °C in the dark. When setting up the assay, add the CPRG substrate at the very end to the reaction tube that already contains the protein lysate and β-gal lysis buffer. The reaction that leads to formation of the colored substrate cannot be terminated; therefore it is important to monitor the reaction and take the absorbance readings as soon as the test sample is different from the blank sample. If the color changes very rapidly, reduce the protein amount added to the lysate and/or perform the incubation at 25 °C rather than at 37 °C.

Acknowledgments

The protocols described here were developed with the help of Jennifer Pasuit, Dr. Denis Chang, and Dr. Thuy Do. Funding was provided by the F.M. Kirby Foundation, the UMDNJ Foundation, and NIH grant (DA0260300)

References

1. Yoo S, van Niekerk EA, Merianda TT, Twiss JL (2010) Dynamics of axonal mRNA transport and implications for peripheral nerve regeneration. Exp Neurol 223(1):19–27
2. Willis DE, van Niekerk EA, Sasaki Y, Mesngon M, Merianda TT, Williams GG, Kendall M, Smith DS, Bassell GJ, Twiss JL (2007) Extracellular stimuli specifically regulate localized levels of individual neuronal mRNAs. J Cell Biol 178(6):965–980
3. Schambra UB, Duncan GE, Breese GR, Fornaretto MG, Caron MG et al (1994) Ontogeny of D1A and D2 dopamine receptor subtypes in rat brain using in situ hybridization and receptor binding. Neuroscience 62(1):65–85
4. Jung AB, Bennett JP (1996) Development of striatal dopaminergic function. I. Pre- and postnatal development of mRNAs and binding sites for striatal D1 (D1a) and D2 (D2a) receptors. Brain Res Dev Brain Res 94(2):109–120
5. Tobón KE, Chang D, Kuzhikandathil EV (2012) MicroRNA 142-3p mediates post-transcriptional regulation of D1 dopamine receptor expression. PLoS One 7(11):e49288

6. Brana C, Caille I, Pellevoisin C, Charron G, Aubert I et al (1996) Ontogeny of the striatal neurons expressing the D1 dopamine receptor in humans. J Comp Neurol 370(1): 23–34
7. Pasuit JB, Li Z, Kuzhikandathil EV (2004) Multimodal regulation of endogenous D1 dopamine receptor expression and function in the CAD catecholaminergic cell line. J Neurochem 89(6): 1508–1519
8. Do T, Kerr B, Kuzhikandathil EV (2007) Brain-derived neurotrophic factor regulates the expression of D1 dopamine receptors. J Neurochem 100(2):416–428

Part II

Computational and Biochemical Methods in the Investigation of Dopamine Receptor Structure, Binding, and Post-translational Regulation

Chapter 3

Computational Approaches in the Structure–Function Studies of Dopamine Receptors

Mayako Michino and Lei Shi

Abstract

In studying the structure–function relationship of dopamine receptors (DARs), computational approaches have played increasingly important roles in integrating the experimental findings and start to lead the discovery process. The sequence conservation among DARs and its homologous receptors provides a framework to deduce the overall topology, to identify the functionally important structural motifs, and to allow the generalization of the findings from other homologous receptors to DARs. The availability of high-resolution structural information of close homologs and dopamine D3 receptor in recent years, in combination with the development of computational algorithms, has promoted detailed characterizations that revealed the structural basis of the subtype-selectivity and efficacy, and led to identification of novel ligands by structure-based virtual screening. Taken together, the accumulated understanding of structure–function relationship of DARs establishes the basis for the structure-based rational drug and ligand discovery for these receptors.

Key words Sequence analysis, Homology modeling, Molecular dynamics, Subtype-selectivity, Virtual screening

1 Introduction

Synthesized in dopaminergic neurons, dopamine is released to stimulate D1-like (D1, D5) and D2-like (D2, D3, D4) dopamine receptors, which are members of the rhodopsin-like G-protein coupled receptor (GPCR) family. While it is generally thought that antagonism of the D2 receptor (D2R) is essential for therapeutic efficacy of all antipsychotic drugs, these agents block D2-like receptors non-selectively, and it remains unclear which downstream signaling processes must be specifically blocked, and in which brain region(s). A significant difficulty in determining the downstream signaling processes and mechanisms of the DARs stems from a lack of appropriate pharmacological tools to selectively target the different receptors and identify relevant effectors [1]. That D3R expression is elevated in response to drugs of abuse, e.g., in the

Mario Tiberi (ed.), *Dopamine Receptor Technologies*, Neuromethods, vol. 96,
DOI 10.1007/978-1-4939-2196-6_3,

brains of cocaine-associated fatalities [2], has prompted efforts towards the development of selective D3R antagonists for the treatment of drug addiction [3]. In addition, expression and imaging studies show that D3Rs are not expressed in the dorsal striatum [4], a region rich in D2R, suggesting that D3R-selective antagonists may be less prone to causing motor side effects that can result from D2R blockade [3]. Growing preclinical evidence indicates that selective D3R antagonists regulate the motivation to self-administer drugs and disrupt drug-associated cue-induced craving. Notably, several D3R-selective agents have been evaluated in animal models of addiction and other neuropsychiatric disorders [3, 5], and clinical trials have been initiated for substance abuse as well as for schizophrenia with several agents targeting D3R [6].

Thus, understanding ligand specificity is one of most critical issues in the studies of structure–function relationship of DARs. Computational approaches have played increasingly important roles in complementing experimental findings, and providing more detailed characterization of the ligand–receptor interactions and the corresponding receptor conformations. Common procedures of molecular modeling and simulation methods are summarized in Appendix 1. However, depending on the particular questions being addressed, the procedures require adjustments according to considerations taking into account of the features that are either shared among other homologous GPCRs or unique to a specific DAR. In this chapter, we review the highlights of computational efforts from the time when there was very little high-resolution structural information of GPCR to the most recent work based on the crystal structure of D3R [7].

2 Methods

2.1 Sequence Alignment and Bioinformatics Analysis

Before the high-resolution structural information of rhodopsin-like GPCRs was available, cryomicroscopy studies of rhodopsin that indicated the existence of seven transmembrane segments gave initial clues of the relative disposition of these segments [8]. Based on this low resolution information and sequence alignments, inferences of residues conservation and physicochemical properties revealed important structure–function insights for rhodopsin-like GPCRs, regarding the binding site residues and critical functional motifs [9, 10]. These insights provided a framework to interpret indirect structural measurements on DARs obtained from mutagenesis experiments, and especially with the substituted-cysteine accessibility method (SCAM) [11–15].

In particular, although the extracellular portion of transmembrane segment 4 (TM4) in homologous amine GPCRs shows high degree of sequence variation, which is more characteristic of a loop domain, a more in-depth sequence analysis indicated this region appeared to be rather conserved within functionally related receptors as well as within species variants of the same receptor subtype [15]. Thus, based on the SCAM results, it was deduced that the extracellular region 4.59–4.62 (Ballesteros and Weinstein indexing system [9], *see* **Note 1**) may be in an α-helical conformation, even though a proline residue, which is more often observed at the end of an α-helix, is located at position 4.59 [15]. Such an insight was later corroborated by the crystal structures of D3R [7] and βARs [16, 17].

Comparative sequence analysis of the conservation of residues has also revealed the divergence of the tilting of transmembrane segments. For example, in the extracellular portion of TM1, it was found that opsins show a pattern consistent with α-helical structure with a conserved face. In contrast, this region in catecholamine receptors is poorly conserved, suggesting a lack of critical contacts. Thus, in catecholamine receptors in the absence of $Pro^{1.48}$, TM1 may be straighter and therefore further from the helix bundle, consistent with the apparent lack of conserved contact residues [18]. Such a divergence between rhodopsin and DARs would have an impact on the tilting of TM1 in the dimer interfaces and was later experimentally validated [19].

2.2 Ligand-Binding Mode Prediction Based on Homology Models

Prior to the determination of the high-resolution crystal structure of the D3R in 2010, homology models of DARs, built based on the template structures of rhodopsin or β_2-adrenergic receptor (β_2AR), were extensively used to understand the receptor–ligand recognition mechanisms for DARs. Ligands were commonly docked to homology models, and then from the predicted ligand binding modes, the key receptor–ligand interactions were characterized and pharmacophore models were generated [20–23].

Specifically, Hobrath and Wang constructed homology models of the D3R based on the 11-*cis*-retinal-bound rhodopsin structure (PDB ID: 1F88), and identified 13 ligand-interacting residues, all of which were later confirmed to be ligand-binding in the eticlopride-bound D3R structure [7] (PDB ID: 3PBL) [20, 21]. Ehrlich et al. constructed homology models for D2-like DARs based on the carazolol-bound β_2AR structure (PDB ID: 2RH1), and identified $Asp^{3.32}$ and $His^{6.55}$ as key residues interacting with the phenylpiperazine moiety of the ligand, and six positions in TM2, TM3, and extracellular loop 2 (ECL2) to be involved in subtype receptor-specific recognition [22].

The successes in the use of homology models were further underscored by the results of the community-wide competition of protein structure modeling and ligand docking for the D3R, the so-called GPCRDock 2010 [24]. This competition was held before the crystallographic coordinates of D3R/eticlopride structure were published, to evaluate how well various GPCR modeling and docking methods performed in identifying key ligand–receptor interactions, by comparing these predictions against the crystal structure. The best models approached the accuracy of the crystallographically determined details, especially in revealing those in the conserved orthosteric binding site (OBS), *see* **Note 2**. Twenty-three out of 117 submitted models correctly predicted the ligand position with their ligand RMSDs <2.5 Å; the best model had a ligand RMSD of only 0.96 Å [24]. Most models were predicted by building a homology model of the receptor based on the structure of β_2AR (PDB: 2RH1), and docking the ligand into the OBS. The top-ranking models were built by modeling methods that made use of additional steps such as refinement of the homology models, pharmacophore-guided ligand docking (e.g., taking into account of the characteristic interaction between the protonated amine of the ligand with Asp$^{3.32}$), and rescoring of poses based on consensus-scoring function [24]. The prediction of the best model (from Selent group) also used a criteria to select for poses with a quasi-planar conformation of eticlopride that is in agreement with experimental structures of the ligand, possessing two intramolecular hydrogen bonds [25]. Another systematic assessment of cognate- and cross-docking to GPCR structures and homology models showed that for D3R, accurate pose prediction can be achieved if the homology models are built based on highly homologous template structures of β_2AR or β_1AR (~ 35 % sequence identity in the TM region and < 1.3 Å Cα-RMSD for binding site residues, with respect to those of D3R) [26].

In addition to predicting the binding mode, homology models of DARs have been used to study the binding of sodium ion (Na^+) to the D2R [27, 28]. The binding property of some classes of ligands is known to be sensitive to Na^+ in dopamine D2-like receptors [29, 30]. Weinstein and colleagues modeled the Na^+-induced conformations of D2R using homology models built based on rhodopsin and β_2AR, and carried out normal mode analysis to rationalize the role of Na^+ binding in the enhanced affinities of substituted benzamides and 1,4-disubstituted piperidines/piperazines (1,4-DAPs) [28]. Selent and colleagues modeled the entry path and allosteric effects of Na^+ in D2R by carrying out a microsecond scale all-atom unconstrained molecular dynamics (MD) simulations (*see* **Note 3**) on a D2R homology model, which was built based on the structure of β_2AR (PDB: 2RH1), and then inserted into a hydrated and equilibrated POPC lipid bilayer membrane. The overall system was stable during the >1 μs simulation (the RMSD of Cα atoms in TM region is <2 Å). The results of the simulations showed that Na^+ enters the receptor from the extracellular side, and traverses a two-step pathway to bind at the energetically stable site near Asp$^{2.50}$, to affect rotamer toggle switch Trp$^{6.48}$ [27].

2.3 Ligand-Binding Mode Prediction Based on the D3R Crystal Structure

The high-resolution structural information provided by the crystal structure of D3R [7] allows more detailed characterizations of ligand–receptor recognition mechanisms, such as the structural basis for the subtype-selectivity and efficacy. To understand the molecular determinants of D3R over D2R selectivity as well as efficacy, we built and refined D3R and D2R models in both inactive and active states. In addition to the D3R structure stabilized in an inactive conformation by eticlopride (a D3R/D2R-selective inverse agonist), we used the differences in the inactive and active structures of β_1AR and β_2AR [16, 31, 32] to guide the construction of a D3R model with the OBS in the active conformation. For comparative investigations of D3R vs. D2R, we built and refined homology models of D2R based on the D3R structure as the template [7, 33]. Using these models, the results of our molecular docking and dynamics studies suggested that the primary pharmacophore (PP) of D3R-selective compounds is bound in the OBS, while the arylamide secondary pharmacophore (SP) is accommodated in a secondary binding pocket (SBP) formed by residues from TMs1, 2, 3, 7 and divergent ECL1 and ECL2 [7, 33] (*see* **Note 4**). We also modeled the binding mode of "synthons," which are deconstructed pharmacophoric elements of full-length D3R-selective compounds (R-22 and its analogs), and found that efficacy of compounds depends on the binding mode of the PP in the OBS [33].

We further investigated the conformational dynamics of the SBP and its role in subtype-selectivity in D3R and D2R by carrying out MD simulations with enhanced-sampling technique, namely replica exchange MD (REMD) [34]. REMD enhances conformational sampling compared to conventional MD by simulating multiple copies (replicas) of the same system at different temperatures and allowing exchanges of replicas between neighboring temperatures according to the Metropolis criterion [35]. This advanced sampling method has been applied to studying protein folding mechanisms [36, 37] and predicting protein–ligand interactions [38]. Thus, in addition to regular 310 K constant-temperature MD, our REMD simulations, which were started from representative docked poses that positioned the SP in different subpockets within the SBP, demonstrated that a converged equilibrium distribution of poses can be obtained from different initial configurations.

Based on simulation results of the D3R and D2R models complexed with D3R-selective 4-phenylpiperazine compounds, we characterized the differences in the size and shape of the SBP in D3R vs D2R, and found that the full-length D3R-selective compounds induces a change in the Prokink wobble angles [39] (*see* **Note 5**) at the conserved $Pro^{2.59}$ in D3R. Such changes correspond to an outward rearrangement of the extracellular segment of TM2, and were not observed in D2R simulations.

Our computational modeling and simulations, in combination with receptor chimera and site-directed mutagenesis experiments, suggested that the non-conserved ECL1 could influence the orientation of conserved TM2, and render different shapes and sizes of the SBP in D3R vs D2R, thereby allowing the SP of the D3R-selective compound to be optimally accommodated by a sub-pocket within the SBP in D3R [34].

2.4 Virtual Screening for the Ligands of DARs

Early efforts in the structure-based virtual high-throughput screening (VHTS) of large libraries of compounds for DARs depended on rhodopsin-based homology models or de novo models, yet showed remarkable results in finding novel active compounds [21, 40]. Varady et al. screened ~250,000 compounds from the NCI 3D database by a step-wise pharmacophore and structure-based approach using multiple homology models of D3R, and identified 4 (out of 20 tested) novel compounds with $K_i < 100$ nM in a D3R binding assay [21]. Becker et al. screened ~120,000 compounds against a de novo model of D2R generated by the PREDICT structure prediction algorithm, and identified 7 (out of 42 tested) compounds with $K_i < 5$ μM [40]. Levoin and colleagues recently demonstrated that the binding site in their "historical" rhodopsin-based D3R homology model was as well suited for lead optimization as the crystal structure of the D3R, which explained the successful utilization of their model in the drug discovery process [41].

Applying VHTS to an ensemble of receptor structures is a promising approach to increase diversity among identified hits and to incorporate protein flexibility into the exploration of optimal ligand–receptor interactions [38]. More recently, VHTS for DARs have been carried out using β_2AR-based homology models and the D3R crystal structure [42]. Carlsson et al. screened over three million "lead-like" molecules from the ZINC library (*see* **Note 6**) and compared the results from screening against a β_2AR-based D3R homology model vs. the D3R crystal structure. Both trials identified hits with μM affinities, and with comparable hit rates of ~20 %. Interestingly, slightly different scaffolds were enriched in the homology model screen compared to the crystal structure screen, due to a more open and larger OBS in the homology model. In the homology model screen, the phenylpiperazine chemotype characteristic of many known dopaminergic ligands was enriched, while in the crystal structure screen, the aryl-amide-aminergic chemotype characteristic of eticlopride was enriched. In terms of efficacy and selectivity properties, however, all of the novel hits identified in this study were found to be antagonists, and showed little D3R/D2R-selectivity [42].

With the general success in the structure-based approach to identifying novel hits, the goal of most recent VHTS for DARs has focused on identifying hits with predefined efficacy and selectivity profiles. Lane et al. has screened ~4 million drug-like compounds using optimized models of D3R generated from the inactive-state structure of D3R, against the OBS and an allosteric binding pocket outside of the OBS [43]. All of the hit compounds identified were antagonists and negative allosteric modulators. The VHTS campaigns for another aminergic GPCR, β_2AR, have investigated the effect of binding site conformation on virtual screening results using the inactive-state and the active-state crystal structures [44, 45]. In these studies, the Shoichet group found that the efficacy of the hits identified by VHTS recapitulates the efficacy of the co-crystallized ligand in the crystal structure used in the screening, despite the relatively small changes in the OBS between the inactive and active state structures of β_2AR. Thus the VHTS using the agonist-bound β_2AR structure only returned potent agonists, while the VHTS using the inverse-agonist-bound β_2AR structure only returned potent inverse agonists [45]. The identification of agonists and biased agonists by VHTS for DARs will likely follow when the active-state structure of a DAR becomes available.

3 Notes

1. Ballesteros and Weinstein indexing system is a residue numbering scheme widely used for the class A GPCRs. The residues in TM helices are assigned generalized index numbers associated with the position of the residues relative to the most conserved residues in each TM helix. These most conserved residues are denoted as x.50, where x is the TM helix number.
2. Orthosteric Binding Site (OBS) is the primary binding site in the receptor where the endogenous agonist binds. In contrast, Secondary Binding Pocket (SBP) is a binding pocket in the receptor that is topographically distinct from the OBS [46].
3. Molecular Dynamics (MD) simulation is a computational technique that simulates the interactions and movements of the atoms in a molecular system according to Newtonian mechanics [47].
4. Primary Pharmacophore (PP) is the moiety within a ligand that binds within the OBS. Secondary Pharmacophore (SP) refers to a ligand moiety extended from PP that binds a secondary binding pocket (SBP).
5. The local conformational flexibility at a proline residue in a proline-containing helix can be quantified by three measures: the bend angle, the wobble angle, and the face shift.
6. The ZINC library is a free public database of over 20 million commercially available molecules that are downloadable in ready-to-dock formats (i.e., with biologically relevant protonation states, and partial charges) and are filtered by physicochemical properties into subsets, such as the “lead-like” and “fragment-like” sets [48].

4 Conclusion

Based on the wealth of information from previous findings regarding the overall topology and the functionally important motifs, recent computational studies that take advantage of the high-resolution structures of close homologs and D3R have revealed the structural basis of the subtype-selectivity and efficacy, and identified novel DAR ligands. However, new questions, such as the structural basis of tailed and/or biased efficacy, arise [49]. The answers to these questions would further our understanding of the signaling and pharmacological properties and lead to more effective and selective therapeutic interventions against drug addiction and other neuropsychiatric disorders.

Appendix 1: Common Procedures in the Molecular Modeling and Simulations of DARs

Homology Modeling and Refinement

In the absence of an experimentally determined high-resolution structure, homology models can be built to generate hypothesis about the structure–function relationships of DARs.

- Select the template structure (preferably with sequence identity >~35–40 % to the target) and align the sequence of DAR being modeled to the sequence of the template;
- Generate initial homology models using the template structure and the sequence alignment;
- Refine the initial homology models to optimize the sidechain and loop conformations;
- Validate models against the restraints derived from indirect structural determinations, such as residue–residue distances and residue accessibilities.

Commonly used software: Modeler [50], MOE (Chemical Computing Group, Inc.), Prime (Schrödinger, Inc.), and SCWRL [51].

Ligand Docking and Virtual Screening

Docking can be used to predict the binding pose of a ligand in the receptor, and is used in virtual high-throughput screening (VHTS) to search a large library of compounds (e.g., ZINC [48]) for novel ligands.

Pose prediction:

- Prepare and dock ligands into defined binding site(s), usually the OBS;
- Select poses among the top-scoring poses based on agreement with experimental data, e.g., mutagenesis data, SAR, ligand structure.

Virtual screening:

- Screen against one or more receptor conformations and binding pockets—receptor conformations can be snapshots obtained from MD simulations [38], and selected by its ability to enrich known ligands in small-scale virtual screen [52];
- Rank the ligands with docking score, which represents the approximate binding energy to distinguish binders from non-binders. Consensus-scoring approach using multiple-scoring functions can improve predictions over any one function by compensating for deficiencies of each function [25].

Commonly used software: DOCK [53], GOLD [54], Glide (Schrödinger, Inc.), and AutoDock [55].

Molecular Dynamics (MD) Simulations

MD simulations can be performed to relax homology models and evaluate the stability of predicted ligand poses, as well as to capture functionally relevant motions of the receptor–ligand complex in the lipid bilayer environment (e.g., conformational changes involved in the activation of receptor in response to ligand-binding) with temporal and spatial resolution.

- Set up the simulation system consisting of the receptor–ligand complex with the explicit water–lipid bilayer–water solvent;
- Equilibrate the system with decreasing restraints on the backbone and Cα atoms of receptor;
- Perform MD at constant temperature and pressure, typically at 310 K and 1 atm;
- Analyze the simulation trajectory to identify functionally relevant changes in conformations.

Commonly used software: NAMD [56], Desmond (D.E. Shaw Research) [47], CHARMM [57], AMBER [58], and Gromacs [59]. MD can be performed with enhanced conformational sampling techniques, such as replica exchange MD (REMD) [34] or accelerated MD (aMD) [60], to study large-scale conformational changes and achieve better convergence.

References

1. Beaulieu JM, Gainetdinov RR (2011) The physiology, signaling, and pharmacology of dopamine receptors. Pharmacol Rev 63(1):182–217
2. Staley JK, Mash DC (1996) Adaptive increase in D3 dopamine receptors in the brain reward circuits of human cocaine fatalities. J Neurosci 16(19):6100–6106
3. Heidbreder CA, Newman AH (2010) Current perspectives on selective dopamine D(3) receptor antagonists as pharmacotherapeutics for addictions and related disorders. Ann N Y Acad Sci 1187:4–34
4. Choi JK, Mandeville JB, Chen YI et al (2010) Imaging brain regional and cortical laminar effects of selective D3 agonists and antagonists. Psychopharmacology (Berl) 212(1):59–72
5. Micheli F (2011) Recent advances in the development of dopamine D3 receptor

antagonists: a medicinal chemistry perspective. ChemMedChem 6(7):1152–1162

6. Lober S, Hubner H, Tschammer N et al (2011) Recent advances in the search for D(3)- and D(4)-selective drugs: probes, models and candidates. Trends Pharmacol Sci 32(3):148–157
7. Chien EY, Liu W, Zhao Q et al (2010) Structure of the human dopamine D3 receptor in complex with a D2/D3 selective antagonist. Science 330(6007):1091–1095
8. Unger VM, Hargrave PA, Baldwin JM et al (1997) Arrangement of rhodopsin transmembrane alpha-helices. Nature 389(6647): 203–206
9. Ballesteros J, Weinstein H (1995) Integrated methods for the construction of three-dimensional models of structure-function relations in G protein-coupled receptors. Meth Neurosci 25:366–428
10. Baldwin JM, Schertler GF, Unger VM (1997) An alpha-carbon template for the transmembrane helices in the rhodopsin family of G-protein-coupled receptors. J Mol Biol 272(1):144–164
11. Fu D, Ballesteros JA, Weinstein H et al (1996) Residues in the seventh membrane-spanning segment of the dopamine D2 receptor accessible in the binding-site crevice. Biochemistry 35(35):11278–11285
12. Javitch JA, Ballesteros JA, Weinstein H et al (1998) A cluster of aromatic residues in the sixth membrane-spanning segment of the dopamine D2 receptor is accessible in the binding-site crevice. Biochemistry 37(4): 998–1006
13. Javitch JA, Ballesteros JA, Chen J et al (1999) Electrostatic and aromatic microdomains within the binding-site crevice of the D2 receptor: contributions of the second membrane-spanning segment. Biochemistry 38:7961–7968
14. Simpson MM, Ballesteros JA, Chiappa V et al (1999) Dopamine D4/D2 receptor selectivity is determined by A divergent aromatic microdomain contained within the second, third, and seventh membrane-spanning segments. Mol Pharmacol 56(6):1116–1126
15. Javitch JA, Shi L, Simpson MM et al (2000) The fourth transmembrane segment of the dopamine D2 receptor: accessibility in the binding-site crevice and position in the transmembrane bundle. Biochemistry 39(40): 12190–12199
16. Cherezov V, Rosenbaum DM, Hanson MA et al (2007) High-resolution crystal structure of an engineered human beta2-adrenergic G protein-coupled receptor. Science 318(5854):1258–1265
17. Warne T, Serrano-Vega MJ, Baker JG et al (2008) Structure of a beta1-adrenergic G-protein-coupled receptor. Nature 454(7203):486–491
18. Shi L, Simpson MM, Ballesteros JA et al (2001) The first transmembrane segment of the dopamine D2 receptor: accessibility in the binding-site crevice and position in the transmembrane bundle. Biochemistry 40(41): 12339–12348
19. Guo W, Urizar E, Kralikova M et al (2008) Dopamine D2 receptors form higher order oligomers at physiological expression levels. EMBO J 27(17):2293–2304
20. Hobrath JV, Wang S (2006) Computational elucidation of the structural basis of ligand binding to the dopamine 3 receptor through docking and homology modeling. J Med Chem 49(15):4470–4476
21. Varady J, Wu X, Fang X et al (2003) Molecular modeling of the three-dimensional structure of dopamine 3 (D3) subtype receptor: discovery of novel and potent D3 ligands through a hybrid pharmacophore- and structure-based database searching approach. J Med Chem 46(21):4377–4392
22. Ehrlich K, Gotz A, Bollinger S et al (2009) Dopamine D2, D3, and D4 selective phenylpiperazines as molecular probes to explore the origins of subtype specific receptor binding. J Med Chem 52(15):4923–4935
23. Sanders MP, Verhoeven S, de Graaf C et al (2011) Snooker: a structure-based pharmacophore generation tool applied to class A GPCRs. J Chem Inf Model 51(9):2277–2292
24. Kufareva I, Rueda M, Katritch V et al (2011) Status of GPCR modeling and docking as reflected by community-wide GPCR Dock 2010 assessment. Structure 19(8):1108–1126
25. Obiol-Pardo C, Lopez L, Pastor M et al (2011) Progress in the structural prediction of G protein-coupled receptors: D3 receptor in complex with eticlopride. Proteins 79(6):1695–1703
26. Beuming T, Sherman W (2012) Current assessment of docking into GPCR crystal structures and homology models: successes, challenges, and guidelines. J Chem Inf Model 52(12):3263–3277
27. Selent J, Sanz F, Pastor M et al. (2010) Induced effects of sodium ions on dopaminergic G-protein coupled receptors. PLoS Comput Biol 6(8):e1000884
28. Ericksen SS, Cummings DF, Weinstein H et al (2009) Ligand selectivity of D2 dopamine

receptors is modulated by changes in local dynamics produced by sodium binding. J Pharmacol Exp Ther 328(1):40–54
29. Neve KA (1991) Regulation of dopamine D2 receptors by sodium and pH. Mol Pharmacol 39(4):570–578
30. Schetz JA, Sibley DR (2001) The binding-site crevice of the D4 dopamine receptor is coupled to three distinct sites of allosteric modulation. J Pharmacol Exp Ther 296(2):359–363
31. Rasmussen SG, DeVree BT, Zou Y et al (2011) Crystal structure of the beta2 adrenergic receptor-Gs protein complex. Nature 477(7366):549–555
32. Warne T, Moukhametzianov R, Baker JG et al (2011) The structural basis for agonist and partial agonist action on a beta(1)-adrenergic receptor. Nature 469(7329):241–244
33. Newman AH, Beuming T, Banala AK et al (2012) Molecular determinants of selectivity and efficacy at the dopamine D3 receptor. J Med Chem 55(15):6689–6699
34. Michino M, Donthamsetti P, Beuming T et al (2013) A single glycine in extracellular loop 1 is the critical determinant for pharmacological specificity of dopamine D2 and D3 receptors. Mol Pharmacol 84(6):854–864
35. Sugita Y, Okamoto Y (1999) Replica-exchange molecular dynamics method for protein folding. Chem Phys Lett 314:141–151
36. Felts AK, Harano Y, Gallicchio E et al (2004) Free energy surfaces of beta-hairpin and alpha-helical peptides generated by replica exchange molecular dynamics with the AGBNP implicit solvent model. Proteins 56(2):310–321
37. Zhou R, Berne BJ, Germain R (2001) The free energy landscape for beta hairpin folding in explicit water. Proc Natl Acad Sci U S A 98(26):14931–14936
38. Osguthorpe DJ, Sherman W, Hagler AT (2012) Exploring protein flexibility: incorporating structural ensembles from crystal structures and simulation into virtual screening protocols. J Phys Chem B 116(23):6952–6959
39. Visiers I, Braunheim BB, Weinstein H (2000) Prokink: a protocol for numerical evaluation of helix distortions by proline. Protein Eng 13(9):603–606
40. Becker OM, Marantz Y, Shacham S et al (2004) G protein-coupled receptors: in silico drug discovery in 3D. Proc Natl Acad Sci U S A 101(31):11304–11309
41. Levoin N, Calmels T, Krief S et al (2011) Homology model versus X-ray structure in receptor-based drug design: a retrospective analysis with the dopamine D3 receptor. ACS Med Chem Lett 2(4):293–297
42. Carlsson J, Coleman RG, Setola V et al (2011) Ligand discovery from a dopamine D3 receptor homology model and crystal structure. Nat Chem Biol 7(11):769–778
43. Lane JR, Chubukov P, Liu W et al (2013) Structure-based ligand discovery targeting orthosteric and allosteric pockets of dopamine receptors. Mol Pharmacol 84(6):794–807
44. Kolb P, Rosenbaum DM, Irwin JJ et al (2009) Structure-based discovery of beta2-adrenergic receptor ligands. Proc Natl Acad Sci U S A 106(16):6843–6848
45. Weiss DR, Ahn S, Sassano MF et al (2013) Conformation guides molecular efficacy in docking screens of activated beta-2 adrenergic G protein coupled receptor. ACS Chem Biol 8(5):1018–1026
46. May LT, Leach K, Sexton PM et al (2007) Allosteric modulation of G protein-coupled receptors. Annu Rev Pharmacol Toxicol 47:1–51
47. Shaw DE, Maragakis P, Lindorff-Larsen K et al (2010) Atomic-level characterization of the structural dynamics of proteins. Science 330(6002):341–346
48. Irwin JJ, Sterling T, Mysinger MM et al (2012) ZINC: a free tool to discover chemistry for biology. J Chem Inf Model 52(7):1757–1768
49. Allen JA, Yost JM, Setola V et al (2011) Discovery of beta-arrestin-biased dopamine D2 ligands for probing signal transduction pathways essential for antipsychotic efficacy. Proc Natl Acad Sci U S A 108(45):18488–18493
50. Eswar N, Webb B, Marti-Renom MA et al. (2006) Comparative protein structure modeling using modeller. In: Curr Protoc Bioinformatics Suppl 15, Wiley, pp 5.6.1–5.6.30
51. Krivov GG, Shapovalov MV, Dunbrack RL Jr (2009) Improved prediction of protein side-chain conformations with SCWRL4. Proteins 77(4):778–795
52. Cavasotto CN, Orry AJ, Murgolo NJ et al (2008) Discovery of novel chemotypes to a G-protein-coupled receptor through ligand-steered homology modeling and structure-based virtual screening. J Med Chem 51(3):581–588
53. Kuntz ID, Blaney JM, Oatley SJ et al (1982) A geometric approach to macromolecule-ligand interactions. J Mol Biol 161(2):269–288
54. Verdonk ML, Cole JC, Hartshorn MJ et al (2003) Improved protein-ligand docking using GOLD. Proteins 52(4):609–623
55. Morris GM, Huey R, Lindstrom W et al (2009) AutoDock4 and AutoDockTools4: automated docking with selective receptor flexibility. J Comput Chem 30(16): 2785–2791

56. Phillips JC, Braun R, Wang W et al (2005) Scalable molecular dynamics with NAMD. J Comput Chem 26(16):1781–1802
57. Brooks BR, Brooks CL 3rd, Mackerell AD Jr et al (2009) CHARMM: the biomolecular simulation program. J Comput Chem 30(10):1545–1614
58. Case DA, Cheatham TE 3rd, Darden T et al (2005) The Amber biomolecular simulation programs. J Comput Chem 26(16):1668–1688
59. Pronk S, Pall S, Schulz R et al (2013) GROMACS 4.5: a high-throughput and highly parallel open source molecular simulation toolkit. Bioinformatics 29(7): 845–854
60. Tikhonova IG, Selvam B, Ivetac A et al (2013) Simulations of biased agonists in the beta(2) adrenergic receptor with accelerated molecular dynamics. Biochemistry 52(33): 5593–5603

Chapter 4

Cell-Free Protein Synthesis and Purification of the Dopamine D2 Receptor

Dipannita Basu, Ritesh Daya, Christal D.R. Sookram, and Ram K. Mishra

Abstract

The dopamine D2 receptor is considered one of the most important neurotransmitter receptors relevant to behavioral and clinical effects of antipsychotic drugs. Its expression and purification however is met with several challenges. This chapter provides a detailed methodology on the cell-free synthesis of the dopamine $D2_L$ receptor, using *Escherichia coli (E. coli)* lysate in a regenerative dialysis membrane system. This cell-free technique utilizes protein synthesis machinery and exogenous dopamine $D2_L$ DNA to synthesize functional protein outside of intact cells. The cell-free system offers various advantages specifically for the expression of transmembrane proteins, like G-protein-coupled receptors, which typically present a significant challenge. Transmembrane protein synthesis via more conventional approaches exhibit a number of innate limitations including protein aggregation, misfolding, and low yield due to cellular toxicity. The cell-free protein synthesis systems allow for the continuous replenishment of depleting precursors and removal of toxic buildup through a size-regulated porous dialysis membrane. As such this system facilitates higher yields of G-protein-coupled receptors when compared to conventional cell-based methods. Furthermore, this method provides the capability to modify the protein product, as it can be designed to incorporate radiolabeled isotopes, unnatural amino acids, solubilizing agents, cofactors, and inhibitors as is relevant for more innovative and specific research questions. Finally, an optimized cell-free system can synthesize high levels of this G-protein-coupled receptor within a few hours of incubation, providing an efficient solution to the challenge of characterizing the dopamine D2 receptor.

Key words Cell-free protein synthesis, *E. coli*, Dopamine D2 receptor, GPCR, Transmembrane protein

Abbreviations

AMPA and NMDA	Ionotropic glutamate receptors
ATP	Adenosine triphosphate
cAMP	Cyclic AMP
cDNA	Complementary DNA
CREB	cAMP-response element-binding protein
CTP	Cytidine triphosphate
dNTP	deoxyribonucleotide triphosphate
$D2_L$	Dopamine D2 receptor long isoform

Mario Tiberi (ed.), *Dopamine Receptor Technologies*, Neuromethods, vol. 96,
DOI 10.1007/978-1-4939-2196-6_4, © Springer Science+Business Media New York 2015

$D2_S$	Dopamine D2 receptor short isoform
DARPP-32	32kDa dopamine and cAMP-regulated phosphoprotein
DNA	Deoxyribonucleic acid
E. coli	Escherichia coli
ECL	Enhanced chemiluminescence
EDTA	Ethylenediaminetetraacetic acid
GPCR	G-protein-coupled receptor
Gαs	Stimulatory G-protein
GTP	Guanosine triphosphate
HRP	Horseradish peroxidase
LB	Lysogeny broth
MOBIX	McMaster Institute for Molecular Biology and Biotechnology
Ni–NTA	Nickel–nitrilotriacetic acid
NPA	Norpropylapomorphine
PAGE	Polyacrylamide gel electrophoresis
PBS	Phosphate buffered saline
PCR	Polymerase chain reaction
PKA	Protein kinase A
PVDF	Polyvinylidene fluoride
RNA	Ribonucleic acid
SDS	Sodium dodecyl sulfate
TAE	Tris-acetate-EDTA buffer
UTP	Uridine triphosphate

1 Introduction

1.1 The Dopamine Receptors

The dopamine receptors were first identified in the 1970s when it was determined that dopamine could have an effect on adenylyl cyclase activity [1–3]. P. F. Spano's laboratory was among the first to determine that there were at least two families of dopamine receptors: the D1 and D2, which could be differentiated based on their capacity to positively or negatively modulate the activity of adenylyl cyclase [4]. The introduction of molecular cloning techniques facilitated further distinctions between the dopamine receptors, and today five dopamine receptors are commonly recognized: D1 and D5, which compose the D1 receptor family, and the D2, D3, and D4 which are classified under the D2 receptor family [1, 5].

Upon stimulation of a D1 receptor family, the stimulatory G-protein, Gαs, translocates to and activates adenylyl cyclase, which in turn catalyzes the conversion of adenosine triphosphate (ATP) into cyclic AMP (cAMP) [1, 6]. cAMP in turn activates protein kinase A (PKA), and thus, D1 stimulation indirectly modulates the targets of PKA including the ion channels, the cAMP-response element-binding protein (CREB), ionotropic glutamate receptors (AMPA and NMDA), and the 32 kDa dopamine- and cAMP-regulated phosphoprotein (DARPP-32) [1, 6]. Stimulation of the D2 receptor family activates the inhibitory G-protein, Gαi/o, to inhibit adenylyl cyclase activity and therein prevents cAMP production [6].

While preliminary distinctions between the dopamine receptors were made via observations of dopamine's effect on adenylyl cyclase activity, there are currently numerous other downstream effects of dopamine receptor signaling that are recognized. Some of the signaling pathways of the dopamine receptors include the effects on calcium signaling and potassium channels, modulation of arachidonic acid synthesis, and other indirect effects [1, 6].

1.2 The Dopamine D2 Receptor

The dopamine D2 receptor is a Type A 7-transmembrane G-protein-coupled receptor (GPCR), which localizes primarily to the striatum, olfactory tubercle, and the nucleus accumbens (Fig. 1) [7]. Molecular cloning techniques have distinguished two splice variants of the dopamine D2 receptor, namely, the presynaptic short isoform ($D2_S$, 415 amino acids) and the postsynaptic long isoform ($D2_L$, 444 amino acids). The dopamine $D2_L$ receptor is structurally distinguished from its dopamine $D2_S$ counterpart as it has 29 more amino acids in its third cytoplasmic loop [8]. The dopamine $D2_S$ receptor is considered predominantly to function as a presynaptic autoreceptor responsible for the detection and regulation of synaptic dopamine release [1, 6]. In contrast, the dopamine $D2_L$ receptor, which is particularly highly expressed in the striatum, facilitates downstream signaling in the postsynaptic neuron [1, 6].

The dopamine D2 receptor regulates a variety of crucial physiological functions and is implicated in reward, motivation, working memory, and higher-order cognitive and executive functions such as

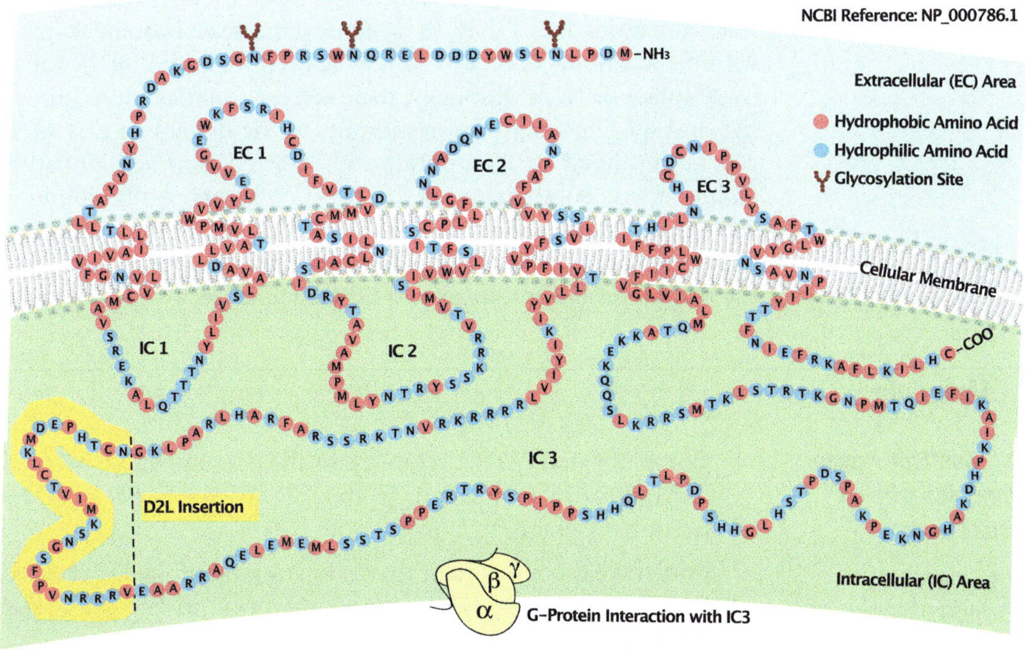

Fig. 1 The amino acid sequence of the human dopamine D2 receptor

attentional and organizational behaviors [6, 9–11]. Due to its important role in these responses, changes in the D2 receptor functions have been linked to the etiology of a number of devastating illnesses, such as schizophrenia, Parkinson's disease, Tourette's syndrome, and addiction [6, 9, 12–18]. As such, protein expression, purification as well as structural and functional characterization of the dopamine D2 receptor, is fundamental to our understanding of a number of severe mental illnesses as well as physiological function and behavior.

1.3 Expression of Dopamine D2 Receptor

Studies characterizing the dopamine D2 receptor have been exceptionally limited, due to the challenges presented in expressing and purifying this hydrophobic protein [19–24]. Similar to other GPCRs, overexpression of the dopamine D2 receptor in heterologous cell-based systems often causes cell toxicity, protein aggregation, and misfolding [25]. Additionally, overexpression of functional receptors in cellular systems is often hindered by shortage of membrane space and inadequate cellular machinery required for membrane insertion and posttranslational modifications [26–28]. These limitations are not typically observed in the expression of the more extensively characterized soluble cytosolic proteins. This helps to account for the relative paucity in our current understanding of the structure and function of hydrophobic GPCRs, such as the D2 receptor.

Herein, we describe the protocols and applications of a cell-free technique for the synthesis of the dopamine $D2_L$ receptor, using *Escherichia coli (E. coli)* lysate in a regenerative dialysis membrane system that circumvents the challenges of conventional cell-based systems. This technique provides excess protein synthesis machinery, a regenerative dialysis membrane system and exogenous dopamine $D2_L$ DNA, to facilitate large-scale protein synthesis within a few hours of incubation, which is not limited by the capacity of intact cells. Additionally, this method facilities more innovative and specific scientific questioning, as the dopamine $D2_L$ DNA can be modified to incorporate radiolabeled isotopes, unnatural amino acids, solubilizing detergents, cofactors, and inhibitors, among other desired components. Consequently, this technique holds significant potential to facilitate a greater understanding of the dopamine $D2_L$ receptor and GPCRs at large.

2 Materials

2.1 Cell-Free Protein Expression with an E. coli Lysate

2.1.1 Construction of Expression Vectors

1. Complementary DNA (cDNA) for the dopamine $D2_L$ receptor (NM_000795) obtained from the Missouri S&T cDNA Resource Center
2. Primers that complement the coding region of the D2 receptor and incorporate a CACC sequence prior to the ATG start codon of the $D2_L$ gene (Table 1)

Table 1
Primer sequences for *E. coli* lysate-based cell-free systems

Primer sequences
Primer for E. coli lysate-based cell-free system
Fwd: CACCATGGATCCACTGAA
Rev: AGACTCGAGTCAGCAGTGGA

3. Polymerase chain reaction (PCR) kit (Invitrogen, cat # 10966-018):
   (a) 10× PCR buffer
   (b) 10 mM dNTP
   (c) 50 mM $MgCl_2$
   (d) Polymerase Taq
4. pET100/D-TOPO vector (Invitrogen, CA)
5. Agarose (BioShop Canada Inc.)
6. 50× TAE:
   (a) 2 M tris.
   (b) 1 M acetic acid.
   (c) 0.05 M ethylenediaminetetraacetic acid (EDTA).
   (d) Make up to a total of 400 mL with distilled water.
7. Bio-Rad Mini-Sub Cell GT rig (Bio-Rad, ON)
8. UV transilluminator (Fotodyne Inc.)
9. One shot TOP10 competent cells with SOC media (Invitrogen, CA)
10. Kit for small-scale DNA extraction (QIAprep Spin Miniprep kit, cat# 27104)
11. Kit for large-scale DNA extraction (QIAGEN Plasmid Plus Maxi kit, cat# 12963)
12. Ethidium bromide (Bio-Rad, ON)
13. 6× DNA loading dye (Fermentas, ON)
14. 1 kB DNA ladder (Fermentas, ON)
15. Freeze-n-Squeeze tubes (Bio-Rad, ON)
16. Nuclease-free water
17. Water bath maintained at 42 °C
18. Bunsen burner
19. Incubator shaker which can shake at 200 rpm, 37 °C
20. 100 mm sterile plates

21. Lysogeny broth (LB), autoclaved:
    (a) 1 % bactotryptone
    (b) 0.5 % bactoyeast
    (c) 1 % NaCl
    (d) pH 7.0
22. LB media agar, autoclaved:
    (a) 1.5 g trypto agar
    (b) 100 mL of LB media
23. Ampicillin (Life Technologies, ON)
24. Spreader
25. Conical flasks
26. Culture tubes: 3 and 10 mL
27. DNA sequencing facility (use a facility that provides molecular biology expertise, e.g., the McMaster Institute for Molecular Biology and Biotechnology (MOBIX))

2.1.2 Cell-Free Protein Synthesis Using the E. coli Lysate

1. $D2_L$ expression vector prepared as outlined in Sect. 3.1.1
2. Float-A-Lyzer 25 kDa dialysis membrane (Spectra/Por, CA)
3. All components from Table 2.

Table 2
Reagents for *E. coli* cell-free protein synthesis system

Chemicals	Volume	Final concentration
(A) Reaction mixture		
1.6 M HEPES-KOH (pH 7.5)	32.63 μL	58 mM
1 M dithiothreitol	2.07 μL	2.3 mM
100 mM ATP[a]	10.8 μL	1.2 mM
100 mM CTP, GTP, UTP[a]	8.1 μL each	0.9 mM each
1 M creatine phosphate[a]	72.9 μL	81 mM
20 mg/mL creatine kinase	12.5 μL	250 μg/mL
40 % polyethylene glycol 8000	90 μL	4 %
5 mg/mL 3′–5′ cyclic AMP	37.7 μL	0.64 mM
1 mg/mL formyl-5,6,7, 8-tetrahydrofolic acid	31.5 μL	35 μg/mL
10 mg/mL *E. coli* tRNA	15.3 μL	170 μg/mL
3.33 M potassium glutamate	54 μL	200 mM
1 M ammonium acetate	24.93 μL	27.7 mM
1 M magnesium acetate	9.63 μL	10.7 mM
10 % sodium azide	10 μL	0.05 %
50 mM each amino acid (-Met)[b]	27 μL	1.5 mM each
75 mM methionine[b]	18 μL	1.5 mM
0.5 mg/mL D2 receptor DNA	46 μL	10 μg/mL
T7 RNA polymerase[b]	20 μL	66.7 μg/mL
E. coli S30 extract[b]	300 μL	–
5 % Brij 35	36 μL	0.2 %
Final volume with nuclease-free water	900 μL	

(continued)

Table 2 (continued)

Chemicals	Volume	Final concentration
(B) Feeding mixture		
1.6 M HEPES-KOH (pH 7.5)	362.5 μL	58 mM
1 M DTT	23 μL	2.3 mM
100 mM ATP[a]	120 μL	1.2 mM
100 mM CTP, GTP, UTP[a]	90 μL/each	0.9 mM/each
1 M creatine phosphate	810 μL	81 mM
40 % polyethylene glycol 8000	1 mL	4 %
5 mg/mL 3′–5′ cyclic AMP	420 μL	0.64 mM
1 mg/mL formyl-5,6,7, 8-tetrahydrofolic acid	350 μL	35 μg/mL
10 mg/mL *E. coli* tRNA	170 μL	170 μg/mL
3.33 M potassium glutamate	600 μL	200 mM
1 M ammonium acetate	277 μl	27.7 mM
1 M magnesium acetate	149 μL	14.9 mM
10 % sodium azide	111 μL	0.05 %
50 mM each amino acid (-Met)[b]	300 μL	1.5 mM each
75 mM methionine[b]	200 μL	1.5 mM
5 % Brij 35	400 μL	0.2 %
Final volume with nuclease-free water	10 mL	

[a]Items purchased from Roche Diagnostics Canada (Mississauga, ON)
[b]Items purchased from Invitrogen (Carlsbad, CA). All other items were purchased from Sigma-Aldrich (St. Louis, MO)

4. 2 mL tubes (Eppendorf, ON)
5. 15 mL tubes (BD Biosciences, ON)
6. 2 L beaker
7. Very small stir bar
8. Thermometer
9. Stir/hot plate

2.1.3 Protein Purification

1. *N*-Lauryl sarcosine (Sigma-Aldrich, MO)
2. ProBond purification system (Invitrogen, CA):
   (a) Nickel–nitrilotriacetic acid (Ni-NTA) affinity resin
   (b) Denaturing binding buffer (8 M urea, 20 mM sodium phosphate, 500 mM NaCl, pH 7.8)
   (c) Denaturing wash buffer 1 (8 M urea, 20 mM sodium phosphate, 500 mM NaCl, pH 6.0)
   (d) Denaturing wash buffer 2 (8 M urea, 20 mM sodium phosphate, 500 mM NaCl, pH 5.3)
   (e) Denaturing elution buffer (8 M urea, 20 mM sodium phosphate, 500 mM NaCl, pH 4.0)

3. Sterile water
4. Rotating wheel
5. Collection tubes

2.1.4 Reconstitution of Protein in Phospholipid Vesicles

1. 25 kDa dialysis membrane (Spectra/Por, CA)
2. Phospholipid mix (Sigma-Aldrich, MO):
   (a) Phosphatidylcholine
   (b) Phosphatidylserine
   (c) Phosphatidylethanolamine
   (d) Dipalmitoylphosphatidylcholine
3. 1 % Carbiosorb (Calbiochem, ON)
4. 1× phosphate-buffered saline (PBS)
5. Mini-extruder (Avanti Polar Lipids Inc., AL)

2.1.5 Immunoblotting

1. Sodium dodecyl sulfate (SDS) buffer:
   (a) 50 mM tris–HCl pH 6.8
   (b) 2 % SDS
   (c) 10 % glycerol
   (d) 1 % β-mercaptoethanol
   (e) 0.02 % bromophenol blue
2. Polyvinylidene fluoride (PVDF) membrane (Bio-Rad, ON)
3. 5 % milk blocking solution
4. Hexahistidine primary antibody raised in rabbit (Invitrogen, CA)
5. Anti-rabbit IgG horseradish peroxidase (HRP)-linked whole secondary antibody raised in donkey (Sigma-Aldrich, MO)
6. Enhanced chemiluminescence (ECL) reagents (GE Healthcare Bio-Sciences Corp., NJ)
7. Kodak XAR film

2.1.6 Radioligand-Binding Assay

1. 1 nM [^{3}H] norpropylapomorphine (NPA) (Perkin Elma, CA)
2. Assay buffer:
   (a) 50 mM tris–HCl pH 7.4
   (b) 5 mM $MgCl_2$
   (c) 1 mM EDTA
   (d) 0.1 mM dithiothreitol
   (e) 0.1 mM phenylmethylsulfonyl fluoride
   (f) Bacitracin 100 μg/mL
   (g) Soybean trypsin 5 μg/mL

3. 1 μM dopamine (Sigma-Aldrich, MO)
4. 5 mL glass assay tubes
5. Incubator with shaker
6. 0.375 % γ-globulin (Sigma-Aldrich, MO)
7. 40 % polyethylene glycol (Sigma-Aldrich, MO)
8. Brandel cell harvester (Brandel, Hertfordshire, UK)
9. 10 % polyethylene glycol filtration buffer:
    (a) 10 % polyethylene glycol.
    (b) 2.5 mM tris–HCl pH 7.4.
    (c) 2 μM EDTA.
    (d) Bring to concentration with distilled water.
10. Filter paper disks
11. Plastic scintillation vials
12. Scintillation fluid (Beckman Coulter, CA)
13. Beckman LS5000 liquid scintillation counter Model LS5KTA (Beckman Coulter, CA)

3 Methods

3.1 Cell-Free Protein Expression with an *E. coli* Lysate

3.1.1 Construction of Expression Vectors

Exogenous DNA coding for the dopamine $D2_L$ receptor is supplied to the cell-free system using expression vectors. These vectors are synthesized using complementary DNA (cDNA) for the $D2_L$ receptor and primers designed to complement the coding region of the D2 receptor and incorporate a CACC sequence prior to the ATG start codon of the $D2_L$ gene (Table 1). This sequence will facilitate ligation of the gene into a vector containing the complementary GTGG sequence. The vector of choice was the linear vector pET100/D-TOPO with an integrated hexahistidine tag selected for effective transcription using an *E. coli*-based system, due to the presence of the T7 promoter (Fig. 2) (*see* **Note 1**):

1. Prepare PCR mixture in an Eppendorf Tube to amplify the $D2_L$ cDNA:
    (a) 5 μL 10× PCR buffer
    (b) 1 μL 10 mM dNTP
    (c) 1.5 μL 50 mM $MgCl_2$
    (d) 1 μL 10 μM forward primer
    (e) 1 μL 10 μM reverse primer
    (f) 1 μL cDNA for the dopamine $D2_L$ receptor
    (g) 1 μL Platinum Taq polymerase enzyme
    (h) 38.5 μL NF H_2O

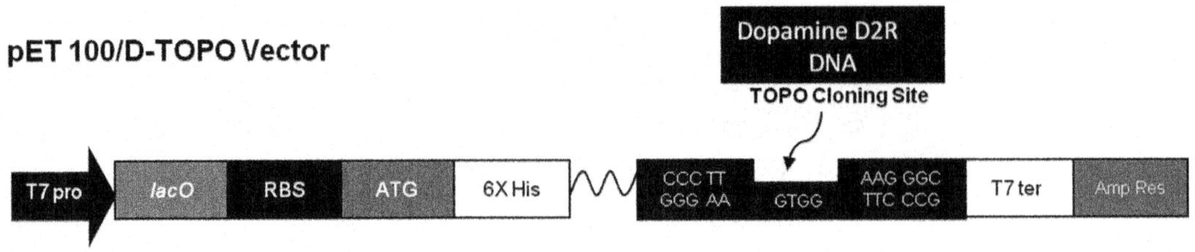

Fig. 2 Vector diagram for constructs used in cell-free protein synthesis systems. Hexahistidine-tagged pET100/D-TOPO vector used in *E. coli* lysate-based cell-free system (Invitrogen, CA) (Basu et al. [18]). Reproduced with permission from the publisher, license #: 3356560998119)

2. Gently pipette up and down three times to mix components properly.
3. Set up tube in PCR machine cycling at:
   (a) 1 cycle of:
   - 95 °C for 2 min and 30 s

   (b) 40 cycles of:
   - 95 °C for 1 min
   - 51 °C for 45 s
   - 72 °C for 1 min

   (c) 1 cycle of:
   - 72 °C for 1 min
4. While the PCR reaction is running, prepare a 1 % agarose gel:
   (a) Dilute the 50× TAE buffer to prepare 40 mL of a 1× stock in distilled water.

   (b) Weigh 0.4 g agarose into 40 mL 1× TAE buffer, in a 250 mL conical flask.

   (c) Loosely cover flask with a paper towel, and microwave mixture for 1 min.

   (d) Use gloves to swirl mixture, and microwave for another 1 min (*see* **Note 2**).

   (e) Allow mixture to cool down so it can comfortably be touched (about 10 min) and add 3 μL ethidium bromide to the mixture and swirl mixture (*see* **Note 3**).

   (f) Pour into Bio-Rad Mini-Sub Cell GT rig, and set up with a tray, black end pieces, and a comb (*see* **Note 4**).

   (g) Let gel set for 30 min.

   (h) Remove comb from set gel, and pour 1× TAE buffer into rig so that the gel is completely covered (~300 mL).
5. Prepare your samples by adding 10 μL of 6× loading dye to 50 μL of the amplified DNA from PCR reaction.

6. Load onto the agarose gel:
   (a) 1 kB ladder into one lane
   (b) Prepared samples from the PCR reaction into adjacent lanes
7. Run at 100 V for approximately 30 min, allowing the DNA to separate by electrophoresis until the dye front approaches ¾ of the length of the gel (*see* **Note 5**).
8. View gel under UV transilluminator to confirm the presence of a positive band at 1,400 bp.
9. Extract the DNA from the gel using a Freeze-n-Squeeze tube:
   (a) Place gel on saran wrap in the transilluminator, and cut out the band with a sharp blade to extract positive DNA (*see* **Note 6**).
   (b) Chop up sliced section into smaller pieces and place into Freeze-n-Squeeze tube for 5 min at −20 °C.
   (c) Spin in a centrifuge for 3 min at 16,000 × *g*.
10. Quantify DNA concentration using a spectrophotometer.
11. Confirm the accuracy of the $D2_L$ receptor amplification by sequencing at MOBIX.
12. Ligate the amplified $D2_L$ cDNA with pET 100/D-TOPO vector by incubating:
    (a) 1.6 ng PCR products
    (b) 1 μL buffer solution
    (c) 20 ng vector pET 100/D-TOPO
    (d) 3.3 μL sterile water
    Determine the amount of DNA to be used according to the formula below:

$$\text{Amount of DNA} = (1,400\ \text{bp} \times \text{amount of vector}) / \text{size of vector} = (1,400 \times 20) / 5,764\ \text{bp} = 4.9\ \text{ng}$$

$$\text{Incubate the Amount of DNA : Amount of Vector at a ratio of } 3:1 = 4.9\ \text{ng} / 3 = 1.6\ \text{ng of amplified } D2_L\ \text{DNA}$$

13. Gently pipette up and down three times to mix components properly.
14. Incubate at room temperature for 30 min to allow for ligation.
15. Transform TOP10 One Shot chemically competent *E. coli* cells with $D2_L$/pET 100 plasmid vector:
    (a) Add 3 μL of the $D2_L$/pET 100 vector DNA to one vial of TOP10 One Shot chemically competent *E. coli* cells and incubate on ice for 30 min.

(b) Heat shock cells by partially immersing vial into 42 °C water bath for 1 min and place vial back on ice.

(c) Recover heat-shocked cells by adding 250 μL room temperature SOC media to the vial using aseptic technique and incubate for 1 h, 37 °C, 250 rpm.

16. Prepare two 100 mm petri dishes of LB agar:

(a) Add 1.5 % agar to LB media, autoclave for sterility, and allow mixture to cool down to comfortably touch.

(b) Add ampicillin (1 mg/mL) to sterile LB agar media.

(c) Pour into 100 mm plates and allow agar to solidify.

(d) Place plates in 37 °C incubator to prewarm.

17. Pipette *E. coli* transformed with $D2_L$/pET100 onto LB agar plates. 50 μL onto one LB agar plate and 200 μL onto another LB agar plate are the most common amounts but can be adjusted to more or less as required to get the growth of individual bacterial colonies.

18. Spread culture with sterile spreader, and incubate at 37 °C, overnight, with the agar side facing up.

19. The next day, use a sterile pipette tip to scrape an individual bacterial colony and place entire tip into 3 mL liquid LB media (1 mg/mL ampicillin). Repeat, to collect several other individual colonies and culture in individual 3 mL tubes (*see* **Note 7**).

20. Grow colonies at 37 °C, overnight, on a shaker at 250 rpm.

21. Spin down 1 mL of each overnight-grown bacterial culture at 800 × *g* for 10 min.

22. Extract DNA from each culture using QiaPrep Spin Miniprep kit (according to the company's outlined protocols).

23. Prepare a 1 % agarose gel and set up as described in step 4 above.

24. Prepare your samples by adding 2 μL of 6× loading dye to 10 μL of DNA extracted from each clone.

25. Load onto the agarose gel:

(a) 1 kB ladder into one lane

(b) Prepared samples from the PCR reaction into adjacent lanes

26. Run at 100 V for approximately 30 min until bands have separated.

27. View gel under UV transilluminator to identify the size of the bacterial clones transformed with $D2_L$/pET100/D-TOPO vector. The size for the positive clones are expected to be size of the D2L DNA, 1,400 bp, plus size of the pET100 vector DNA, 5,764 bp, for a total of 7,164 bp.

28. Identify the positive clone of size 7,164 bp, and confirm the sequence (using MOBIX).
29. Inoculate 1.5 mL of the positive clone expressing $D2_L$/pET 100/D-TOPO into 500 mL liquid LB media with 1 mg/mL ampicillin.
30. Grow clones at 37 °C, overnight on a shaker at 250 rpm.
31. Extract DNA from 500 mL of overnight culture using QIAGEN Plasmid Plus Maxi kit (according to the company's outlined protocols).
32. Quantify amount of $D2_L$/pET100 DNA extracted using a spectrophotometer and store DNA at −80 °C until use.

3.1.2 Cell-Free Protein Synthesis

The cell-free synthesis system (Fig. 3) is made up of a reaction mixture and a feeding mixture separated by a dialysis membrane (with a molecular weight cutoff of 25 kDa). Two mixtures are prepared and exist in balance with each other in this system. The reaction mixture contains high- and low-molecular-weight compounds that facilitate protein synthesis. The previously prepared $D2_L$ plasmid DNA will be included in this reaction mixture. The feeding mixture functions to replenish the reaction mixture and as such contains only of low-molecular-weight precursors. This system facilitates a continuous supply of fresh substrates from the feeding mixture to the reaction mixture for effective protein synthesis. Simultaneously, effective concentration gradients that build up across the dialysis membrane allow for the removal of toxic wastes:

1. Prepare the reaction mixture in a 1.5 mL Eppendorf Tube (Table 2).
2. Prepare the feeding mixture in a 15 mL falcon tube (Table 2).

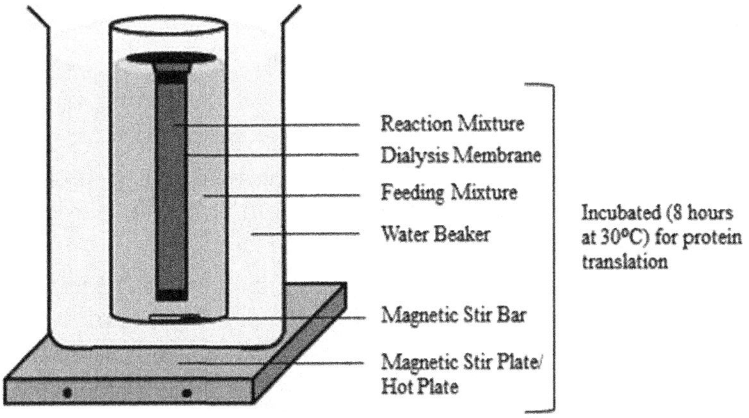

Fig. 3 Cell-free synthesis setup diagram for the *E. coli* lysate-based cell-free system setup (Basu et al. [18]). Reproduced with permission from the publisher, license #: 3356560998119)

3. Pour reaction mixture into Float-A-Lyzer dialysis membrane (*see* **Note 8**).
4. Pour feeding mixture into a round-bottomed BD tube, and put a small magnet into the tube to facilitate mixing.
5. Place this apparatus in a 2 L beaker of distilled water, and incubate at 30 °C for 8 h with consistent stirring.
6. To assess by western blotting:
   (a) Remove a 25 μL sample of the sample at 8 h.
   (b) To precipitate synthesized protein, add four volumes cold acetone (−20 °C), and incubate for 30 min at−20 °C.
   (c) Spin for 5 min at 13,000×*g* at room temperature to precipitate protein.
   (d) Resuspend $D2_L$ protein in 1× SDS buffer and use for western analysis.

3.1.3 Protein Purification

Protein purification of the solubilized hexahistidine-tagged recombinant dopamine $D2_L$ receptor was performed using a ProBond purification system (Invitrogen, CA), which included a Ni–NTA affinity resin selected for its affinity to the histidine tag associated with the dopamine $D2_L$ receptor using specified vector design:

1. Solubilize 250 μL of protein product with 27.5 μL of 20 % *N*-lauryl sarcosine, and rotate mixture for 1 h, at 4 °C.
2. Prepare 250 μL of the Ni–NTA column:
   (a) Invert and tap bottle with resin to resuspend it.
   (b) Take a p1000 tip, snip the end to widen opening, and use it to pipette 250 μL resin into an Eppendorf Tube.
   (c) Centrifuge (800×*g*, 1 min), and discard supernatant.
   (d) Add 750 μL sterile water to the resin, and invert/tap gently to mix properly.
   (e) Centrifuge (800×*g*, 1 min) and discard supernatant.
   (f) Add 750 μL denaturing binding buffer, mix, centrifuge, and discard supernatant as written above.
3. Add 500 μL protein lysate to prepared resin, and allow binding for 1 h, at room temperature with gentle agitation on a rotating wheel.
4. Following binding of the crude protein lysate to the resin, perform two rounds of washes with 1 mL of the following buffers, with centrifugation and removal of supernatant between each wash:
   (a) Denaturing binding buffer (*see* **Note 9**)
   (b) Denaturing wash buffer 1 (*see* **Note 9**)
   (c) Denaturing wash buffer 2 (*see* **Note 9**)

Elute the bound protein by adding 250 μL denaturing elution buffer, and centrifuge (800 × *g*, 1min) to collect purified protein.

3.1.4 Reconstitution of Protein in Phospholipid Vesicles

A phospholipid vesicle is necessary for proper folding of the purified receptor. As such the final, purified, protein product should be incubated with a phospholipid mix that will accommodate this folding. Additionally, detergents must be removed from the protein products since these can interfere with effective folding. 1 % Carbiosorb facilitates removal of these detergents. A mini-extruder is utilized to accommodate the formation of smaller vesicles:

1. Prepare the phospholipid mix in a ratio of 1:1:1:1 using 500 μL of each component:
   (a) Phosphatidylcholine
   (b) Phosphatidylserine
   (c) Phosphatidylethanolamine
   (d) Dipalmitoylphosphatidylcholine
2. Mix thoroughly and dry in tube with constant rotation using nitrogen in a vacuum desiccator for 2 h, until only a film remains.
3. To create phospholipid vesicles, use following protocol:
   (a) Hydrate lipid mixture by resuspending the lyophilized solution in 500 μL PBS.
   (b) Load the hydrated sample into one of the gas-tight syringes, and place into one end of the mini-extruder, while placing the other empty syringe with plunger on the other end (*see* **Note 10**).
   (c) Insert the assembled extruder into the extruder stand, and push the filled syringe slowly to transfer the solution through the extruder membrane and into the empty syringe.
   (d) Push the plunger on the filled syringe to send solution back to the original syringe, and repeat this step 4 more times (total of ten passes through the extruder membrane).
   (e) Remove the filled syringe from the extruder and inject the homogenous phospholipid solution into a clean Eppendorf Tube.
4. Add the protein product to the phospholipid mixture passed through the mini-extruder.
5. Resuspend and vortex well.
6. Transfer the mixture to the 25 kDa dialysis membrane.
7. Place dialysis membrane 1 % Carbiosorb dissolved in 1× PBS for 3 days at 4 °C.

8. Maintain the solution by adding fresh Carbiosorb solution at 18, 32, 40, and 56 h.
9. Collect final folded protein product and store at −20 °C.

3.1.5 Immunoblotting

SDS-polyacrylamide gel electrophoresis (SDS-PAGE) as previously described by our lab should be performed using cell-free synthesized D2 receptor protein to confirm the positive synthesis of this protein. (For a detailed description of an immunoblotting protocol, please refer to [29].)

Western blotting parameters:

1. Membrane type: PVDF activated in methanol according to manufacturer's instructions.
2. Blocking solution: 5 % milk block.
3. Prepare incubations with the primary antibodies at 1:5,000 dilution, overnight at 4 °C:
   (a) Anti-rabbit hexahistidine primary antibody
4. Prepare incubations with the secondary antibodies at 1:5,000 dilution, for 1.5 h at room temperature:
   (a) For the anti-rabbit-exposed membranes, incubate with anti-rabbit IgG HRP-linked whole secondary antibody from donkey (Sigma-Aldrich, MO).
5. Develop membranes by 1 min incubation with enhanced chemiluminescence (ECL) reagents.
6. Expose to Kodak XAR film.
7. Developed films were used for qualitative protein expression analysis.

3.1.6 Radioligand-Binding Assay

An assessment of the radioligand-binding capabilities of the dopamine $D2_L$ receptor product produced in the cell-free system will characterize effective binding capacities of these receptors, as a reflection of effective folding of the protein. The maximal amount of synthesized protein previously used for the assays was 19 μg for crude and 0.15 μg for pure fraction: (For a detailed description of a radioligand-binding assay protocol, please refer to [30].)

1. Prepare glass assay tubes with:
   (a) 10 μL of sample:
      - Crude protein product
      - Pure protein product
   (b) 25 μL of 10 mM dopamine.
   (c) 25 μL of 1 nM [^{3}H] NPA.
   (d) Make up this solution to a total volume of 250 μL with assay buffer.

2. Incubate tubes for 20 h, at 4 °C with gentle rocking.
3. Add 100 μL of 0.375 % γ-globulin and 150 μL of 40 % polyethylene glycol.
4. Incubate the tubes at 4 °C for 15 min.
5. Filter samples through a Brandel cell harvester using 10 % polyethylene glycol filtration buffer.
6. Assess the bound radioactivity on filter paper disks by placing them in plastic scintillation vials filled with scintillation fluid.
7. Measure radioactivity counts using a liquid scintillation counter such as the Beckman LS5000 liquid scintillation counter Model LS5KTA.
8. Calculate the percent displacement by calculating the ratio in radioactivity binding in the presence and absence of 1 μM dopamine and expressed per μg of D2 protein.

4 Typical Expected Results

4.1 Immunoblot Detection of Bacterial Cell-Free Synthesized Dopamine $D2_L$ Receptor

The accuracy and effectiveness of synthesis of the dopamine $D2_L$ receptor with the *E. coli* cell-free technique was characterized using SDS-PAGE and immunoblotting assessments with an anti-hexahistidine antibody (Fig. 4). These assessments confirmed the presence of the dopamine $D2_L$ receptor in both the crude and purified lysates. The bands identified were at a molecular weight of 47 kDa, slightly smaller than the anticipated molecular weight of 50.65 kDa for the dopamine $D2_L$ receptor. The differences in the observed and calculated molecular weight may reflect secondary and tertiary protein structure. The highest concentrations of dopamine $D2_L$ receptors were observed in the crude lysate. The protein expression levels observed following protein purification were lower than the crude lysate due to anticipated losses which are innate to purification processing. The final concentration of dopamine $D2_L$ protein synthesized was 17 μg/mL reaction volume in the purified lysate. As such, the cell-free system is an effective approach to accurately synthesize and prepare dopamine $D2_L$ receptors.

4.2 Ligand-Binding Properties of Bacterial Cell-Free Synthesized Dopamine $D2_L$ Receptor

Radioligand competition assays were utilized to confirm the correct functional synthesis and purification of the dopamine $D2_L$ receptor protein (Fig. 5). These ligand-binding assays confirmed that both the crude and the purified bacterial cell-free generated dopamine $D2_L$ protein were capable of binding the receptor-specific agonist [^{3}H]-NPA, which was displaced by unlabeled dopamine. These radioligand-binding assays demonstrate the ability of the receptor to clearly bind and displace ligands.

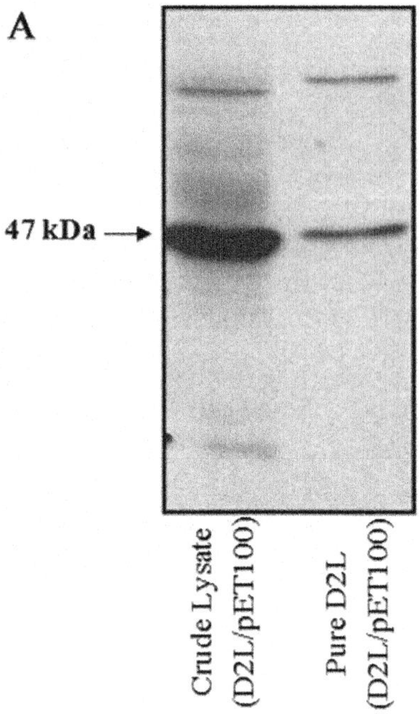

Fig. 4 Confirmation of dopamine D2L receptor expression in the *E. coli* lysate-based cell-free system using immunoblot with an anti-hexahistidine antibody to demonstrate the expression of tagged dopamine $D2_L$ receptor in crude and purified lysate (Basu et al. [18]). Reproduced with permission from the publisher, license #: 3356560998119)

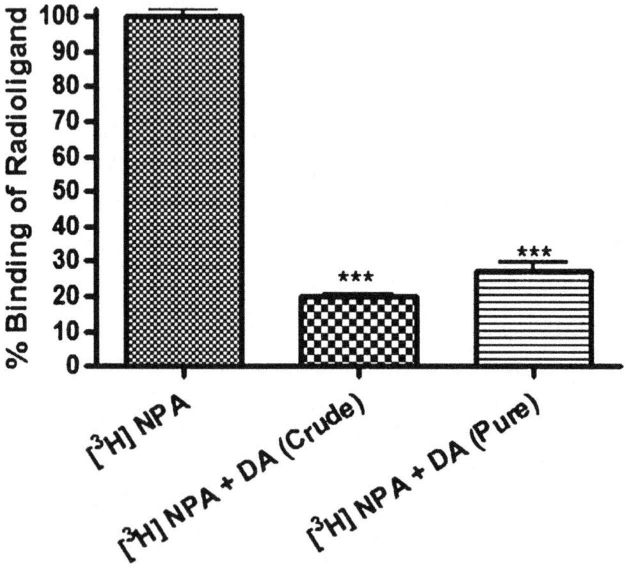

Fig. 5 Confirmation of dopamine D2L receptor expression in the *E. coli* lysate-based cell-free system using a dopamine (DA) radioligand ([^{3}H]NPA) competition assay of crude and pure lysate (Basu et al. [18]). Reproduced with permission from the publisher, license #: 3356560998119)

This displacement can be used to estimate a yield of 0.83 μg/mL (15 pmol D2L/mg of total protein) for fully folded and functional dopamine $D2_L$ using the *E. coli* cell-free system.

5 Conclusion

The dopamine D2 receptor is a GPCR with substantial implications in central nervous system diseases including schizophrenia, Parkinson's disease, attention-deficit–hyperactivity disorder, Tourette's syndrome and various addictions. Furthermore, dopamine $D2_L$ receptor targeting has been the therapeutic mechanism of action for a number of currently available therapeutics including antipsychotic drugs (dopamine $D2_L$ antagonists) and l-DOPA (a precursor to the dopamine $D2_L$ receptor agonist dopamine). Consequently, the three-dimensional structural characterization of the dopamine $D2_L$ receptor would significantly contribute to our understanding of ligand-binding sites and allow for the design of highly discriminating and, therefore, safer therapeutics, targeting various central nervous system disorders. Unfortunately, structural investigations of these receptors are largely unfeasible with insufficient quantities of pure protein. The *E. coli* lysate-based cell-free system provides a partial resolution to this predicament by presenting a novel method to synthesize and purify the human dopamine $D2_L$ receptor. The expression and purification procedure reported here appears to be rapid and highly specific with respect to both quantity and fold purification. The results of this study can be used as an avenue toward better research and improved drug design for a receptor highly implicated in physiological and mental health.

6 Notes

1. Prior to conducting any DNA work, make sure to clean all working surfaces with nuclease-free detergent to prevent contamination.
2. Mixture is very hot, so handle with heavy duty gloves.
3. Ethidium bromide is a mutagen, so wear a lab coat and gloves when handling it.
4. Pour gel slowly to avoid any bubbles. If there are bubbles, push to the side using clean pipette tip.
5. DNA is negatively charged and will run from the positive electrode (black) to the negative electrode (red); therefore, set gel up accordingly.
6. Make sure to cut as close to the band as possible, removing any unlabeled agarose, thereby preventing dilution of the DNA concentration.

7. Make sure to scrape single colonies only, to prevent heterogeneity in the DNA content of grown clones.
8. Dialysis membranes are very fragile and can easily rip. Reinforce strength of attachment between dialysis membrane and floating top of Float-A-Lyzer using parafilm.
9. The pH of each buffer should be checked before usage as these can change with time and are crucial to proper binding, washing, and elution.
10. To reduce loss of volume during the extrusion process, pre-wet the extruder parts by passing a syringe full of PBS buffer through the extruder and discarding before placing phospholipid mixture in the apparatus.

Acknowledgments

The study was funded by the National Institutes of Health, Canadian Institutes of Health Research, and the Ontario Mental Health Foundation.

References

1. Missale C, Nash SR, Robinson SW, Jaber M, Caron MG (1998) Dopamine receptors: from structure to function. Physiol Rev 78:189–225
2. Spano PF, Govoni S, Trabucchi M (1978) Studies on the pharmacological properties of dopamine receptors in various areas of the central nervous system. Adv Biochem Psychopharmacol 19:155–165
3. Kebabian JW, Calne DB (1979) Multiple receptors for dopamine. Nature 277:93–96
4. Garau L, Govoni S, Stefanini E, Trabucchi M, Spano PF (1978) Dopamine receptors: pharmacological and anatomical evidences indicate that two distinct dopamine receptor populations are present in rat striatum. Life Sci 23:1745–1750
5. Sibley DR, Monsma FJ (1992) Molecular biology of dopamine receptors. Trends Pharmacol Sci 13:61–69
6. Beaulieu JM, Gainetdinov RR (2011) The physiology, signaling, and pharmacology of dopamine receptors. Pharmacol Rev 63:182–217
7. Levey AI, Hersch SM, Rye DB, Sunahara RK, Niznik HB, Kitt CA, Price DL, Maggio R, Brann MR, Ciliax BJ (1993) Localization of D1 and D2 dopamine receptors in brain with subtype-specific antibodies. Proc Natl Acad Sci U S A 90:8861–8865
8. Malek D, Munch G, Palm D (1993) Two sites in the third inner loop of the dopamine D2 receptor are involved in functional G protein-mediated coupling to adenylate cyclase. FEBS Lett 325:215–219
9. Noble EP (2003) D2 dopamine receptor gene in psychiatric and neurologic disorders and its phenotypes. Am J Med Genet B Neuropsychiatr Genet 116B:103–125
10. Pivonello R, Ferone D, Lombardi G, Colao A, Lamberts SW, Hofland LJ (2007) Novel insights in dopamine receptor physiology. Eur J Endocrinol 156(Suppl 1):S13–S21
11. van Holstein M, Aarts E, van der Schaaf ME, Geurts DE, Verkes RJ, Franke B, van Schouwenburg MR, Cools R (2011) Human cognitive flexibility depends on dopamine D2 receptor signaling. Psychopharmacology (Berl) 218:567–578
12. Vallone D, Picetti R, Borrelli E (2000) Structure and function of dopamine receptors. Neurosci Biobehav Rev 24:125–132
13. Seeman P (2001) Antipsychotic drugs, dopamine receptors, and schizophrenia. Clin Neurosci Res 1:53–60

14. Seeman P (2006) Targeting the dopamine D2 receptor in schizophrenia. Expert Opin Ther Targets 10:515–531
15. Howes OD, Kapur S (2009) The dopamine hypothesis of schizophrenia: version III–the final common pathway. Schizophr Bull 35:549–562
16. Steeves TD, Ko JH, Kideckel DM, Rusjan P, Houle S, Sandor P, Lang AE, Strafella AP (2010) Extrastriatal dopaminergic dysfunction in tourette syndrome. Ann Neurol 67: 170–181
17. Beyaert MG, Daya RP, Dyck BA, Johnson RL, Mishra RK (2013) PAOPA, a potent dopamine D2 receptor allosteric modulator, prevents and reverses behavioral and biochemical abnormalities in an amphetamine-sensitized preclinical animal model of schizophrenia. Eur Neuropsychopharmacol 23:253–262
18. Basu D, Tian Y, Bhandari J, Jiang JR, Hui P, Johnson RL, Mishra RK (2013) Effects of the dopamine D2 allosteric modulator, PAOPA, on the expression of GRK2, arrestin-3, ERK1/2, and on receptor internalization. PLoS One 8:e70736
19. Ramwani J, Mishra RK (1986) Purification of bovine striatal dopamine D-2 receptor by affinity chromatography. J Biol Chem 261: 8894–8898
20. Senogles SE, Amlaiky N, Falardeau P, Caron MG (1988) Purification and characterization of the D2-dopamine receptor from bovine anterior pituitary. J Biol Chem 263: 18996–19002
21. Clagett-Dame M, Schoenleber R, Chung C, McKelvy JF (1989) Preparation of an affinity chromatography matrix for the selective purification of the dopamine D2 receptor from bovine striatal membranes. Biochim Biophys Acta 986:271–280
22. Usui H, Takahashi Y, Maeda N, Mitui H, Isobe T, Okuyama T, Nishizawa Y, Hayashi S (1990) Purification of D2 dopamine receptor by photoaffinity labelling, high-performance liquid chromatography and preparative sodium dodecyl sulphate polyacrylamide gel electrophoresis. J Chromatogr 515:375–384
23. Chazot PL, Strange PG (1992) Molecular characterization of D2 dopamine-like receptors from brain and from the pituitary gland. Neurochem Int 21:159–169
24. Strange PG (1993) Dopamine receptors in the basal ganglia: relevance to Parkinson's disease. Mov Disord 8:263–270
25. Sarramegna V, Talmont F, Demange P, Milon A (2003) Heterologous expression of G-protein-coupled receptors: comparison of expression systems from the standpoint of large-scale production and purification. Cell Mol Life Sci 60:1529–1546
26. Grisshammer R (2006) Understanding recombinant expression of membrane proteins. Curr Opin Biotechnol 17:337–340
27. Wagner S, Bader ML, Drew D, de Gier JW (2006) Rationalizing membrane protein overexpression. Trends Biotechnol 24:364–371
28. Katzen F, Peterson TC, Kudlicki W (2009) Membrane protein expression: no cells required. Trends Biotechnol 27:455–460
29. Chong VZ, Young LT, Mishra RK (2002) cDNA array reveals differential gene expression following chronic neuroleptic administration: implications of synapsin II in haloperidol treatment. J Neurochem 82:1533–1539
30. Bhagwanth S, Mishra S, Daya R, Mah J, Mishra RK, Johnson RL (2012) Transformation of Pro-Leu-Gly-NH(2) peptidomimetic positive allosteric modulators of the dopamine D(2) receptor into negative modulators. ACS Chem Neurosci 3:274–284

Chapter 5

Wnt Ligand Binding to and Regulation of Dopamine D2 Receptors

Sehyoun Yoon, Mi-Hyun Choi, and Ja-Hyun Baik

Abstract

The dopamine D2 receptor (D2R) plays an important role in mesencephalic dopaminergic neuronal development, and we have previously reported that Wnt5a interacts with D2R to promote the differentiation of dopaminergic neurons. Co-immunoprecipitation of Wnt5a and D2R, displacement of the D2R receptor antagonist [^{3}H]spiperone from D2R by Wnt5a, as well as glutathione S-transferase (GST) pull-down assays demonstrated that Wnt5a binds D2R. In this chapter, we present the co-immunoprecipitation, pull-down, and competition-binding assays used in the analysis of the D2R–Wnt5a interaction.

Key words Wnt5a, Dopamine D2 receptor, Radioligand binding, Co-immunoprecipitation, GST pull-down assay

1 Introduction

Dopamine (DA) is the predominant catecholamine neurotransmitter in the brain, and dopaminergic pathways in the midbrain have been closely associated with serious neurological disorders [1–3]. DA-producing cells are generated within the embryonic ventral midbrain. The developmental program of the ventral midbrain has been shown to require a complex network of transcription factors and signaling pathways [4–6].

DA binds to membrane receptors belonging to the family of seven transmembrane domain G protein-coupled receptors, leading to the formation of second messengers and the activation or inhibition of specific signaling pathways. To date, five different subtypes of DA receptors have been identified, and based on their structural and pharmacological properties, a general subdivision into two groups has been made: the D1-like and D2-like receptors. D1-like receptors (D1 and D5) are coupled to heterotrimeric G proteins such as $G\alpha_s$ and $G\alpha_{olf}$, with activation leading to increased adenylyl cyclase (AC) activity and increased cyclic adenosine monophosphate (cAMP) production [7–10]. In contrast, D2-like

Mario Tiberi (ed.), *Dopamine Receptor Technologies*, Neuromethods, vol. 96,
DOI 10.1007/978-1-4939-2196-6_5, © Springer Science+Business Media New York 2015

receptors (D2, D3, and D4) are coupled to $G\alpha_i$ and $G\alpha_o$ proteins and negatively regulate the production of cAMP, resulting in decreased protein kinase A activity, activation of K^+ channels, and the modulation of numerous other ion channels [11–14].

We have previously reported that the dopamine D2 receptor (D2R) as an autoreceptor expressed in midbrain plays a crucial role in the development of dopaminergic neurons [15–18], which involves D2R-mediated extracellular signal-regulated kinase (ERK) activation [15–18]. It is well known that the members of the Wnt protein family are involved in differentiation, organogenesis, and cell migration [19–21], and it has been suggested that Wnt5a regulates dopaminergic neuron differentiation [22, 23]. The Wnt ligands are extracellular proteins that bind to other cell types through primarily distinct classes of receptors, the Frizzled G protein-coupled receptors, and single transmembrane co-receptors such as LRP5/6 and Ryk [24–27]. It has been reported that Wnt5a can bind to the Frizzled 3, 4, and 5 receptors [28–30].

Wnt5a protein is also known to regulate the development of dopaminergic neurons. In a previous study, we analyzed the effect of Wnt5a on dopaminergic neuron development in mesencephalic primary cultures from wild-type (WT) and D2R knockout (D2R−/−) mice [17]. Treatment with Wnt5a increased the neurite number and length of dopamine neurons in primary mesencephalic neuronal cultures from WT mice, but not from D2R−/− mice [17]. Since the effect of Wnt5a on dopaminergic neurons was absent in D2R−/− mice, we investigated whether Wnt5a can interact with D2R. We found that Wnt5a interacts with D2R to promote the differentiation of dopaminergic neurons through the regulation of ERK signaling [17]. Wnt5a-mediated ERK activation is regulated not only by D2R but also by EGF receptor (EGFR) signaling suggesting that stimulation of D2Rs by Wnt5a activates ERK phosphorylation via EGFR, thereby promoting dopaminergic neuron development [17].

Recently, we have demonstrated that D2R activation induces shedding of heparin-binding EGF by activating a disintegrin and metalloproteinase (ADAM) 10 or 17, causing EGFR transactivation in mesencephalic neurons [18]. It has been also suggested that Wnt binding to its receptor, Frizzled (Fz), transactivates ErbB1, probably by matrix metalloproteinases (MMP)-mediated release of soluble ErbB1 ligands and then Wnt-transactivated ErbB1 stimulates MAPK activation [31]. Therefore, it is possible to imagine that Wnt5a binding to D2Rs promotes MMP-mediated EGF activation. However, further studies are needed to clarify how Wnt signaling can transactivate the EGFR–ERK signaling and whether Wnt binding to D2R mediates inhibition of AC.

In order to investigate whether Wnt5a can interact with D2R, co-immunoprecipitation of Wnt5a and D2R was conducted

(Fig. 1), and displacement of [^{3}H]spiperone, a D2R receptor antagonist, from D2R by Wnt5a was also performed, demonstrating that Wnt5a binds D2R (Fig. 2) [17]. This interaction was confirmed by glutathione S-transferase (GST) pull-down assays demonstrating that the domain including transmembrane domain 4, the second extracellular loop, and transmembrane domain 5 of D2R binds to Wnt5a (Fig. 1) [17]. In this chapter, we describe the co-immunoprecipitation, pull-down, and competition-binding assays used to analyze the D2R–Wnt5a interaction as briefly discussed below.

First, co-immunoprecipitation has allowed determining if an interaction between D2R and Wnt5a is detectable in HEK293T cells transfected with FLAG-tagged D2R and HA-tagged Wnt5a. Lysates of the transfected cells were subjected to immunoprecipitation with an anti-FLAG antibody, followed by Western blotting to detect HA (Fig. 1a). Second, GST pull-down assays were performed to examine in more detail the interaction between D2R and Wnt5a using five domain fragments of D2R (comprising amino acids 1–71, 72–151, 131–210, 211–374, and 375–444) which were expressed as GST fusion proteins. The interaction of each purified GST fusion protein with recombinant His–Wnt5a was analyzed by the GST pull-down assay (Fig. 1b, c). Lastly, radioligand-binding assays were done to determine the capacity of Wnt5a to bind D2R by measuring the ability of Wnt5a to compete with [^{3}H]spiperone, an antagonist of D2R. To validate the specificity of Wnt5a binding to D2R, competition-binding experiments were conducted using haloperidol, another D2R antagonist. The specificity of Wnt5a binding to D2R was further tested by comparing the D2R binding capacity to another related Wnt ligand, Wnt9b, using [^{3}H]spiperone. Our results suggest Wnt9a has a very low affinity relative to that of Wnt5a (Fig. 2).

2 Materials

2.1 Co-immunoprecipitation

1. HEK 293 T cells (CRL-3216, ATCC).
2. Constructs: p3XFLAG-myc CMV-26-mouse D2R (constructed from our lab) and pLNCX-HA-Wnt5a (provided by Dr. Jang-Soo Chun (Gwangju Institute of Science and Technology, Gwangju, South Korea).
3. Cell lysis buffer: 20 mM Tris (pH 7.5), 150 mM NaCl, 1 mM ethylenediaminetetraacetic acid (EDTA), 1 mM ethyleneglycoltetra-acetic acid (EGTA), 2.5 mM sodium pyrophosphate, 1 mM β-glycerophosphate, 1 % Triton X-100, 50 mM dithiothreitol (DTT), 1 μg/ml leupeptin, 1 μg/ml aprotinin, 1 mM Na_3VO_4, and 1 mM phenylmethylsulfonyl fluoride (PMSF) (*see* **Note 1**).

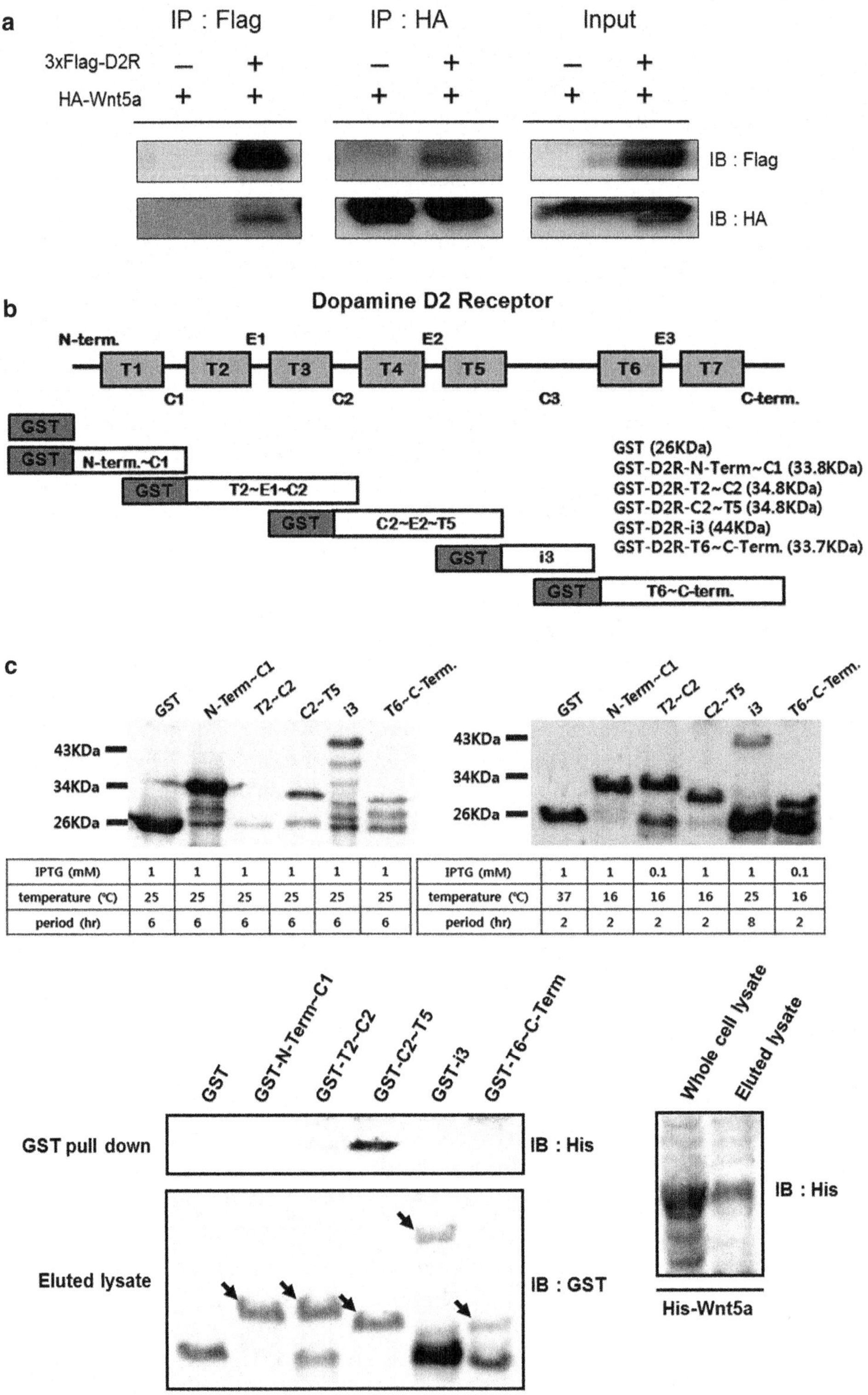

IPTG (mM)	1	1	1	1	1	1
temperature (℃)	25	25	25	25	25	25
period (hr)	6	6	6	6	6	6

IPTG (mM)	1	1	0.1	1	1	0.1
temperature (℃)	37	16	16	16	25	16
period (hr)	2	2	2	2	8	2

Fig. 1 Wnt5a interacts with D2R. (**a**) Immunoprecipitation shows that D2R is associated with Wnt5a. Protein samples were obtained from lysates of HEK293T cells transfected with p3xFLAG-myc-CMV-26-D2R and pLNCX-HA-Wnt5a. (**b**) Schematic diagram of five domain fragments of D2R (comprising amino acids 1–71,

4. 0.1 M phosphate buffered saline (PBS) (pH 7.4).
5. Anti-FLAG M2 monoclonal antibody (F3165, Sigma).
6. Anti-HA mouse monoclonal antibody (sc-7392, Santa Cruz Biotechnology).
7. Protein G Sepharose 4 Fast Flow (17-0618-01, GE Healthcare), Protein A Sepharose 4 Fast Flow (17-5208-01, GE Healthcare) (*see* **Note 2**).
8. Binding buffer: 50 mM Tris–HCl (pH 7.5), 150 mM NaCl, 2 mM EDTA, 2.5 mM sodium pyrophosphate, 0.5 % sodium deoxycholate, 1 % Triton™ X-100, 1 mM β-glycerophosphate, 1 μg/ml leupeptin, 1 μg/ml aprotinin, 1 mM Na_3VO_4, and 1 mM PMSF (*see* **Note 1**).
9. 5× SDS loading buffer: 250 mM Tris (pH 6.8), 50 % glycerol, 10 % SDS, 500 mM DTT, and 0.3 % bromophenol blue.

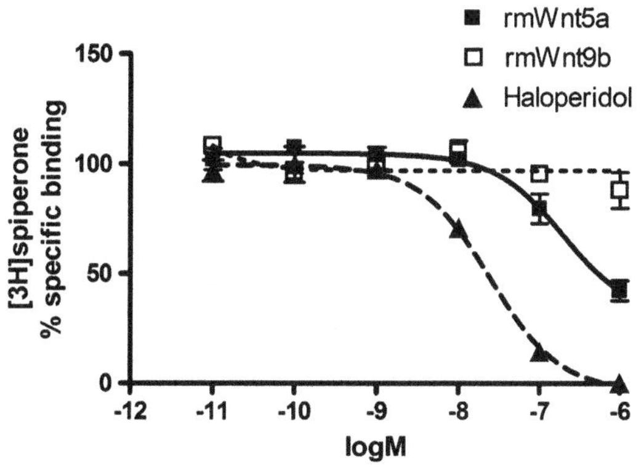

Fig. 2 Wnt5a binds to D2R. The binding capacities of rmWnt5a, rmWnt9b, and haloperidol were measured by inhibition of [^{3}H]spiperone binding to D2R following treatment with rmWnt5a and haloperidol. Nonspecific binding was measured in the presence of 5 μm butaclamol. Values are mean ± S.E. of four independent experiments, each conducted in duplicate. In each of the experiments, the curves were fitted via nonlinear regression analysis using the one-site competition model (GraphPad Prism program). Reprinted with permission from ref. [17]

Fig. 1 (continued) 72–151, 131–210, 211–374, and 375–444). Domain fragments are represented as GST-N-TermC1, GST-T2C2, GST-C2T5, GST-i3, and GST-T6C-Term, respectively. Expression conditions were construct dependent. Soluble proteins were obtained by modulating the concentration of IPTG, temperature, and expression duration. (**c**) GST pull-down assays used bacterially expressed GST-D2R mutants and His–Wnt5a. Pulled-down proteins were analyzed by SDS-PAGE and immunoblotting with His antibody. The *left bottom panel* shows GST and GST-D2R mutants eluted with glutathione Sepharose 4B and immunoblotted with GST antibody. The *right bottom panel* shows His–Wnt5a purified on NTA agarose and immunoblotted with His antibody. Reprinted with permission from ref. [17]

2.2 GST Pull Down

1. GST-tagged recombinant mouse D2R clone, His-tagged recombinant mouse Wnt5a clone.
2. Luria Bertani (LB) broth.
3. Ampicillin, kanamycin.
4. Isopropylthio-β-D-galactoside (IPTG, I1401, Duchefa).
5. PBS (pH 7.4).
6. Glutathione Sepharose 4B (17-0756-01, GE Healthcare) for GST-tagged protein, Ni–NTA agarose (30210, Qiagen) for His-tagged protein.
7. GST lysis buffer: 0.1 M PBS (pH 7.4), 1 mM DTT, 1 μg/ml leupeptin, 1 μg/ml aprotinin, 1 mM Na_3VO_4, 1 mM PMSF, and 10 μg/ml lysozyme. The buffer is prepared immediately before use.
8. 10 % Triton X-100.
9. His lysis and washing buffer: 50 mM Tris–HCl (pH 8.0), 0.3 M NaCl, 20 mM imidazole, 1 μg/ml leupeptin, 1 μg/ml aprotinin, 1 mM Na_3VO_4, 1 mM PMSF, and 10 μg/ml lysozyme.
10. His bead-binding buffer: 50 mM Tris–HCl (pH 8.0), 500 mM NaCl, 5 mM imidazole, and 5 mM β-mercaptoethanol.
11. GST washing buffer: 0.1 M PBS (pH 7.4), 1 μg/ml leupeptin, 1 μg/ml aprotinin, 1 mM Na_3VO_4, and 1 mM PMSF.
12. GST elution buffer: 50 mM Tris–HCl (pH 8.5), 20 mM glutathione (GSH), 0.1 % Triton X-100, 100 mM NaCl, 1 μg/ml leupeptin, 1 μg/ml aprotinin, 1 mM Na_3VO_4, and 1 mM PMSF.
13. His elution buffer: 50 mM Tris–HCl (pH 8.0), 500 mM NaCl, 250 mM imidazole, 1 μg/ml leupeptin, 1 μg/ml aprotinin, 1 mM Na_3VO_4, and 1 mM PMSF.
14. GST pull-down buffer: 20 mM Tris (pH 7.5), 1 mM EDTA (pH 8.0), 100 mM NaCl, 1 % Triton X-100, 1 % NP-40, 1 μg/ml leupeptin, 1 μg/ml aprotinin, 1 mM Na_3VO_4, and 1 mM PMSF.
15. 5× SDS loading buffer.

2.3 Competition Ligand-Binding Assay

1. Cell culture medium: Dulbecco's modified Eagle medium (DMEM)-high glucose, fetal bovine serum (FBS), and antibiotics (penicillin–streptomycin). Store at 4 °C.
2. Sterilized PBS by autoclaving: 137 mM NaCl, 2.7 mM KCl, 10 mM Na_2HPO_4, and 2 mM KH_2PO_4 in distilled water. Store at 4 °C.
3. 0.1 % trypsin–EDTA. Store at 4 °C.
4. Transfection reagent: jetPEI (Qbiogene). Store at 4 °C. Protect from light.

5. 100-mm culture dishes.
6. Construct p3xFLAG-myc-CMV-26-mouse D2R.
7. 15-ml and 50-ml conical tubes.
8. 10 mM Tris–Cl (pH 7.5), 5 mM EDTA. Store at 4 °C.
9. Polytron homogenizer (Ultra-Turrax T25, IKA).
10. Optima MAX-XP Ultracentrifuge (Beckman Coulter) with 100 Ti rotor.
11. 50 mM Tris–Cl (pH 7.7). Store at 4 °C.
12. [^{3}H]Spiperone (NET-856, PerkinElmer). Dissolve in binding buffer. Store at −80 °C.
13. (+)-Butaclamol (D033, Sigma), dissolve in ethanol. Solutions should be freshly prepared. Haloperidol (H1512, Sigma). Dissolve in DMSO. Store at −20 °C.
14. L-ascorbic acid (A4544, Sigma). Prepare solution immediately before use because it is easily oxidized. Store at 4 °C.
15. Chloride solution (30× stock): 3 M NaCl, 0.125 M KCl, 0.05 M $CaCl_2$, and 0.025 M $MgCl_2$ in distilled water. Store at room temperature (RT).
16. Reaction buffer: 1× chloride solution, 62.5 mM Tris–Cl (pH 7.7), and 0.125 % ascorbic acid in distilled water. Store at 4 °C.
17. Recombinant human/mouse Wnt5a (645-WN-010, R&D systems) and recombinant mouse Wnt-9b (3669-WN-025, R&D systems). Store at −80 °C (Stability and storage: 12 months from date of receipt). Prepare the solution immediately before use.
18. High-density polyethylene plastic 20-ml liquid scintillation vials (#986701, Wheaton).
19. Scintillation liquid (BetaplateScint, 1205-440, PerkinElmer). Store at RT.
20. Glass microfiber filters, grade GF/B (1821-915, Whatman).
21. Harvesters (M-48, Brandel).
22. Scintillation counter (LS6500, Beckman).

3 Methods

3.1 Co-immunoprecipitation Assay

3.1.1 DNA Preparation and Transfection

1. Transfect HEK293T cells with p3XFLAG-myc-CMV-26-mouse D2R and pLNCX-HA-Wnt5a using PEI transfection reagent (1 μg/μl) in 100-mm dishes.
2. Incubate the transfected cells at 37 °C in an atmosphere of 5 % CO_2 for 48 h.

3.1.2 Harvest and Lysis of Cells

1. Carefully wash cultured cells twice with prechilled PBS.
2. Scrape cells off dishes with a clean, cold scraper and transfer to new tubes.
3. Centrifuge at 1,000 × *g* at 4 °C for 5 min, and discard the supernatants.
4. Add cold cell lysis buffer described in item 3 of Sect. 2.1 (200 μl/100-mm dish).
5. Incubate on ice for 30 min, vortex three times.
6. Centrifuge at 16,000 × *g* at 4 °C for 30 min.
7. Transfer the supernatant to new tubes immediately.
8. Measure protein concentration by Bradford assay.

3.1.3 Preclearing

1. Wash Protein A/G Sepharose beads twice with cold PBS and prepare a 30 % slurry in PBS.
2. Mix the slurry and cell lysate at a ratio of 20 μl slurry to 1 mg cell lysate, and add cold PBS to get a final volume of 500 μl.
3. Incubate with gentle agitation for 2 h at 4 °C.
4. Centrifuge at 16,000 × *g* at 4 °C for 20 s. Transfer the supernatant to new tubes and discard the beads.

3.1.4 Immunoprecipitation

1. Add 7.5 μg primary antibody (anti-FLAG or anti-HA antibody) to precleared lysate. Add cold binding buffer up to a total volume of 500 μl.
2. Incubate lysate–antibody mixture with gentle agitation at 4 °C overnight.
3. Add 20 μl 30 % Protein A/G Sepharose slurry and incubate on an orbital shaker at 60 rpm for 4 h at 4 °C.
4. Centrifuge at 16,000 × *g* at 4 °C for 20 s. Discard the supernatant and wash the pellet with 1 ml cold binding buffer 3–5 times (*see* **Note 3**).
5. Centrifuge at 16,000 × *g* at 4 °C for 1 min, and discard the supernatant.
6. Add 5× SDS loading buffer to the beads and boil for 5 min.
7. Separate immunoprecipitated proteins by SDS-PAGE using 10 % polyacrylamide gel.

3.2 GST Pull Down

3.2.1 Growth and Induction of Cells

1. Seed cultures by inoculating separately D2R and Wnt5a clones in 3 ml of LB medium containing ampicillin.
2. Grow the inoculum for 12–16 h at 37 °C with vigorous shaking.
3. Dilute the culture 1:50 in desired volume of LB medium with ampicillin.
4. Grow liquid cultures at 37 °C with vigorous agitation to an A_{600} of 0.5–0.6 (approximately 2–3 h) (*see* **Note 4**).

5. Induce fusion protein expression by addition of IPTG to the culture to a final concentration of 0.1–1 mM.
6. Continue incubation for an additional 2 h at optimized temperature (16–37 °C).
7. Transfer the liquid cultures to labeled centrifuge tubes.
8. Centrifuge cells at 1,500 × *g* for 10 min at 4 °C and discard the supernatants. Drain the pellets thoroughly and place tubes on ice.
9. Resuspend each pellet in 50 μl of ice-cold lysis buffer for each milliliter of culture centrifuged (20 ml for 400 ml cultures). Save 20-μl aliquots for analysis by SDS-PAGE.

3.2.2 Cell Lysis

1. Lyse the cells using a sonicator. Sonicate 5–6 times for 30 s each with 30 s intervals (*see* **Note 6**).
2. Add 10 % Triton X-100 to a final concentration of 1 %.
3. Centrifuge at 16,000 × *g* at 4 °C for 10 min to remove insoluble material. Transfer the supernatants to fresh tubes. Save 20-μl aliquots for analysis by SDS-PAGE.

3.2.3 Equilibration of Beads

1. Use glutathione Sepharose 4B beads for GST-tagged protein and Ni–NTA (nitriloacetic acid) agarose for His-tagged protein.
2. Gently shake the bottle of beads to resuspend the matrix.
3. Use a pipette to remove sufficient slurry for use and transfer to a 15-ml Falcon tube.
4. Sediment the matrix by centrifugation at 1,000 × *g* for 30 s. Decant carefully the supernatant.
5. Wash the beads by adding cold PBS to GST-tagged protein and adding distilled water to His-tagged protein. Invert to mix.
6. Repeat steps 4 and 5 twice.
7. Decant carefully the supernatant and save only the beads. Add bead washing buffer to get a 50 % slurry.

3.2.4 Purification of Fusion Proteins

1. Add 200 μl of 50 % equilibrated bead slurry to 20 ml lysates and incubate in an orbital shaker at 60 rpm for 2 h at 4 °C.
2. Centrifuge at 1,000 × *g* for 30 s to sediment the beads and discard the supernatants.
3. Add 1 ml cold washing buffer, mix gently, and centrifuge at 1,000 × *g* for 30 s. Repeat twice for a total of three washes.
4. Elute fusion protein by the addition of 200 μl elution buffer. Suspend the beads and incubate by an orbital shaker at 60 rpm for 90 min at 4 °C.

5. Centrifuge at 16,000 × *g* for 1 min, and transfer the supernatants to fresh tubes.
6. Measure protein concentration by Bradford assay.

3.2.5 Protein Binding and Pull Down

1. Add 60 μl GST beads to 40 μg domain fragments of purified GST-tagged D2R (comprising amino acids 1–71, 72–151, 131–210, 211–374, and 375–444) and incubate at 4 °C for 2 h [17].
2. Add 1 ml of cold GST washing buffer, mix gently, and centrifuge at 1,000 × *g* for 30 s. Repeat once for a total of two washes.
3. Add 40 μg purified His-tagged Wnt5a to GST-tagged protein-bead complex, and add GST pull-down buffer to get a final volume of 1 ml and incubate at 4 °C overnight.
4. Centrifuge at 1,000 × *g* for 30 s and remove the supernatants.
5. Wash the beads with GST pull-down buffer three times.
6. Add 5× SDS loading buffer to the beads and boil for 5 min.
7. Separate fusion proteins by SDS-PAGE using 10 % polyacrylamide gels.
8. Perform Western blotting.
9. Detect specific bands by enhanced chemiluminescence (ECL; Amersham Biosciences) and analyze using a LAS3000 image analysis system (Fuji).

3.3 Competition Ligand-Binding Assay

3.3.1 DNA Preparation and Transfection

1. Prepare plasmid DNA (p3xFLAG-myc-CMV-26-mouse D2R) with Plasmid Midi Kit (QIAGEN).
2. Subculture HEK293T cells in 100-mm dishes (2×10^6 cells/dish) in 10 ml of growth medium.
3. Grow HEK293T cells for 18–22 h to 60–70 % confluence.
4. Transfect HEK293T cells with 14 μg p3xFLAG-myc-CMV-26-mouse D2R and 24 μl of 1 μg/μl PEI transfection reagent.
5. After 4 h incubation with plasmid DNA, replace transfection mix with fresh cell culture medium.
6. Incubate transfected HEK293T cells at 37 °C in a humidified atmosphere of 5 % CO_2 for 48 h.

3.3.2 Cell Harvest and Membrane Preparation

1. Remove medium and wash the transfected cells with 3 ml of prechilled PBS two times.
2. Harvest cells using a scraper and transfer into conical tubes.
3. Centrifuge at 1,000 × *g* for 5 min at 4 °C.
4. Discard supernatant and add 2 ml of 10 mM Tris–Cl (pH 7.5) containing 5 mM EDTA to each dish.

5. Homogenize with an Ultra-Turrax T25 at 11,000 rpm 10 times for 10 s each with 10 s intervals. Isolate membranes by centrifugation at 45,000*g* in an Optima MAX-XP ultracentrifuge for 45 min. Resuspend pellet in 2 ml of the same Tris–Cl–EDTA buffer used in step 4 and repeat centrifugation. Resuspend final pellet in 50 mM Tris–Cl (pH 7.7).

3.3.3 Competition Binding

1. Use 40 μg of membrane protein and 0.5 nM [^{3}H]spiperone dissolved in binding buffer (specific activity 79.6 Ci/mmole) for ligand-binding assays (*see* **Note 7**).
2. Use 5 μM (+)-butaclamol to measure nonspecific binding. (+)-Butaclamol (D033, Sigma) is dissolved in ethanol. Solutions should be freshly prepared.
3. For the competition experiments, use haloperidol and Wnt5a or Wnt9b at concentrations ranging from 10^{-11} to 10^{-6} M. Incubate all samples with 0.5 nM [^{3}H]spiperone (Table 1). All reagents are described in Sect. 2.3 (steps 15 and 16) (*see* **Note 8**).
4. Incubate at 37 °C for 1 h.
5. Rapidly filter through Whatman GF/B filters using a harvester (M-48, Brandel Harvester).
6. Wash twice with 4 ml of 50 mM ice-cold Tris–Cl (pH 7.7).
7. Put Whatman GF/B filters into scintillation vials, add 5 ml of scintillation cocktail and incubate overnight at 4 °C.
8. Count radioactivity on the filter with a scintillation counter.

3.3.4 Analysis

Binding data can be analyzed to one-site binding model with GraphPad Prism for Windows or Mac (GraphPad Software, La Jolla California USA, www.graphpad.com).

Table 1
Composition of reaction mixture for competition-binding assay

Membrane protein	0 μg	40 μg	40 μg					
Wnt5a or Wnt9b	0	0	10^{-11} M	10^{-10} M	10^{-9} M	10^{-8} M	10^{-7} M	10^{-6} M
(+)-Butaclamol	Final 5 μM							
[^{3}H]Spiperone	Final 0.5 nM							
Binding buffer	Up to 100 μl							

Membrane proteins are prepared in 50 mM Tris–Cl (pH 7.7). Composition of binding buffer is described in Sect. 2.3 (steps 15 and16). The condition with "0 μg" membrane protein and "0 μg" of Wnt competitor is used as a negative control (optional)

4 Notes

1. All solutions should be freshly prepared before each use. All aqueous solutions are prepared using deionized distilled water.
2. Select protein Sepharose beads based on antibody isotypes. For instance, anti-FLAG M2 monoclonal antibody (F3165, Sigma): IgG1→Protein G (+++; extremely recommended) or Protein A (+; recommended). Anti-HA mouse monoclonal antibody (sc-7392, Santa Cruz Biotechnology): IgG2a→Protein A (+++; extremely recommended) or Protein G (+++; extremely recommended).
3. After immunoprecipitation, the NaCl concentration (up to 200 mM) can be increased to reduce the level of nonspecific binding.
4. The production of soluble and undergradable (stable) proteins requires optimization of several parameters, including host strain, optical density of cells at time of induction, induction temperature, and induction duration. A greater yield of soluble, full-length fusion protein is usually obtained using lower growth temperature, shorter induction time, higher cell density, and lower IPTG concentration.
5. For the host strain, we tested the *Escherichia coli* expression hosts, BL21 (DE3), BL21 (DE3) pLysS, and SoluBL21. Under our experimental conditions, the best yields were obtained with BL21 (DE3).
6. Excessive sonication can result in co-purification of *E. coli* host proteins and may also lead to denaturation and break down of the fusion protein. The frequency and intensity of sonication should be adjusted such that complete lysis is achieved without frothing (which can denature proteins). After sonication, add Triton X-100 to facilitate the solubilization of the fusion protein.
7. Reactions are carried out in a volume of 100 μl. Estimate the concentration of [^{3}H]spiperone (79.6 Ci/mmol) by measuring the cpm in 1 μl of [^{3}H]spiperone (*see* Table 1).
8. Recombinant human/mouse Wnt5a (645-WN-010, R&D systems) and recombinant mouse Wnt-9b (3669-WN-025, R&D systems) are diluted to adequate concentrations for competition-binding assay. Prepare recombinant mouse (rm) Wnt5a: the molecular weight of rmWnt5a is 38 kDa (MW: 4.18×10^6). Dissolve 10 μg of Wnt5a in 26.8 μl of reaction mixture (1×10^{-5} M) and serially dilute the solution. Prepare recombinant mouse (rm) Wnt9b: the molecular weight of rmWnt9b is 36.8 kDa (MW: 4.05×10^6). Dissolve 25 μg of Wnt9b in 67.0 μl of reaction mixture (1×10^{-5} M) and serially dilute the solution.

Acknowledgements

This work was supported in part by research grants from the Korea Healthcare Technology R&D Project A084139and the National Research Foundation of Korea (2011-0015678), funded by the Korean government.

References

1. Beaulieu JM, Gainetdinov RR (2011) The physiology, signaling, and pharmacology of dopamine receptors. Pharmacol Rev 63:182–217
2. Tritsch NX, Sabatini BL (2012) Dopaminergic modulation of synaptic transmission in cortex and striatum. Neuron 76:33–50
3. Baik JH (2013) Dopamine signaling in reward-related behaviors. Front Neural Circuits 7:152
4. Simon HH, Bhatt L, Gherbassi D, Sgadó P, Alberí L (2003) Midbrain dopaminergic neurons: determination of their developmental fate by transcription factors. Ann N Y Acad Sci 991:36–47
5. Smidt MP, Burbach JP (2007) How to make a mesodiencephalic dopaminergic neuron. Nat Rev Neurosci 8:21–32
6. Lindvall O, Kokaia Z (2009) Prospects of stem cell therapy for replacing dopamine neurons in Parkinson's disease. Trends Pharmacol Sci 30:260–267
7. Dearry A, Gingrich JA, Falardeau P, Fremeau RT Jr, Bates MD, Caron MG (1990) Molecular cloning and expression of the gene for a human D1 dopamine receptor. Nature 347:72–76
8. Zhou QY, Grandy DK, Thambi L, Kushner JA, Van Tol HH, Cone R, Pribnow D, Salon J, Bunzow JR, Civelli O (1990) Cloning and expression of human and rat D1 dopamine receptors. Nature 347:76–80
9. Sunahara RK, Guan HC, O'Dowd BF, Seeman P, Laurier LG, Ng G, George SR, Torchia J, Van Tol HH, Niznik HB (1991) Cloning of the gene for a human dopamine D5 receptor with higher affinity for dopamine than D1. Nature 350:614–619
10. Grandy DK, Zhang YA, Bouvier C, Zhou QY, Johnson RA, Allen L, Buck K, Bunzow JR, Salon J, Civelli O (1991) Multiple human D5 dopamine receptor genes: a functional receptor and two pseudogenes. Proc Natl Acad Sci U S A 88:9175–9179
11. Bunzow JR, Van Tol HH, Grandy DK, Albert P, Salon J, Christie M, Machida CA, Neve KA, Civelli O (1988) Cloning and expression of a rat D2 dopamine receptor cDNA. Nature 336:783–787
12. Dal TR, Sommer B, Ewert M, Herb A, Pritchett DB, Bach A, Shivers BD, Seeburg PH (1989) The dopamine D2 receptor: two molecular forms generated by alternative splicing. EMBO J 8:4025–4034
13. Sokoloff P, Giros B, Martres MP, Bouthenet ML, Schwartz JC (1990) Molecular cloning and characterization of a novel dopamine receptor (D3) as a target for neuroleptics. Nature 347:146–151
14. Van Tol HH, Bunzow JR, Guan HC, Sunahara RK, Seeman P, Niznik HB, Civelli O (1991) Cloning of the gene for a human dopamine D4 receptor with high affinity for the antipsychotic clozapine. Nature 350:610–614
15. Kim SY, Choi KC, Chang MS, Kim MH, Kim SY, Na YS, Lee JE, Jin BK, Lee BH, Baik JH (2006) The dopamine D2 receptor regulates the development of dopaminergic neurons via extracellular signal-regulated kinase and Nurr1 activation. J Neurosci 26:4567–4576
16. Kim SY, Lee HJ, Kim YN, Yoon S, Lee JE, Sun W, Choi EJ, Baik JH (2008) Striatal-enriched protein tyrosine phosphatase regulates dopaminergic neuronal development via extracellular signal-regulated kinase signaling. Exp Neurol 214:69–77
17. Yoon S, Choi MH, Chang MS, Baik JH (2011) Wnt5a-dopamine D2 receptor interactions regulate dopamine neuron development via extracellular signal-regulated kinase (ERK) activation. J Biol Chem 286:15641–15651
18. Yoon S, Baik JH (2013) Dopamine D2 receptor-mediated epidermal growth factor receptor transactivation through a disintegrin and metalloprotease regulates dopaminergic neuron development via extracellular signal-related kinase activation. J Biol Chem 288:28435–28446
19. Wodarz A, Nusse R (1998) Mechanisms of Wnt signaling in development. Annu Rev Cell Dev Biol 14:59–88
20. Moon RT, Brown JD, Torres M (1997) WNTs modulate cell fate and behavior during vertebrate development. Trends Genet 13:157–162
21. Budnik V, Salinas PC (2011) Wnt signaling during synaptic development and plasticity. Curr Opin Neurobiol 21:151–159

22. Arenas E (2005) Engineering a dopaminergic phenotype in stem/precursor cells: role of Nurrl, glia-derived signals, and Wnts. Ann N Y Acad Sci 1049:51–66
23. Castelo-Branco G, Wagner J, Rodriguez FJ, Kele J, Sousa K, Rawal N, Pasolli HA, Fuchs E, Kitajewski J, Arenas E (2003) Differential regulation of midbrain dopaminergic neuron development by Wnt-1, Wnt-3a, and Wnt-5a. Proc Natl Acad Sci U S A 100:12747–127529
24. Bhanot P, Brink M, Samos CH, Hsieh JC, Wang Y, Macke JP, Andrew D, Nathans J, Nusse R (1996) A new member of the frizzled family from Drosophila functions as a Wingless receptor. Nature 382:225–230
25. Tamai K, Semenov M, Kato Y, Spokony R, Liu C, Katsuyama Y, Hess F, Saint-Jeannet JP, He X (2000) LDL-receptor related proteins in Wnt signal transduction. Nature 407: 530–535
26. Wehrli M, Dougan ST, Caldwell K, O'Keefe L, Schwartz S, Vaizel-Ohayon D, Schejter E, Tomlinson A, DiNardo S (2000) *arrow* encodes an LDL-receptor-related protein essential for Wingless signalling. Nature 407:527–530
27. Lu W, Yamamoto V, Ortega B, Baltimore D (2004) Mammalian Ryk is a Wnt coreceptor required for stimulation of neurite outgrowth. Cell 119:97–108
28. He X, Saint-Jeannet JP, Wang Y, Nathans J, Dawid I, Varmus H (1997) A member of the Frizzled protein family mediating axis induction by Wnt-5A. Science 275:1652–165
29. Mikels AJ, Nusse R (2006) Purified Wnt5a protein activates or inhibits β-catenin–TCF signaling depending on receptor context. PLoS Biol 4:e115
30. Kawasaki A, Torii K, Yamashita Y, Nishizawa K, Kanekura K, Katada M, Ito M, Nishimoto I, Terashita K, Aiso S, Matsuoka M (2007) Wnt5a promotes adhesion of human dermal fibroblasts by triggering a phosphatidylinositol-3 kinase/Akt signal. Cell Signal 19: 2498–2506
31. Civenni G, Holbro T, Hynes NE (2003) Wnt1 and Wnt5a induce cyclin D1 expression through ErbB1 transactivation in HC11 mammary epithelial cells. EMBO Rep 4:166–171

Chapter 6

Regulation of Pre- and Postsynaptic Protein Phosphorylation by Dopamine D2 Receptors

Alessandra Bonito-Oliva, Giada Spigolon, and Gilberto Fisone

Abstract

Dopamine D2-type receptors are coupled to $G_{i/o}$ proteins, which inhibit adenylyl cyclase, the enzyme responsible for the synthesis of 3′,5′-cyclic monophosphate (cAMP). Therefore, a considerable proportion of the effects produced on protein phosphorylation by pharmacological manipulations of these receptors is mediated by changes in the activity of cAMP-dependent protein kinase. Studies performed in the striatum, a brain region particularly enriched in D2-type receptors, have led to the identification of several downstream target phosphoproteins regulated by D2-type receptor agonists and antagonists. This chapter provides a short introductory summary of the mechanisms involved in such regulations and describes two standard methodologies, Western blotting and immunohistochemistry, which can be employed to investigate changes in the state of phosphorylation of D2-type receptor targets localized at presynaptic and postsynaptic level.

Key words cAMP-dependent protein kinase, Haloperidol, Immunofluorescence, Quinpirole, Striatum, Western blotting

1 Introduction

1.1 Dopamine Receptors and Signaling

Dopamine exerts its effects through activation of metabotropic, heptahelical receptors, coupled to heterotrimeric, guanosine triphosphate binding proteins (G-proteins). The current criterion for classification of dopamine receptors is based on their ability to interact with specific G-proteins, which once activated stimulate or inhibit the activity of adenylyl cyclase (AC), the enzyme responsible for the synthesis of 3′,5′-cyclic monophosphate (cAMP). Dopamine D1 and D5 receptors are coupled to G_s and G_{olf} proteins, which stimulate AC, and are referred to as D1-type receptors. Conversely, dopamine D2, D3, and D4 receptors belong to the D2-type receptors and act through $G_{i/o}$ proteins, which inhibit AC [1].

A large proportion of the current understanding of dopamine signaling derives from studies performed in the striatum, a major component of the basal ganglia innervated by midbrain dopaminergic neurons. D1-type and D2-type receptors are expressed in

Mario Tiberi (ed.), *Dopamine Receptor Technologies*, Neuromethods, vol. 96,
DOI 10.1007/978-1-4939-2196-6_6, © Springer Science+Business Media New York 2015

distinct populations of GABAergic medium spiny neurons (MSNs), which are the principal neuronal type in the striatum [2]. This chapter is centered on the regulation exerted by D2-type receptors on the phosphorylation of pre- and postsynaptic proteins, in striatal MSNs. Focus will be given to targets of the canonical cAMP/cAMP-dependent protein kinases (PKA) signaling cascade. However, D2-type receptors also act via G-protein independent mechanisms, which affect a different set of targets (e.g., Akt and GSK-3), typically localized at the postsynaptic level [3].

The regulation of AC by D1-type and D2-type receptors leads to opposite changes in the activity of PKA, which in turn affects the phosphorylation of a multitude of downstream target proteins, involved in the control of short- and long-term neuronal responses [4]. The characterization of the mechanisms implicated in D1-type receptor-mediated transmission has led to the identification of numerous effectors, which are phosphorylated in response to the robust activation of the cAMP/PKA pathway produced by selective D1-type receptor agonists (e.g., SKF38393 and SKF81297), psychostimulants, such as cocaine and amphetamine, and anti-Parkinsonian drugs, such as l-dopa [4].

In contrast, a clear appreciation of the impact produced on protein phosphorylation by activation of D2-type receptors has been more elusive. This depends in part on the coupling of these receptors to inhibition of cAMP/PKA signaling. Thus, whereas activation of D1-type receptors, through stimulation of PKA, increases by severalfold the levels of a phosphoprotein, a selective D2-type receptor agonist (e.g., quinpirole) would typically produce a more modest effect, limited to the reduction of basal phosphorylation. To circumvent this problem, D2-type receptor-mediated phosphorylation is often studied by examining the effects of specific antagonists. These drugs promote cAMP/PKA signaling by suppressing the inhibitory control exerted by tonic activation of D2-type receptors. In this way, administration of selective D2-type receptor antagonists (e.g., eticlopride and raclopride), or even of typical antipsychotic drugs with preferential affinity for D2-type receptors (e.g., haloperidol), results in a robust increase in PKA-mediated protein phosphorylation [4].

This situation is exemplified by comparing the effects of quinpirole and haloperidol, on the phosphorylation of PKA substrates, such as the GluA1 subunit of the α-amino-3-hydroxy-5-methylisoxazole-4-propionic acid (AMPA) glutamate receptor and the dopamine- and cAMP-regulated phosphoprotein of 32 kDa (DARPP-32). In the mouse striatum, administration of quinpirole (0.2 mg/kg, IP) produces a 25 % reduction in the levels of phospho-Ser845-GluA1 and phospho-Thr34-DARPP-32 (Fig. 1a). Conversely, administration of haloperidol (0.5 mg/kg, IP) results in larger increases of both phosphoproteins (Fig. 1b) [5].

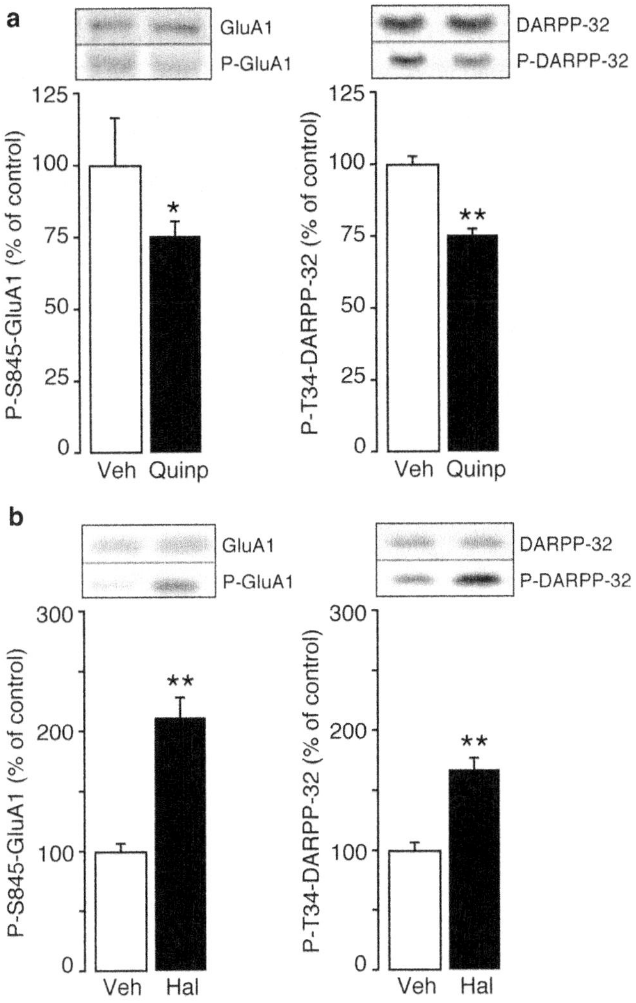

Fig. 1 Effects of quinpirole and haloperidol on the phosphorylation of GluA1 and DARPP-32 measured in the mouse striatum. (**a**) Male C57Bl/6 mice were treated IP with (**a**) quinpirole (0.2 mg/kg) or (**b**) haloperidol (0.5 mg/kg) and killed after 15 min. PhosphoSer845-GluA1, phosphoThr34-DARPP-32, total GluA1, and total DARPP-32 were determined by Western blotting. *Upper panels* show representative autoradiograms. *Lower panels* show summaries of data expressed as means ± SEM (n=6–12). The amount of phospho-Ser845-GluR1 and phospho-Thr34-DARPP-32 are expressed as percentage of those determined after vehicle administration. $p<0.05$, $p<0.01$ versus vehicle (one-way ANOVA followed by Dunnett's test). Modified from [5]

The use of antagonists in the study of D2-type receptor-mediated phosphorylation is often the only alternative when changes in phosphorylation are examined by immunohistochemistry. With this technique, basal levels of phosphoproteins are in many cases below the detection threshold [6, 7], which precludes from appreciating reductions in phosphorylation caused by activation of D2-type receptors. In contrast, administration of D2-type receptor antagonists results in large increases in the state

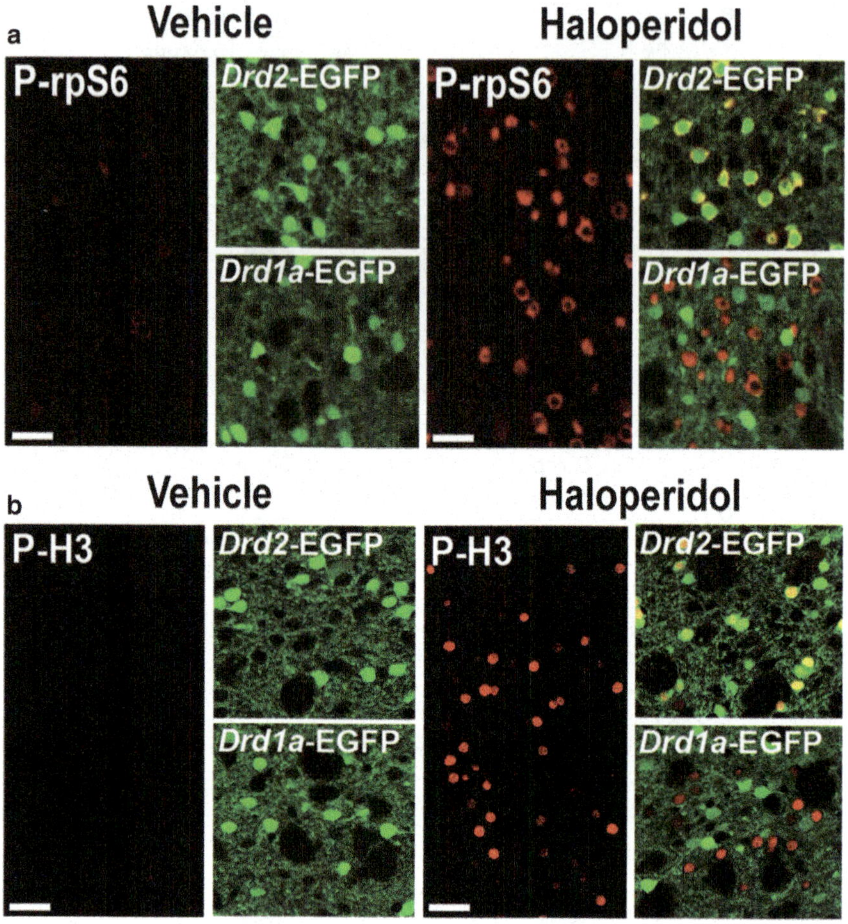

Fig. 2 Effect of haloperidol on the phosphorylation of rpS6 and histone H3 in striatal MSNs. Mice expressing EGFP (*green*) in MSNs containing D2Rs (*Drd2*-EGFP) or D1Rs (*Drd1a*-EGFP) were treated with vehicle or haloperidol (0.5 mg/kg) and perfused 15 min later. Confocal sections of the dorsal striatum show immunofluorescence (*red*) for (**a**) phospho-Ser235/236-rpS6 (P-rpS6) or (**b**) phospho-Ser10-histone H3 (P-H3). As mentioned in the text, immunohistochemistry does not allow for a clear detection of basal levels of phosphoproteins (cf. absence of P-rpS6- and P-H3-immunoreactivity in the striata of mice treated with vehicle). Haloperidol produces a robust increase in both P-rpS6 and P-H3. Note the distinct localization of P-rpS6 and P-H3 in the cytoplasm and nucleus, respectively. Upper small panels show co-localization (*yellow*) of (**a**) P-rpS6 and (**b**) P-H3 with EGFP, in *Drd2*-EGFP mice. Conversely, in *Drd1a*-EGFP mice (lower small panels), no co-localization is observed between phosphoproteins and EGFP. Scale bars: 40 μm. Modified from [6, 7]

of phosphorylation of cytoplasmic (e.g., ribosomal protein S6; rpS6) [7, 8] and nuclear (e.g., histone H3) [6] targets (Fig. 2). Immunohistochemistry is particularly useful when applied to mice expressing fluorescent probes in specific groups of neurons, such as bacterial artificial chromosome transgenic mice in which EGFP is selectively expressed in MSNs containing dopamine D1 or D2 receptors [9]. Using this approach the effects produced by raclopride and haloperidol on histone H3 and rpS6 have been localized to dopamine D2 receptor-expressing MSNs (Fig. 2) [6–8].

The increases in protein phosphorylation produced by haloperidol, raclopride, and eticlopride depend on the ability of these drugs to antagonize D2-type receptors. Since antagonists are devoid of intrinsic activity, these effects are likely exerted by suppressing the inhibition of cAMP signaling exerted by dopamine-mediated activation of D2-type receptors. Thus, augmented protein phosphorylation produced by blockade of D2-type receptors relies on the presence of basal activation of AC. Indeed, dopamine D2-type receptor-expressing MSNs are highly enriched in adenosine A_{2A} receptors (A2ARs), which are coupled to G_{olf}-mediated activation of AC [10, 11]. A2ARs are tonically activated by endogenous adenosine, which is generated as a product of nucleotides breakdown [12]. Therefore, in striatal MSNs, cAMP/PKA signaling is controlled by the opposite actions of D2-type receptors and A2ARs. In line with this interaction, pharmacological or genetic inactivation of A2ARs prevents the increase in phosphorylation induced by D2-type receptor antagonists at the level of several downstream targets [5–7, 13].

One important issue concerning the study of D2-type receptor signaling regards the distinction between effects produced at pre- versus postsynaptic level. Dopamine D2 receptors are highly expressed in the soma and terminals of midbrain dopaminergic neurons, where they function as inhibitory autoreceptors. In this regard, the regulation of the phosphorylation of tyrosine hydroxylase (TH), the rate limiting enzyme in the synthesis of catecholamines, has emerged as an ideal parameter to determine the effects produced presynaptically by pharmacological and genetic manipulation of D2-type receptors. The utility of this approach is exemplified by the study of the two isoforms of the dopamine D2 receptor, which are generated by alternative splicing and referred to as short (D2S) and long D2 (D2L) receptor. Initial studies indicated that D2S receptors are involved in presynaptic dopamine transmission [14]. Using transgenic mice lacking both D2S and D2L, or only D2L receptors, it was subsequently shown that the D2S receptor is implicated in the regulation of TH phosphorylation [15]. As in the case of postsynaptic targets, effects produced on TH phosphorylation by D2-type receptor antagonists are much larger in comparison to those observed with agonists. Thus, administration of eticlopride or haloperidol results in severalfold increases in the phosphorylation of TH at Ser40, a site regulated by PKA [16].

In the following sections we will describe basic methodologies to determine changes in the state of phosphorylation of pre- and postsynaptic proteins, following systemic administration of agonists or antagonists at D2-type receptors. Description will be centered on two standard techniques, Western blotting and immunohistochemistry, combined with the use of the antibodies listed in Table 1.

Table 1
Summary of the antibodies used to determine the effects produced by D2R agonists and antagonists on pre- and postsynaptic phosphoproteins

Phosphoprotein	Localization	I Ab-Western blotting (ECL detection)	II Ab-Western blotting (ECL detection)	I Ab-IHC	II Ab-IHC	References
P-Ser31-TH	Presynaptic	PhosphoSolutions cat. #p1580-31 (1:10,000)	Polyclonal goat anti-rabbit (Dako, cat. #P0448) 1:15,000			[16]
P-Ser40-TH	Presynaptic	PhosphoSolutions cat. #p1580-40 (1:10,000)	Polyclonal goat anti-rabbit (Dako, cat. #P0448) 1:15,000			[16]
P-Ser845-GluA1	Postsynaptic	PhosphoSolutions cat. #p1160-845 (1:1,000)	Polyclonal goat anti-rabbit (Dako, cat. #P0448) 1:30,000			[5]
P-Ser235/236-rpS6	Postsynaptic	Cell signaling cat. #2211S (1:1,000)	Polyclonal goat anti-rabbit (Dako, cat. #P0448) 1:10,000	Cell signaling cat. #2211S (1:500)	Jackson ImmunoResearch Goat anti-rabbit Cy3-conj. cat. #111-165-003 (1:400) Goat anti-rabbit-Cy2 cat. #111-225-003 (1:400)	[7]
P-Ser240/244-rpS6	Postsynaptic	Cell signaling cat. #2215S (1:1,000)	Polyclonal goat anti-rabbit (Dako, cat. #P0448) 1:10,000	Cell signaling cat. #2215S (1:500)	Jackson ImmunoResearch Goat anti-rabbit Cy3-conj. cat. #111-165-003 (1:400) Goat anti-rabbit-Cy2 cat. #111-225-003 (1:400)	[8]
P-Ser10-H3	Postsynaptic			Millipore cat. #06-570 (1:500)	Jackson ImmunoResearch Goat anti-rabbit Cy3-conj. cat. #111-165-003 (1:400) Goat anti-rabbit-Cy2 cat. #111-225-003 (1:400)	[6]
P-Thr34-DARPP32	Postsynaptic	PhosphoSolutions cat. #P-1025-34 (1:1,000)	Polyclonal goat anti-rabbit (Dako, cat. #P0448) 1:15,000			[5]

1.2 Western Blotting

Western blotting is an immunochemical technique commonly used to detect and quantify proteins in biological samples. Cell or tissue lysates are prepared in a cocktail of detergents, salts, and buffers, and proteins are separated by sodium dodecyl sulfate (SDS)-polyacrylamide gel electrophoresis (PAGE). During this step, denatured, charged molecules migrate in the gel through an electrical field. Using this technique, it is possible to separate proteins ranging from about 10 to 200 kDa in molecular weight. Following separation by SDS-PAGE, proteins are electrophoretically transferred and immobilized onto membranes, which are incubated with primary antibodies. Secondary antibodies are then used to bind the complexes formed by the primary antibodies with specific proteins or phosphoproteins.

In this chapter we will refer to secondary antibodies linked to horseradish peroxidase (HRP). In the enhanced chemiluminescence (ECL) assay, HRP catalyzes the oxidation of luminol, which generates light detected by exposing an X-ray film to the membranes. The film can then be scanned to produce an image file in which the intensity of the bands corresponding to the proteins, or phosphoproteins of interest, is measured by an image analysis software (i.e., Bio-Rad).

An alternative detection can be performed using the LI-COR Odissey System (LI-COR, Lincoln, NE, USA). In this case, the membrane is incubated with a fluorophore-conjugated secondary antibody, scanned to visualize protein-antibody complexes, and the intensity of the bands corresponding to the proteins or phosphoproteins of interest can be analyzed with appropriate software (i.e., LI-COR or Bio Rad). Notably, with the LI-COR Odyssey System, it is possible to detect, on the same membrane, a target protein in its total (phosphorylated and non-phosphorylated) and phosphorylated form. In this case, the primary antibodies against the total and the phosphorylated protein must have been generated in different species (e.g., mouse vs. goat). This will make it possible the combined use of two species-specific secondary antibodies conjugated with different fluorophores, which can be detected in the same blot.

The levels of phosphorylated proteins must be normalized according to the amount of total (i.e., non-phosphorylated plus phosphorylated) protein present in the sample. Moreover, as an additional control, sample loading can be controlled by measuring the amount of a housekeeping protein (e.g., actin).

1.3 Immunohistochemistry

Immunohistochemistry is a common laboratory technique used to visualize the distribution of specific antigens in tissues. In addition to qualitative information regarding protein localization, immunohistochemical methods and analysis have been adapted to provide semiquantitative as well as quantitative information. However, Western blotting remains the preferable method for quantitative analysis, particularly when studying the effects of D2-type receptor agonists on protein phosphorylation (cf. Sect. 1). The use of immunohistochemistry is particularly advantageous in combination with transgenic mice in which specific groups of neurons are labeled with fluorescent probes (e.g., EGFP; *see* Sect. 1 and Fig. 2) [9].

2 Materials

2.1 Drug Treatment and Brain Dissection

1. Quinpirole (Sigma-Aldrich, cat. #Q111), eticlopride (Sigma-Aldrich, cat. #E101), raclopride (Sigma-Aldrich, cat. #R121), haloperidol (Sigma-Aldrich, cat. #H1512), and appropriate vehicles.
2. Acetic acid 100 % (MERCK, cat. #1.00063.2500).
3. 1 ml syringes and 22 G syringe needles.
4. Small animal decapitator.
5. Liquid nitrogen and appropriate containers for dissection and storage of brain samples.
6. Mouse brain matrix (ASI instruments, RBM cat. #2000C) and feather blades (Agar Scientific, Stansted, UK; cat. # T5115).
7. Surgical tools (operating scissors, micro-dissecting forceps, spatula), Sample Corer (internal diameter 1 mm) for tissue punches (Fine Science Tools, Heidelberg, Germany; cat. # 18035-01).
8. Eppendorf tubes (1.5 and 2 ml) for collection of brain samples.

2.2 Sample Preparation for Western Blotting

1. 1 % SDS (Sigma-Aldrich, cat. #L3771).
2. UP50H-Compact Lab Homogenizer (Hielscher Ultrasound Technology).
3. Techne Dri-block DB-3.

2.3 Protein Quantification

1. Pierce BCA Protein Assay Reagent A (Thermo Scientific, cat. #23222).
2. Pierce BCA Protein Assay Reagent B (Thermo Scientific, cat. #23224).
3. Disposable 12 × 75 mm glass tubes (Corning 99445-12 PYREX).
4. Albumin standard (Thermo Scientific, cat. #23209).
5. Sample buffer 4×: 0.6 g Tris base, 8 ml glycerol, 1.6 g SDS, 4 ml β-mercaptoethanol, 500 μl 0.1 % bromophenol blue, adjust pH to 6.75, and add ddH_2O to 20 ml.

2.4 SDS-PAGE and Electrophoretic Transfer

1. Upper buffer, pH 6.8: 30.25 g Tris base, 2 g SDS, adjust pH to 6.8, and add ddH_2O to 500 ml.
2. Lower buffer, pH 8.8: 90.75 g Tris base, 2 g SDS, adjust pH to 8.8, and add ddH_2O to 500 ml.

3. 40 % acrylamide/bis solution (Bio-Rad, cat. #161-0148).
4. 10 % ammonium persulfate (APS) (Sigma-Aldrich, cat. #A3678-25G).
5. N,N,N′,N′-Tetramethylethylenediamine (TEMED) (Sigma-Aldrich, cat. #T9281-25G).
6. Dual gel caster (Hoefer, cat. #SE245) including casting cradle, casting clamp assembly, silicone rubber gaskets, cams, and clamps.
7. Two (10×10 cm) glasses, two alumina plates (10×10 cm), and two spacers (GE Healthcare Life Science).
8. Two 18-wells combs (GE Healthcare Life Science).
9. Gel electrophoresis unit (Hoefer SE260) including lower buffer chamber, upper buffer chamber, color-coded lead, safety lead, and spring clamps.
10. Running buffer (10×): 144 g glycine, 30 g Tris base, 10 g sodium dodecyl sulfate; add ddH_2O to 1 l.
11. Prestained Protein Ladder (Thermo Scientific, cat. #26616).
12. Electrophoresis power supply (Amersham Pharmacia Biotech).
13. Transfer buffer (10×): 30.3 g Tris base, 144 g glycine; add ddH_2O to 1 l.
14. Mighty Small Transfer Tank (Hoefer, cat. #TE22) including lower chamber w/heat exchanger, safety lid w/high voltage leads, electrode panels, cassettes, foam sponges, and cassette hook. Blotter papers.
15. PVDF Immobilon-P transfer membranes (Millipore, cat. #IPVH00010).
16. Methanol (Millipore, cat. #1.06009.2500).

2.5 Antibody Incubation

1. Methanol (Millipore, cat. #1.06009.2500).
2. Incubation boxes (9.5×3 cm).
3. PBS-Tween (0.1 % Tween 20 in phosphate buffered saline).
4. Blocking solution: 5 % nonfat milk powder in PBS-Tween (for ECL). For LI-COR System use the Odyssey Blocking Buffer (LI-COR, cat. # 927-40000).
5. Primary antibodies (*see* Table 1).
6. Secondary antibodies (*see* Table 1).

2.6 Signal Detection and Acquisition

1. ECL detection kit (Amersham, GE Healthcare, cat. #RPN2232).
2. Fixer and developer solutions (Kodak, cat. #1901859).
3. Amersham Hyperfilm ECL (GE Healthcare, cat. #28906837).
4. LI-COR Odyssey Scanner System (LI-COR).

2.7 Materials for Immunohistochemistry

1. Pentobarbital (APL, Stockholm, Sweden, cat. #338327); dilute 1:1 with saline.
2. Syringes (1 ml) and 22 G syringe needles.
3. Phosphate buffer 0.2 M, pH 7.5 (PB). For 500 ml, mix 95 ml of 0.2 M NaH_2PO_4 with 405 ml of 0.2 M Na_2HPO_4.
4. Phosphate-buffered saline pH 7.5 (PBS): 137 mM NaCl, 2.7 mM KCl, 10 mM Na_2HPO_4, and 1.8 mM K_2PO_4.
5. 10× Tris-buffered saline (TBS) (10×): 0.25 M Trizma base (Sigma-Aldrich, cat. #T1503) and 5 M NaCl (Merck Millipore, cat. #567440). Adjust to pH 7.5 with HCl. This stock solution can be stored at room temperature for up to 2 months. The 1× working solution is stable for up to 2 weeks at room temperature (readjust pH to 7.5 if necessary).
6. Blocking buffer (TBS 0.3 %; Triton 5 % BSA): 1× TBS, 0.3 % Triton X-100 (Merck Millipore, cat. #648466) and 5 % BSA (Sigma-Aldrich, cat. #A3059).
7. Tris buffer (TB): 0.25 M Trizma base (adjust to pH 7.5 with HCl).
8. 4 % paraformaldehyde (PFA) (*see* **Note 1**) (wt/vol). For 1 l dissolve 40 g of PFA powder (Merck Millipore, cat. #1040051000) in 400 ml of ddH_2O. Heat to 60 °C (*see* **Note 2**) on a magnetic heating plate and gradually add some drops of 5 M NaOH to keep the pH below 7.5 (preferably use pH strip). Continue to stir until most of the powder is dissolved. Add 500 ml of 0.2 M PB and continue to stir for about 10 min. Let the solution cool to below 40 °C and filter (e.g., using a Nalgene filter unit). Store at 4 °C (*see* **Note 3**).
9. Cryoprotective solution: 30 % (vol/vol) ethylene glycol (Sigma-Aldrich, cat. #03750), 30 % glycerol (Acros Organics, cat. #295600010) in 0.1 M PB pH 7.4.
10. Primary antibodies (*see* Table 1).
11. Fluorophore-conjugated secondary antibodies (*see* Table 1).
12. Mounting medium: 2.5 % DABCO (Merck Millipore, cat. #8034560100) in glycerol (Acros Organics, cat. #295600010).
13. Poly-l-lysine-coated glasses (Sigma-Aldrich, cat. #P0425-72EA).
14. Microscopy coverslips (Fisher Scientific, cat. #S175212).
15. Perfusion tools: 300 ml syringe, butterfly needles (23 g, w/12 in. tubing), hemostatic forceps, cork pad, microdissecting forceps, spring scissors (approx. 2.5–3 in. blade length), blunt-ended forceps (approx. 6 in.), and spatula.

3 Methods

3.1 Drug Treatment and Brain Dissection

1. Handle and inject the animals (*see* **Note 4**) with saline during the 3 days preceding the experiment to familiarize them with the procedure and reduce stress on the test day.
2. Prepare the drug as follows: dissolve haloperidol in a drop of acetic acid, and add ddH_2O (final concentration 0.5 mg/10 ml), adjusting the pH to 6.0 with 1.0 M NaOH (*see* **Note 5**). Eticlopride (0.2 mg/10 ml), quinpirole (0.2 mg/10 ml), and raclopride (0.3 mg/10 ml) are dissolved in 0.9 % saline.
3. Move the animals to the test room and let them acclimatize for 1 h. Then inject them IP with drugs or vehicle according to their body weight, in a volume of 10 ml/kg.
4. To detect phosphoproteins, the animals are sacrificed 15–60 min after drug administration. *For immunohistochemistry go to Sects. 3.6 and 3.7 for perfusion procedure and immunolabeling.* For Western blotting decapitate the mouse and immediately place the head in liquid nitrogen for 5 s.
5. Carefully extract the brain and insert it into the matrix (*see* **Note 6**).
6. To dissect out the striatum, insert the blades in the matrix to obtain two slices of 1 mm each, ranging from −1.34 to 0.70 mm from the bregma [17]. Place the slices on the border of the matrix, and with a cold Sample Corer (1 mm internal diameter), collect a sample of striatum from each slice (*see* **Note 7**). The corpus callosum represents a good anatomical reference to localize the striatum (*see* **Note 8**).
7. Place the punches in an Eppendorf tube, which will be kept in liquid nitrogen until the end of the experiment. At this stage, the tissue samples can be stored at −80 °C, up to several months. Alternatively, they can be immediately processed according to the next step.
8. Place the Eppendorf tubes on an ice-cold surface, and add 180 μl 1 % SDS to each sample. Homogenize the tissue with a sonicator and incubate 10 min at 100 °C. At this stage, the lysates can be stored at −20 °C, up to few months.

3.2 Protein Quantification and Sample Preparation for Western Blotting

1. Prepare the tubes for the standard curve by pipetting 0, 2.5, 5, 10, or 20 μl of bovine serum albumin (BSA; 2 mg/ml) (0, 5, 10, 20, or 40 μg of protein) in each tube. Each point is measured in duplicate, which are averaged.
2. Pipette 5 μl from each sample in duplicate.

3. Prepare the BCA solution (50:1 reagent A over reagent B), and immediately add 2 ml in each tube. Incubate 30 min at 37 °C.
4. Remove the tubes from the water bath, measure absorbance at 562 wave length, and calculate the protein concentration of the samples.
5. Add 4× sample buffer (*see* **Note 9**) to each sample in order to obtain a final 1× concentration and incubate them 10 min at 100 °C. At this stage, the samples can be stored at −20 °C, for up to few months, or immediately process according to the next step.

3.3 SDS-PAGE and Electrophoretic Transfer

1. Prepare the gel sandwich stack (*see* **Note 10**). For each sandwich, assemble one notched alumina plate, one rectangular glass plate, and two spacers so that the long flat side of the T-shaped spacer fits between both plates. Align bottom edges of plates and spacers carefully. While holding the sandwich in place, secure it into the casting clamp by tightening all screws until they are finger tight. Place the gasket into the bottom of the casting cradle with the screws of the casting clamp facing out. Push down the casting assembly, insert a cam into each hole on both sides of the casting cradle, and seal the sandwich by turning both cams 180°.
2. Fill the glass sandwich with water to check for proper sealing.
3. Prepare 10 % acrylamide (*see* **Note 11**) running solution for two gels. In a falcon tube mix in the following order: 7.5 ml ddH_2O, 3.7 ml lower buffer, 3.7 ml acrylamide 40 %, 150 μl 10 % APS, and 15 μl TEMED (*see* **Note 12**).
4. Pour 7 ml of running gel into each sandwich, fill the remaining space with ddH_2O, and let polymerize for approximately 30 min (room temperature).
5. Prepare stacking solution for two gels. In a falcon tube mix in the following order: 4.5 ml ddH_2O, 1.9 ml upper buffer, 750 μl acrylamide 40 %, 74 μl 10 % APS, and 15 μl TEMED.
6. Remove the excess of water from the sandwich and pour the stacking solution up to the edge of the glasses (4 ml should be sufficient). Immediately insert the multi-wells comb inside the sandwich, use spring clamps to secure the top section of the plates, and let it polymerize for about 15 min (room temperature).
7. Remove the gel sandwiches from the casting cradle by loosening the pressure bar screws (at this stage, the sandwich can be stored overnight at 4 °C, wrapped in ddH_2O wet paper into a plastic bag, to avoid dehydration).
8. Gently remove the combs. Place the gel sandwiches into the lower buffer chamber of the electrophoresis unit (*see* **Note**

13), with the alumina plate facing the core of the buffer chamber, and secure them with two clamps for each gel sandwich.

9. Pour 1× running buffer into the lower buffer chamber. The space between each sandwich and the buffer chamber core has to be completely filled (500 ml should be sufficient).
10. Connect the coolant ports of the electrophoresis unit with tubes attached to cold water supply, to control the temperature during the electrophoresis (optional).
11. Defrost the samples and the protein ladder at room temperature.
12. Vortex the protein ladder and load 0.8 μl in the first well of the first gel sandwich (*see* **Note 14**). Vortex each sample and load 10 μg of protein (*see* **Note 15**) in the first gel (*see* **Note 16**), and then load 0.8 μl of ladder in the last well.
13. Repeat for the second gel.
14. Place the safety lid on the unit. Plug the color-coded leads into the outlets of the electrophoresis power supply.
15. Run the electrophoresis at 60 V until the dye reaches the end of the stacking gel (*see* **Note 17**) (this should take approximately 30 min). Then increase the voltage to 150 V. The electrophoresis can be terminated when the dye reaches the bottom of the gel (this should take approximately 3 h) (*see* **Note 18**).
16. Turn off the power supply, disconnect the leads and the cooler tubes, and remove the safety lid. Remove the clamps and lift away gel sandwiches from the upper buffer chamber core (*see* **Note 19**).
17. Prepare one transfer cassette at the time. Place the first sandwich into a tray filled with cold 1× transfer buffer, and then slip an extra spacer between the glass and the alumina plates and carefully lift up the glass plate.
18. Cut a PVDF membrane (8.5×7 cm) (*see* **Note 20**), and activate it by placing it in 100 % methanol for 2 min at room temperature.
19. Place in the tray filled with transfer buffer one activated PVDF membrane, one transfer cassette (opened), two sponges, and two blotter papers. Assemble the transfer sandwich by placing on the gray side of the cassette one sponge, one blotter paper, the activated membrane, and then the gel (*see* **Note 21**). Complete the sandwich by placing on the gel a blotter paper and a sponge, and then close the cassette (*see* **Note 22**). Keep in mind that the membrane must face the gray side of the cassette, while the gel is placed toward the black side.
20. Assemble the second cassette following the same procedure.

21. Fill the transfer chamber with cold transfer buffer, insert a small magnet, and slide the cassettes in. The cassettes must be oriented so that the hinge side is facing up and all black panels of the cassettes are facing the same side of the transfer unit (*see* **Note 23**).
22. Connect the coolant outlets to the water supply and place the unit on a magnetic plate, to ensure homogeneous cooling in the transfer chamber.
23. Place the transfer chamber lid and connect to the electrophoresis power supply (*see* **Note 24**). Important: the cassettes are color coded to match the leads in the lid. To transfer toward the anode, orient the lid so that the gray half of the cassette faces the anode (+), or red lead, and the black half of the cassette faces the cathode (−), or black lead (*see* **Note 25**).
24. Run the electrophoretic transfer at 70 V for 90 min (*see* **Note 26**).
25. Turn off the power and disconnect lid and water tubes. Open each cassette carefully, discard the gel, and lift the membrane to dry (*see* **Note 27**).

3.4 Antibody Incubation

1. Using the protein ladder identify the position of the target protein or phosphoprotein and cut the part of interest (*see* **Note 28**) (do not exceed the 2 cm high). At this stage the membranes can be left to dry, wrapped in foil, and stored at room temperature.
2. Activate the membranes in methanol for 2 min, and then quickly rinse in ddH_2O.
3. Place each membrane in an incubation box filled with PBS-Tween.
4. Incubate each membrane in blocking solution, at room temperature for 50 min, under mild shaking (*see* **Note 29**). Use different blocking solution depending on the detection procedure (ECL vs. LI-COR Odissey System; see Sect. 2.5).
5. Rinse 10 min × 3 with PBS-Tween.
6. Dilute the primary antibody in PBS-Tween (*see* Table 1 for dilutions), and incubate overnight at 4 °C (*see* **Note 30**).
7. Rinse 10 min × 3 with PBS-Tween.
8. Incubate for 50 min at room temperature with secondary antibody (*see* **Note 31**) prepared in PBS-Tween (*see* Table 1 for dilutions).
9. Rinse 10 min × 3 with PBS-Tween.

3.5 Signal Detection and Acquisition

1. Incubate the membranes for 5 min in freshly prepared ECL solution.
2. Place the membranes in the developer cassette and cover it with a transparent plastic sheet.

3. In the dark room, expose an Amersham ECL film to the membrane (*see* **Note 32**).
4. Incubate the film in the developer solution until the bands appear. Rinse in water for 20 s, and then incubate in fixative solution for 2 min and rinse again. Alternatively, this step can be carried out using an automatic developer.
5. Dry the film in the dark room and scan it to obtain an image file.
6. Quantify the signal in the image file with Bio-Rad quantification software. Alternatively, after step 9 in Sect. 3.4, scan the membranes using the LI-COR Odyssey Scanner System.
7. For each gel, the resulting values are then expressed as % of control, vehicle-treated samples.

3.6 Perfusion and Brain Dissection

To detect proteins by immunohistochemistry, mice are first treated with an overdose of pentobarbital and transcardially perfused with fixative. The following protocol has been optimized for the detection of phosphorylated proteins in the brain and is based on the direct perfusion of the mice with PFA (*see* **Note 33**), which should be carried out under a chemical fume hood.

1. Mice are treated as described in Sect. 3.1 (points 1–4) and perfused after 15–60 min (see below).
2. Fill a tube connected to a peristaltic pump with freshly prepared 4 % PFA pH 7.5 (*see* **Note 34**). Eliminate any air bubble.
3. Inject the animal with pentobarbital (200 mg/kg, IP).
4. As soon as the mouse is deeply anesthetized (*see* **Note 35**), place it on its back and fix each limb on a cork pad using butterfly needles.
5. Using the scissors, open the thorax by a midline incision. Pinch the tip of the sternum with the forceps. Cut the diaphragm and make an incision along the lateral edges of the chest until the thorax is open to expose the heart.
6. Flip the chest and clamp it with hemostatic forceps.
7. Gently grasp the heart, make a small cut with the spring scissors in the right atrium (*see* **Note 36**), and place the needle connected to the pump in the apex of the left ventricle (*see* **Note 37**).
8. Turn on the peristaltic pump to deliver the fixative solution and eliminate blood. Use a 20 ml/min rate for 5 min (*see* **Note 38**).
9. Stop the pump and proceed to the brain dissection. Decapitate the mouse using the scissors and open the scalp. Cut off the cerebellum and make an incision along the sagittal suture. Remove the dorsal bones of the skull.

10. Place a small spatula at the level of the olfactory bulbs to lift the brain, cut the optical nerve and carefully remove the brain (*see* **Note 39**).
11. Postfix (*see* **Note 40**) the brains at 4 °C overnight, in tubes containing 4 % PFA.

3.7 Immunolabeling

1. Cut the striatum in 40-μm-thick coronal sections using a vibratome (Leica, Nussloch, Germany) (*see* **Note 41**). With this particular instrument, set the cutting speed to 4–5 and vibration level to 9–10. Use cold PBS to fill the bath where the brain is submerged.
2. Using a soft brush, transfer the sections to a multi-well plate filled with cryoprotective solution. Samples can be stored at -20 °C or otherwise processed according to the next steps.
3. Select floating striatal sections of interest to be processed for the ICH reaction, and place them in a multi-well plate filled with TBS.
4. Rinse the sections 10 min × 3 in TBS under mild shaking (*see* **Note 42**).
5. Incubate the specimen for 60 min in blocking buffer (*see* **Note 43**). This step is necessary as all epitopes on the tissue sample have to be blocked to prevent the nonspecific binding of the antibodies that will be used in the next steps.
6. Rinse the sections 10 min × 3 in TBS.
7. Dilute the primary antibody in TBS (see Table 1 for dilutions) and incubate overnight at 4 °C (*see* **Note 44**). If the reaction is carried out in 24-well plates, a volume of 300 μl is enough for 1 well containing up to four sections (*see* **Note 45**).
8. Rinse the sections 10 min × 3 in TBS.
9. Incubate with fluorophore-conjugated secondary antibodies (Cy3/Cy2-conjugated goat anti-rabbit antibodies; *see* **Note 46**) diluted in TBS for 1 h at room temperature in the dark (*see* **Note 47**).
10. Rinse the sections 10 min × 3 in TBS.
11. Rinse 5 min × 2 in TB.
12. Mount the sections on poly-l-lysine-coated slides using a soft brush and a petri dish filled with TB diluted in ddH_2O (1:1) (*see* **Note 48**).
13. Remove the excess of liquid from the specimen with filter paper, and add a small drop of mounting medium (*see* **Note 49**) on the slices (*see* **Note 50**). Then carefully lower a coverslip onto the medium avoiding bubbles.
14. Seal the slides with nail polish around the edges of the coverslips.

15. For best results examine specimens immediately, or store at 4 °C in the dark.
16. For optimal quantification *Z*-stack (6–8 μm) images of the striatum are generated at the confocal microscope. To acquire double-stained images, it is suggested to use 488 nm argon laser (to excite Cy2 fluorochrome) and 546 nm mercury-arc lamp (to excite Cy3 fluorochrome). Specific filter wavelengths (505–530 nm for Cy2 and 585–615 nm for Cy3) can be used to reduce the background and to optically separate the two fluorescent cyanines. To quantify fluorescent striatal neurons, the 20× objective represents a good compromise between resolution and cell number included in each acquisition. A reliable quantification is obtained by averaging 2–4 20× images from each of three coronal striatal sections. ImageJ software (open source platform) can be used to count the number of fluorescent cells or, alternatively, to measure the level of fluorescence in a given region. Cells can be counted manually by marking them using the Cell Counter plugin or automatically using the analyze particles tool on binary images. In the case of high cell density, or differences in signal intensity rather than cell number, ImageJ can be used to evaluate the level of fluorescence. To this end, total cell fluorescence is calculated using area, integrated density, and mean gray values.

4 Notes

1. PFA is particularly toxic and need to be handled with care. Weigh the powder and prepare the solution in a chemical fume hood. Otherwise, a premade PFA solution (32 %, EM grade) is commercially available.
2. Make sure the temperature does not exceed 70 °C, as PFA tends to crystallize above this temperature.
3. 4 % PFA should not be prepared more than 3 days before use.
4. The use of animals must be approved by the Institutional Animal Care and Use Committee. In our case the experiments were performed in accordance with the guidelines of Research Ethics Committee of Karolinska Institutet, Swedish Animal Welfare Agency, and European Communities Council Directive 86/609/EEC.
5. Care should be taken to keep the pH at 6 or just below this value, to avoid precipitation.
6. The entire dissection procedure has to be as fast as possible (1.5–3 min) to limit modifications in protein phosphorylation. As an alternative, protein phosphorylation can be detected

using focused microwave irradiation, which results in rapid heath inactivation of brain enzymes [18].

7. The striatum can be dissected unilaterally or bilaterally depending on the amount of tissue required for the analysis. The punching procedure (number and thickness of brain sections, coordinates, and diameter of the Sample Corer) can be modified to collect other brain areas.
8. As an alternative, the striatum can be dissected out using a freehand procedure. In this case the brain is placed on an ice-cold surface covered with PBS or saline-wet filter paper. The cortex is removed and the entire striatum pinched out with micro-dissecting forceps.
9. As an alternative, the amount of sample buffer added to the lysates can be adjusted according to the lysate protein concentration, resulting in samples containing the same amount of protein.
10. Wear gloves to keep the caster and plates free of finger marks.
11. The percent of acrylamide in the running solution can be adjusted according to the molecular weight of the protein to detect, using higher percent for lower molecular weight proteins (10 % acrylamide is usually appropriate for proteins between 20 and 200 kDa.
12. After addition of TEMED and APS, the gel polymerizes rapidly, so make this last addition when ready to cast.
13. Electrophoresis should be run with two gels. In case all the samples fit in one gel, a second sandwich (empty) should be inserted in the unit.
14. Remember to label the sides of the electrophoresis chamber, to identify the samples.
15. The amount of protein to load can vary based on the quality and affinity of the primary antibody for the protein of interest.
16. A loading list should be prepared, indicating the sequence of the samples (balanced for treatment group, or for any other variables, so that every gel has at least four samples belonging to the same experimental group) and the volume of lysate to be loaded for each sample.
17. When the bromophenol blue in the sample buffer (which corresponds to the front of the electrophoresis) reaches the end of the stacking gel, it forms a continuous band, and the protein ladder starts separating into the different molecular weight bands.
18. The electrophoresis can be terminated even before the bromophenol blue reaches the bottom of the gel. However, a longer electrophoresis allows a better separation of the proteins.

19. To minimize diffusion of the bands, transfer should be performed as soon as possible after electrophoresis.
20. Identify the membranes (i.e., cut the upper right corners of the membranes to distinguish them and maintain proper orientation).
21. When the gel is lifted up from the electrophoresis sandwich and placed in the transfer cassette, carefully respect the orientation of the gel, to ensure that the order of the samples loaded onto the gel is reproduced on the membrane.
22. Pay attention to trapped air bubbles.
23. If transferring only one or two gels, choose the cassette positions nearest to the center.
24. Control the presence of ascendant air bubbles into the transfer unit chamber, which confirms a correct functioning of the system.
25. Given the position of the PVDF membrane and the gel toward the gray and the back side of the cassette, respectively, this orientation of the lid allows the negative current to hit the (negative charged) proteins in the gel first, moving them toward the membrane.
26. For high molecular weight proteins, transfer can be carried out overnight at 10 V. In this case, cooling is not necessary.
27. Whereas the running buffer is discarded after use, the transfer buffer can be reused a few times. All the other accessories (blotting paper, sponges) can be washed and reused several times.
28. To detect several proteins with different molecular weights, the membrane can be cut in correspondence to different protein sizes.
29. All subsequent incubation and washing steps are done under shaking.
30. Alternatively, the primary incubation can be performed for 2 h, at room temperature. In case of long term storage for reutilization, the primary antibody solution must contain 0.05 % sodium azide to prevent contamination.
31. The secondary antibody has to be specifically directed against the host species of the primary antibody.
32. The optimal exposure time varies from a few seconds to minutes and should be empirically determined for each experiment.
33. The direct use of PFA without any previous washing step with saline is particularly suitable for the detection of phosphoproteins, as the fixation phase occurs as soon as the animal is deeply anesthetized, thereby reducing the risk of modifications in the state of phosphorylation.

34. Keep the PFA solution in an ice-filled styrofoam box.
35. Before proceeding with the perfusion phase, the efficacy of the anesthesia must be checked by observing the loss of the righting and palpebral reflexes, as well as the response to painful stimulation (pinching of the paws).
36. The right atrium is typically darker than the rest of the heart. The incision is made to release the pressure generated by blood and fixative solution.
37. The left ventricle is thicker and lighter than the right ventricle. Be careful to not perforate the wall between the two ventricles when inserting the needle.
38. Blanching of the liver and body stiffness are signs of a good perfusion.
39. This step should be done as soon as possible to avoid brain desiccation.
40. The duration of the postfixation step depends on the specific immunolabeling technique to be carried out. Alternative methods are based on a shorter postfixation step (2–4 h), followed by transfer of the brains to a sucrose solution. Once the tissue has sunk, it can be frozen and sectioned.
41. Following overnight postfixation brains can be stored at 4 °C in PB 0.1 M—0.05 % sodium azide until cutting. For best result it is suggested to section the brain as soon as possible.
42. It is preferable to let this washing step last longer, to ensure complete removal of the cryoprotective solution. All subsequent incubation and washing steps are done under shaking.
43. For best results, this buffer must be freshly prepared. The use of concentrated BSA increases the coating of all proteins, therefore enhancing the specificity of the primary antibody for the protein, or phosphoprotein of interest. An alternative blocking reagent commonly used is normal serum. In this case, the serum must originate from the same host species of the secondary antibody. It should be noted that there is no optimal blocking method for immunohistochemical experiments and that this step needs to be optimized ad hoc.
44. Optimal dilutions of the primary antibodies need to be determined for each experimental system even though the dilutions indicated in Table 1 have been already verified and optimized in our laboratory.
45. In case some wells remain empty, is recommended to fill them with ddH_2O to create a humid chamber during long incubation steps.

46. The secondary antibody should be specifically directed against the host species of the primary antibody. Cy3-/Cy2-conjugated affinity-purified goat anti-rabbit antibodies have been optimized for this protocol at the dilutions indicated in Table 1.
47. All the steps performed after the secondary antibody incubation must be carried out in the dark.
48. The specimen is rinsed with TB and dipped in TB diluted with ddH_20 during the mounting step to get rid of excess salts as they can increase nonspecific fluorescent signals.
49. DABCO is added to glycerol presents in the mounting medium as anti-fading agent.
50. Use as less medium as possible. This allows for the coverslip to come as close to the specimen as possible, thereby reducing focus aberration.

Acknowledgments

Work was supported by Swedish Research Council Grant 13482 (G.F.), StratNeuro at Karolinska Institutet (G.F. and A.B.O.), the Foundation Blanceflor Boncompagni-Ludovisi née Bildt (A.B.O.), Åhlén-stiftelsen (A.B.O.), and C.M. Lerici Foundation (G.S.).

References

1. Missale C, Nash SR, Robinson SW et al (1998) Dopamine receptors: from structure to function. Physiol Rev 78:189–225
2. Gerfen CR (1992) The neostriatal mosaic: multiple levels of compartmental organization in the basal ganglia. Annu Rev Neurosci 15:285–320
3. Beaulieu JM, Gainetdinov RR, Caron MG (2007) The Akt-GSK-3 signaling cascade in the actions of dopamine. Trends Pharmacol Sci 28:166–172
4. Bonito-Oliva A, Feyder M, Fisone G (2011) Deciphering the actions of antiparkinsonian and antipsychotic drugs on cAMP/DARPP-32 signaling. Front Neuroanat 5:38
5. Håkansson K, Galdi S, Hendrick J et al (2006) Regulation of phosphorylation of the GluRl AMPA receptor by dopamine D2 receptors. J Neurochem 96:482–488
6. Bertran-Gonzalez J, Hakansson K, Borgkvist A et al (2009) Histone H3 phosphorylation is under the opposite tonic control of dopamine D2 and adenosine A2A receptors in striatopallidal neurons. Neuropsychopharmacology 34:1710–1720
7. Valjent E, Bertran-Gonzalez J, Bowling H et al (2011) Haloperidol regulates the state of phosphorylation of ribosomal protein S6 via activation of PKA and phosphorylation of DARPP-32. Neuropsychopharmacology 36:2561–2570
8. Bonito-Oliva A, Pallottino S, Bertran-Gonzalez J et al (2013) Haloperidol promotes mTORC1-dependent phosphorylation of ribosomal protein S6 via dopamine- and cAMP-regulated phosphoprotein of 32 kDa and inhibition of protein phosphatase-1. Neuropharmacology 72:197–203
9. Gong S, Zheng C, Doughty ML et al (2003) A gene expression atlas of the central nervous system based on bacterial artificial chromosomes. Nature 425:917–925
10. Herve D, Le Moine C, Corvol JC et al (2001) Galpha(olf) levels are regulated by receptor usage and control dopamine and adenosine action in the striatum. J Neurosci 21:4390–4399

11. Schiffmann SN, Jacobs O, Vanderhaegen JJ (1991) Striatal restricted adenosine A2 receptor (RDC8) is expressed by enkephalin but not by substance P neurons: an in situ hybridization histochemistry study. J Neurochem 57:1062–1067
12. Ballarin M, Fredholm BB, Ambrosio S et al (1991) Extracellular levels of adenosine and its metabolites in the striatum of awake rats: inhibition of uptake. Acta Physiol Scand 142:97–103
13. Svenningsson P, Lindskog M, Ledent C et al (2000) Regulation of the phosphorylation of the dopamine- and cAMP-regulated phosphoprotein of 32 kDa in vivo by dopamine D1, dopamine D2, and adenosine A2A receptors. Proc Natl Acad Sci U S A 97:1856–1860
14. Usiello A, Baik JH, Rouge-Pont F et al (2000) Distinct functions of the two isoforms of dopamine D2 receptors. Nature 408:199–203
15. Lindgren N, Usiello A, Goiny M et al (2003) Distinct roles of dopamine D2L and D2S receptor isoforms in the regulation of protein phosphorylation at presynaptic and postsynaptic sites. Proc Natl Acad Sci U S A 100:4305–4309
16. Håkansson K, Pozzi L, Usiello A et al (2004) Regulation of striatal tyrosine hydroxylase phosphorylation by acute and chronic haloperidol. Eur J Neurosci 20:1108–1112
17. Franklin KBJ, Paxinos G (1997) The Mouse brain in stereotaxic coordinates. Academic, San Diego, CA
18. Bateup HS, Svenningsson P, Kuroiwa M et al (2008) Cell type-specific regulation of DARPP-32 phosphorylation by psychostimulant and antipsychotic drugs. Nat Neurosci 11:932–939

Chapter 7

Study of Dopamine D1 Receptor Regulation by G Protein-Coupled Receptor Kinases Using Whole-Cell Phosphorylation and Cross-Linking Methods

Keyvan Sedaghat, Boyang Zhang, Xiaodi Yang, Caroline Lefebvre, and Mario Tiberi

Abstract

Cells utilize receptor desensitization to prevent sustained receptor signaling and potential cellular injuries through diminishing the receptor's responsiveness toward agonists. One of the key facilitators for desensitization among G protein-coupled receptors (GPCR) is the family of G protein-coupled receptor kinases (GRK). In the agonist-activated state, the receptor is phosphorylated by GRKs to allow binding of cytosolic arrestins, which leads to uncoupling from the G protein. Yet there are many nuances when studying GRK-mediated phosphorylation, which include basal phosphorylation by GRKs, the various GRK isoforms, and structural targets of GRKs on the receptor. To address questions stemming from such examples as well as to pursue other avenues concerning GRK-mediated phosphorylation, we focus on the dopamine D1 receptor subtype (D1R) and detail two assays: whole-cell phosphorylation and coimmunoprecipitation using the cross-linker dithiobis(succinimidyl propionate) (DSP) to demonstrate potential interactions between D1R and GRK isoforms. In addition, we provide an overview of past studies concerning the desensitization properties of D1R and a brief protocol for indirect immunofluorescence confocal microscopy to visualize the co-localization between D1R and GRK isoforms.

Key words Phosphorylation, Cross-linking, GRKs, Dopamine, D1R, Immunoblotting, HEK293 cells, Co-transfection

1 Introduction

Upon agonist exposure, G protein-coupled receptors (GPCR) are classically subjected to a series of cellular processes that function to attenuate their signaling. In the initial phase, homologous desensitization occurs in which the receptor, in its agonist-bound conformation, is phosphorylated by G protein-coupled receptor kinases (GRK1, GRK2, GRK3, GRK4, GRK5, GRK6, GRK7) to increase receptor affinity for the binding of cytosolic arrestins (β-arrestin 1, β-arrestin 2, visual and cone arrestins). Arrestin sterically hinders further G protein coupling and causes the receptor's responsiveness

Mario Tiberi (ed.), *Dopamine Receptor Technologies*, Neuromethods, vol. 96,
DOI 10.1007/978-1-4939-2196-6_7, © Springer Science+Business Media New York 2015

toward agonists to diminish [1, 2]. Receptor phosphorylation can also be facilitated by second messenger-dependent kinases (PKA, PKC), which is sufficient by itself to impair coupling between receptor and G protein. Since this can occur without agonist occupancy to the target receptor, it is referred to as heterologous desensitization [3]. Although events following heterologous desensitization are currently unclear, in the GRK-mediated pathway, the desensitized receptor is subjected to internalization through arrestin recruitment of clathrin and the subsequent formation of clathrin-coated vesicles [3].

The desensitization properties of the dopamine D1 receptor subtype (D1R) have been well studied. Meanwhile, how posttranslational mechanisms such as phosphorylation and protein–protein interactions fine-tune D1R desensitization remains unclear. D1R couples to stimulatory heterotrimeric G proteins ($G_{s/olf}$) to activate adenylyl cyclase (AC), resulting in an increase of cyclic adenosine monophosphate (cAMP) levels [4]. Its intracellular C-terminal (CT) region contains multiple serine and threonine residues that pose as potential phosphorylation sites for GRKs and second messenger-dependent kinases [5]. In human embryonic kidney 293 (HEK293) cells, basal and dopamine-induced D1R phosphorylation by GRK2, GRK3, and GRK5 occur predominantly on D1R-CT serine residues [6–10]. Phosphorylation of D1R by GRK2 and GRK3 induced a significant rightward shift in dose–response curves for dopamine without any change in maximal response, while GRK5 phosphorylation mediated both a rightward shift and a 40 % decrease in maximal response in HEK293 cells expressing high receptor levels (8–9 pmol/mg membrane proteins) [6]. These results suggest that GRK5 produces a more pronounced desensitization on D1R in comparison with GRK2 or GRK3 [6]. Intriguingly, in cells expressing moderate levels of D1R (~3 pmol/mg membrane proteins), GRK3-induced D1R phosphorylation produced a significantly greater rightward shift and maximal response diminution relative to that mediated by GRK2-induced receptor phosphorylation [8]. Moreover, removing the distal-central portion of CT of D1R (Ser380-Leu424) impaired dopamine-induced phosphorylation by GRK2 as well as basal and dopamine-induced phosphorylation by GRK3 [8].

In HEK293T cells, D1R has been shown to be constitutively phosphorylated by GRK4α [11]. This constitutive phosphorylation was demonstrated to be restricted to the distal portion of CT for D1R, specifically to residues Thr428 and Ser431. GRK4α-mediated D1R phosphorylation resulted in a reduction in cAMP accumulation, an increase in receptor internalization, and a decrease in total receptor number. The pathophysiological relevance of GRK-mediated D1R phosphorylation in the brain remains to be fully appreciated. Indeed, results obtained so far with

GRK-isoform-specific gene knockout mice suggest an intricate regulation of D1R function by GRKs in vivo [1, 12–15]. Meanwhile, studies have shown an augmentation in basal phosphorylation of Ser residues of D1R and D1R-Gαs uncoupling caused by increased GRK activity in kidney proximal tubules of hypertensive animal models and human subjects [16–21]. Interestingly, the GRK4 gene has been linked to human essential hypertension [22]. GRK4 activity in isolated renal proximal tubule cells from hypertensive patients was found to be enhanced compared to normotensive patients, and this could be attributed by single nucleotide polymorphisms in the GRK4γ isoform coding sequence [21]. The expression of one GRK4 variant in transgenic mice also induced hypertension [21].

The involvement of PKA in D1R desensitization has been inconsistent. Some groups have reported that PKA contributes to D1R desensitization [23–27], whereas others have obtained opposing results [28, 29]. Surprisingly, there is evidence that PKA may even serve as a positive regulator of D1R signaling rather than a desensitizing factor in HEK293 cells [30]. Four putative PKA phosphorylation sites exist within D1R: Thr135 in intracellular loop 2, Ser229 and Thr268 in intracellular loop 3, and Ser380 in CT [26]. Mutation of Thr268 to valine served as a primary determinant in the reduction of the rate of agonist-induced D1R desensitization when expressed in C6 glioma cells [26]. In addition, this mutation seemed to be specific in regulating the kinetics of desensitization because other pharmacological properties of D1R were not altered including [^{3}H]-SCH23390 ligand binding, receptor expression, and AC activation. Using a fusion protein derived from CT of D1R (residues 372–442) and purified PKA, it was demonstrated that PKA could phosphorylate Ser380 in the CT of D1R [31]. Using NS20Y neuroblastoma cells stably expressing D1R in conjunction with mutagenesis and pharmacological approaches, it was revealed that substitution of Ser380 to Ala had no effect on D1R phosphorylation, sequestration, desensitization, and trafficking to the perinuclear region [32]. Furthermore, PKA-dependent phosphorylation of Thr268 was shown to regulate a late step in the sorting of the receptor to the perinuclear region of the cell, but this phosphorylation was not necessary for D1R sequestration or desensitization of cAMP accumulation [32].

Although desensitization and internalization go hand in hand based on the classical model, for D1R, these events can be dissociated from one another [9]. Mutation of either Glu359 or Thr360 in the proximal portion of the CT led to a complete loss of desensitization of the receptor without any effect on internalization, indicating that these two residues may form a motif necessary for GRK-mediated desensitization. Conversely, mutation of Ser431, Thr439, and Thr446 in the distal portion of the CT culminated in a lack of internalization, whereas desensitization was not impacted [9].

To investigate the processes underlying D1R homologous desensitization, namely, GRK phosphorylation/interaction at the cytoplasmic domains of D1R, this chapter will focus mainly on two assays: whole-cell phosphorylation and protein cross-linking/coimmunoprecipitation (co-IP). Also included is a brief protocol for indirect immunofluorescence confocal microscopy to support the interaction between GRKs and D1R detected by cross-linking experiments. These assays are relatively straightforward and should be utilized as a starting point before advanced techniques such as mass spectrometry or bioluminescence/fluorescence resonance energy transfer techniques (BRET, FRET) are attempted. Although HEK293 cells transfected with D1R are employed in the following procedures, other receptor constructs and heterologous systems can also be used.

In whole-cell phosphorylation assay, co-transfected cells (HA-tagged D1R and GRK isoforms) are first supplemented with [^{32}P]-orthophosphate that are to be incorporated into the intracellular regions of D1R via GRK phosphorylation. D1R is then isolated from the cellular milieu based on principles of immunoprecipitation. The engineered HA epitope on the extracellular N-terminal region of D1R is recognized by anti-HA antibodies conjugated to an agarose matrix, and together they form an insoluble antigen–antibody complex. Although optional, a preclearing step is performed beforehand in which protein A-Sepharose beads are added to remove unwanted lysate proteins that may bind to the bead matrix. The antigen (D1R) is separated from the complex using a disruption buffer containing sodium dodecyl sulfate and 2-mercaptoethanol (sample buffer), and its incorporation of [^{32}P]-orthophosphate is visualized by autoradiography.

The co-IP assay detailed in this chapter involves the application of the cell-permeable and thiol-cleavable homobifunctional cross-linker dithiobis(succinimidyl propionate) (DSP) prior to cell lysis because detection of GPCR interactions with either GRKs or arrestins in cells has been proven difficult without the use of cross-linking agents [33, 34]. DSP has identical reactive groups (N-hydroxysuccinimide (NHS) ester) toward primary amines at both ends of its 12 Å (8-carbon) spacer arm (Fig. 1). It has the advantage of reacting rapidly with any protein possessing primary amines (pH 7.0–9.0) in a one-step chemical cross-linking reaction

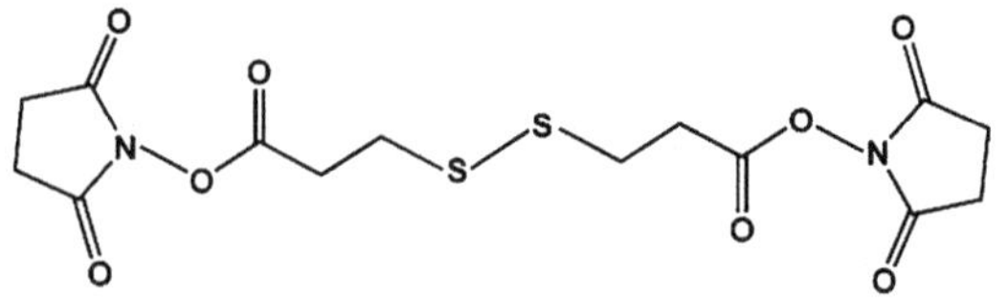

Fig. 1 Chemical structure of DSP. DSP (M.W. 404.42) is also known as Lomant's reagent or DTSP. It has an 8-atom (12 Å) spacer arm

to form stable amide bonds. Direct protein–protein interactions happen on a spatial scale of less than 10 nm or 100 Å, which is within the realm of the DSP spacer arm length (12 Å or 1.2 nm). After DSP treatment, immunoprecipitation is followed, whereby D1R is used as the "bait" protein to be extracted by anti-HA affinity matrix. Co-extracted D1R cross-linked proteins such as GRK isoforms ("prey") that interact with D1R are then probed using specific antibodies via Western blotting. Although co-IP is the most commonly used technique to identify protein–protein interactions, one also needs to consider the possibility of indirect interactions where bait and prey proteins are united by an unidentified protein or the possibility that direct interactions may only occur after cell lysis and therefore may not exist under physiological conditions. However, under cross-linking conditions, the latter is not a major concern as the DSP reaction is quenched by tris(hydroxymethyl) aminomethane (50 mM) contained in the cell lysis buffer.

2 Materials

2.1 Cell Culture and Transfection

1. Human embryonic kidney 293 (HEK293) cells (CRL1573, American Type Culture Collection (ATCC), Manassas, VA).
2. Minimum essential medium (MEM) containing Earle's salts (Invitrogen).
3. Fetal bovine serum (FBS) (*see* **Note 1**).
4. Gentamicin (10 mg/mL stock; Invitrogen).
5. Phosphate-buffered saline (PBS) without Ca^{2+} and Mg^{2+} (Wisent, QC, Canada).
6. Trypsin (0.25 %)-ethylenediaminetetraacetic acid (EDTA; 0.05 % w/v) solution (Invitrogen).
7. Filter-sterilized 4-(2-hydroxyethyl)-1-piperazineethanesulfonic acid (HEPES) (1 M, pH 7.0).
8. Filter-sterilized 2× HEPES-buffered saline (HBS; 0.05 M HEPES, 0.28 M NaCl, and 1.5 mM Na_3PO_4, pH 7.1) (*see* **Note 2**).
9. Filter-sterilized 2.5 M $CaCl_2$. Store in cell culture room at ambient temperature.
10. Autoclaved Milli-Q water. Store in cell culture room at ambient temperature.
11. 0.2 μm vented blue plug seal capped 75 cm^2 polystyrene cell culture flasks (VWR International, Montréal, QC, Canada).
12. 100×20 mm polystyrene cell culture dishes (Sarstedt, Montréal, QC, Canada).

13. 6-well polystyrene cell culture plates (VWR International, Montréal, QC, Canada).
14. Sterile 13 mL (16×100 mm) snap-capped polypropylene tubes (Sarstedt, Montréal, QC, Canada).
15. Sterile 50 mL (28×114 mm) screw-capped polypropylene conical tubes (Sarstedt, Montréal, QC, Canada).
16. Sterile 15 mL (17×120 mm) screw-capped polystyrene conical tubes (VWR International, Montréal, QC, Canada).
17. Expression plasmids (HA-tagged rat D1R (HA-rD1R), rat GRK2, rat GRK3) (*see* **Note 3**).
18. Certified biological safety cabinet (BSC).

2.2 Phosphorylation

1. Phosphate-free Dulbecco's Modified Eagle's Medium (DMEM) (Invitrogen).
2. [^{32}P]-orthophosphate, carrier-free (10 mCi/mL) (PerkinElmer, Boston, MA, USA).
3. Sterile HEPES (1 M, pH 7.4) (Invitrogen).
4. Plexiglas (Lucite) shield (1 cm thick).
5. Plexiglas containers for radioactive waste.
6. Plexiglas boxes for [^{32}P]-orthophosphate storage and metabolic labeling.
7. Geiger–Müller detector.
8. Ascorbic acid (Sigma-Aldrich).
9. Dopamine hydrochloride (Sigma-Aldrich).
10. Nonidet P-40 (Calbiochem, La Jolla, CA, USA).
11. Sodium deoxycholate (Sigma-Aldrich).
12. Sodium dodecyl sulfate (Bio-Rad Laboratories, Hercules, CA, USA).
13. Disodium pyrophosphate (Sigma-Aldrich).
14. Sodium fluoride (NaF) (Sigma-Aldrich).
15. Glycerol (Thermo Fisher Scientific).
16. 2-mercaptoethanol (Sigma-Aldrich).
17. Sterile EDTA solution (0.5 M, pH 8.0). Store at room temperature.
18. Aprotinin (Sigma-Aldrich).
19. Benzamidine (Sigma-Aldrich).
20. Leupeptin (Sigma-Aldrich).
21. Pepstatin A (Sigma-Aldrich).
22. Soybean trypsin inhibitor (Sigma-Aldrich).
23. Phenylmethylsulfonyl fluoride (PMSF) (Sigma-Aldrich).

24. Dimethyl sulfoxide (DMSO) (Sigma-Aldrich).
25. Bromophenol blue (Bio-Rad Laboratories, Hercules, CA, USA).
26. Tris(hydroxymethyl)aminomethane base (Tris) (Sigma-Aldrich).
27. Radioimmunoprecipitation assay plus (RIPA+) buffer (50 mM Tris–HCl (pH 8.0), 150 mM NaCl, 5 mM EDTA, 1 % (v/v) Nonidet P-40, 0.5 % (w/v) sodium deoxycholate, 0.1 % (w/v) SDS, 10 mM NaF, 10 mM disodium pyrophosphate) containing protease inhibitors (2 μg/mL aprotinin, 10 μg/mL benzamidine, 10 μg/mL leupeptin, 1 μg/mL pepstatin A, 10 μg/mL soybean trypsin inhibitor, and 10 μg/mL PMSF) (*see* **Notes 4** and **5**).
28. Teflon cell lifters (Thermo Fisher Scientific).
29. Protein A-Sepharose beads (GE Healthcare Life Sciences, Baie d'Urfé, QC, Canada).
30. 15 mL screw-capped tubes (VWR International, Montréal, QC, Canada).
31. 1.5 mL screw-capped conical tubes (Thermo Fisher Scientific).
32. 1.5 mL Eppendorf tubes (Sarstedt, Montréal, QC, Canada).
33. Parafilm (VWR International, Montréal, QC, Canada).
34. Rotating mixer (Thermo Fisher Scientific).
35. Rat monoclonal anti-hemagglutinin (HA) affinity matrix (Covance Research Products, Richmond, CA, USA).
36. Bovine serum albumin (BSA) (Sigma-Aldrich) (*see* **Note 6**).
37. Bio-Rad DC protein assay kit (Bio-Rad Laboratories, Hercules, CA, USA).
38. Tris-buffered saline with BSA (TBS-BSA) (20 mM Tris–HCl, 137 mM NaCl, 2 % (w/v) BSA) (*see* **Note 7**).
39. 2× SDS sample buffer (25 mM Tris–HCl (pH 6.5), 8 % (v/v) SDS, 5 % (v/v) 2-mercaptoethanol, 10 % (v/v) glycerol, 2 mg/mL bromophenol blue) (*see* **Note 8**).

2.3 Radioligand Receptor-Binding Assay

1. PBS.
2. Lysis Buffer (10 mM Tris–HCl, pH 7.4; 5 mM EDTA, pH 8.0). Store at 4 °C.
3. Resuspension buffer (62.5 mM Tris–HCl, pH 7.4; 1.25 mM EDTA, pH 8.0). Store at 4 °C.
4. Binding buffer (62.5 mM Tris–HCl, pH 7.4; 1.25 mM EDTA, pH 8.0; 200 mM NaCl; 6.7 mM $MgCl_2$; 2.5 mM $CaCl_2$; 8.33 mM KCl). Store at 4 °C.
5. Washing buffer (50 mM Tris–HCl, pH 7.4; 100 mM NaCl). Store at 4 °C.
6. [^{3}H]-SCH23390 (PerkinElmer, Boston, MA, USA).

7. *cis*-Flupenthixol (Sigma-Aldrich). Store at room temperature.
8. Teflon cell lifters (Thermo Fisher Scientific).
9. 12 × 75 mm polystyrene test tubes (VWR International, Montréal, QC, Canada).
10. Three-tier polypropylene racks (Thermo Fisher Scientific).
11. 15 mL (18 × 100 mm) polycarbonate centrifuge tubes (Beckman Coulter Canada LP, Mississauga, ON, Canada).
12. BSA (Sigma-Aldrich) (*see* **Note 6**).
13. Bio-Rad protein assay dye concentrate (Bio-Rad Laboratories, Hercules, CA, USA).
14. 7 mL (17 × 54 mm) polyethylene scintillation vials (VWR International, Montréal, QC, Canada).
15. BioSafe II biodegradable scintillation liquid (Research Products International Corp., Mount Prospect, IL, USA).
16. Brinkmann Polytron (Kinematica, Lucerne, Switzerland).
17. Whatman GF/C glass fiber filter strips (Brandel Inc., Gaithersburg, MD, USA).
18. Brandel semiautomated harvesting system M-48 (Brandel Inc., Gaithersburg, MD, USA).

2.4 Cell Lysates and Preparation of GRK2/3 Input Samples

1. Lysis buffer (10 mM Tris–HCl (pH 7.4), 5 mM EDTA). Store at 4 °C.
2. Cell sonicator.
3. Teflon cell lifters (Thermo Fisher Scientific).
4. 1.5 mL Eppendorf tubes (Sarstedt, Montréal, QC, Canada).
5. Bio-Rad protein assay dye concentrate (Bio-Rad Laboratories, Hercules, CA, USA).
6. BSA (Sigma-Aldrich) (*see* **Note 6**).
7. 2× SDS sample buffer.

2.5 Cross-Linking

1. Dimethyl sulfoxide (DMSO) (Sigma).
2. Dithiobis(succinimidyl propionate) (DSP) (Thermo Fisher Scientific) (*see* **Note 9**). Store desiccated at 4 °C.
3. Dithiothreitol (DTT) (Sigma) (*see* **Note 10**).
4. PBS (pH 7.4).
5. Ascorbic acid (Sigma).
6. Dopamine hydrochloride (Sigma).
7. RIPA+ buffer (*see* Sect. 2.2, item 27).
8. Teflon cell lifters (Thermo Fisher Scientific).

9. Protein A-Sepharose beads (GE Healthcare Life Sciences, Baie d'Urfé, QC, Canada).
10. Rat monoclonal anti-hemagglutinin (HA) affinity matrix (Covance Research Products, Richmond, CA, USA).
11. Rotating mixer (Thermo Fisher Scientific).
12. 15 mL screw-capped tubes (VWR International, Montréal, QC, Canada).
13. 1.5 mL screw-capped conical tubes (Thermo Fisher Scientific).
14. 1.5 mL Eppendorf tubes (Sarstedt, Montréal, QC, Canada).
15. Biotinylated anti-HA antibody (Covance Research Products, Richmond, CA, USA).
16. Bovine serum albumin (BSA) (Sigma).
17. Bio-Rad DC protein assay kit (Bio-Rad Laboratories, Hercules, CA).
18. 2× SDS/DTT sample buffer (25 mM Tris–HCl (pH 6.5), 8 % (v/v) SDS, 5 % (v/v) 2-mercaptoethanol, 10 % (v/v) glycerol, 100 mM DTT, 2 mg/mL bromophenol blue) (*see* **Note 11**).

2.6 SDS-Polyacrylamide Gel Electrophoresis (SDS-PAGE) and Immunoblotting

1. Acrylamide (Bio-Rad Laboratories, Hercules, CA, USA).
2. *N,N*′-methylenebisacrylamide (Bio-Rad Laboratories, Hercules, CA, USA).
3. 30.8 % (w/v) acrylamide stock solution. Add 72.5 g acrylamide and 2.5 g *N,N*′-methylenebisacrylamide to 150 mL Milli-Q water, dissolve by heating at 37 °C, adjust to 250 mL final volume with distilled water, verify that the pH is 7.0 or less, filter sterilize through 0.45 μm filter, and store at room temperature (*see* **Note 12**).
4. Ammonium persulfate (Bio-Rad Laboratories, Hercules, CA, USA).
5. Tetramethylethylenediamine (TEMED) (Bio-Rad Laboratories, Hercules, CA, USA).
6. Glycine (Sigma-Aldrich).
7. SDS (Bio-Rad Laboratories, Hercules, CA, USA).
8. 10× running buffer (288 g of glycine, 60.5 g Tris, 20 g SDS in 2 L of Milli-Q water).
9. Stained protein molecular weight markers (Bio Basic Canada Inc., Markham, ON, Canada).
10. SE640 wide-mini vertical electrophoresis unit (Hoefer, Holliston, MA, USA).
11. Gel fixing solution (10 % (v/v) acetic acid and 40 % methanol (v/v) prepared in Milli-Q water). Store at room temperature (*see* **Note 13**).

12. Semidry transfer buffer (2.9 g of glycine, 5.8 g Trizma Base, 0.37 g SDS in 200 mL of Milli-Q water).
13. Nonfat dry milk.
14. Blotto solution (50 mM Tris (pH 8.0), 2 mM $CaCl_2$, 80 mM NaCl, 5 % (w/v) nonfat dry milk, 0.2 % Nonidet P-40, 0.02 % NaN_3). Store at 4 °C for 1 month.
15. Tween-20 (Bio-Rad Laboratories, Hercules, CA, USA).
16. Trans-Blot® SD semidry electrophoretic transfer cell (Bio-Rad Laboratories, Hercules, CA, USA).
17. 10× Tris-buffered saline with Tween-20 (TBS-T) (200 mM Tris–HCl (pH 7.4), 1.37 mM NaCl and 2 % (v/v) Tween-20).
18. Sodium azide (NaN_3) (Sigma-Aldrich).
19. 150×25 mm polystyrene cell culture dish (Sarstedt, Montréal, QC, Canada).
20. Whatman filter paper (VWR International, Montréal, QC, Canada).
21. Blot paper (Hoefer, Holliston, MA, USA).
22. Stripping buffer (100 mM 2-mercaptoethanol, 0.5 % (w/v) SDS, 500 mM acetic acid).
23. Polyvinylidene difluoride (PVDF) (Thermo Fisher Scientific).
24. Rabbit polyclonal anti-GRK2/GRK3 antibodies (Gift of Dr. Robert J. Lefkowitz, Duke University. Antibodies are described in reference [35]) (*see* **Note 14**).
25. Horseradish peroxidase (HRP)-conjugated goat anti-rabbit antibodies (GE Healthcare Life Sciences, Baie d'Urfé, QC, Canada).
26. HRP-conjugated streptavidin (GE Healthcare Life Sciences, Baie d'Urfé, QC, Canada).
27. Enhanced Chemiluminescent (ECL) Kit (GE Healthcare Life Sciences, Baie d'Urfé, QC, Canada).
28. Kodak BioMax MR films (VWR International, Montréal, QC, Canada).
29. Typhoon phosphorimager (GE Healthcare Life Sciences, Baie d'Urfé, QC, Canada).

2.7 Confocal Microscopy

1. 12 mm diameter coverslips (Bellco Glass Inc., Vineland, NJ, USA)
2. 24-well dishes (VWR International, Montréal, QC, Canada)
3. 150×20 mm polystyrene cell culture dish (VWR International, Montréal, QC, Canada)
4. Paraformaldehyde (32 % w/v) solution, EM grade (Electron Microscopy Sciences, Hatfield, PA, USA) (*see* **Note 15**)

5. Fixation solution (4 % w/v paraformaldehyde in PBS) (*see* **Note 15**)
6. Glycine (Sigma)
7. Ammonium chloride (Sigma)
8. Quenching buffer (0.37 % (w/v) glycine, 0.27 % (w/v) ammonium chloride in PBS)
9. Saponin (Acros Organics, Morris Plains, NJ, USA)
10. Blocking/permeabilization buffer (1 % (w/v) BSA and 0.4 % (w/v) saponin in PBS)
11. Parafilm
12. Sharp-angled microscopy forceps
13. Mouse monoclonal anti-HA (12CA5) antibody (Roche Diagnostics, Laval, QC, Canada)
14. Rabbit polyclonal anti-GRK2/GRK3 antibodies (Gift of Dr. Robert J. Lefkowitz, Duke University. Antibodies are described in reference [35]) (*see* **Note 14**)
15. Alexa 488-conjugated goat anti-mouse antibody (Jackson ImmunoResearch Laboratories, Inc., West Grove, PA, USA)
16. Texas red-conjugated goat anti-rabbit antibody (Jackson ImmunoResearch Laboratories, Inc., West Grove, PA, USA)
17. SlowFade® Gold antifade reagent kit (Life Technologies Inc., Burlington, ON, Canada)
18. Glass slides (Thermo Fisher Scientific)

3 Methods

3.1 Cell Culture and Transfection

1. Maintain stocks of HEK293 cells in 0.2 μm vented blue plug seal capped 75 cm^2 tissue culture flasks in minimal essential medium (MEM) supplemented with 10 % heat-inactivated FBS and gentamicin (40 μg/mL) in a humidified incubator at 37 °C and 5 % CO_2. Detailed procedures about the making, propagation, and transfection of HEK293 cell stocks are discussed elsewhere [36, 37]. Only key steps are described below.
2. Place tissue culture flasks in BSC, aspirate MEM, and wash cells once with 5 mL of PBS.
3. Add 2 mL of trypsin per flask, incubate 1–2 min in BSC, tap briskly once the side of flasks to detach cells, add 20 mL of MEM (supplemented with 10 % heat-inactivated FBS and 20 μg/mL gentamicin), triturate media gently with a 10 mL pipette to resuspend cells without clumps, add an additional 40 mL of MEM, and count cells.

4. For each transfection condition seed four 100 mm dishes with 10 mL of cell resuspension at a density of 2.5×10^6 cells/dish and grow cells for ~24 h before transfection.
5. Put dishes in BSC and transfect cells with a total of 5 μg of plasmid DNA/dish using a modified calcium phosphate precipitation method as described [36, 37] (*see* **Note 16**).
6. Add plasmid DNAs (total amount of 10 μg) to sterile uncapped plastic 13 mL tubes in BSC, complete volume to 900 μL with autoclaved Milli-Q water, add 100 μL of 2.5 M $CaCl_2$, and mix by flicking tubes (*see* **Note 17**).
7. Add dropwise 1 mL of 2× HBS to DNA–calcium solution using a P1000 pipette (*see* **Note 18**).
8. Flick tubes to mix final solution (2 mL), add dropwise 1 mL to one dish (5 μg/dish), and add the last 1 mL to another dish (*see* **Note 19**).
9. Place dishes in a humidified incubator at 37 °C and 5 % CO_2 and incubate cells with DNA-calcium phosphate precipitate overnight (18–24 h).

3.2 Whole-Cell Phosphorylation Studies

3.2.1 Handling and Seeding of Transfected Cells (See Note 20)

1. Aspirate medium following overnight transfection, wash dishes with PBS (5 mL/dish), add fresh complete MEM (MEM, 10 % v/v FBS and 20 μg/mL gentamicin), and put dishes in a humidified incubator at 37 °C and 5 % CO_2.
2. The following day put dishes from the same transfection condition in BSC, aspirate medium, wash dishes with PBS, trypsinize cells using 0.5 mL trypsin/dish, add fresh complete MEM (5 mL/dish), triturate, pool cells in sterile 50 mL screw-capped polypropylene conical tubes, and count cells with a hemacytometer and fluorescence microscope (*see* **Note 21**).
3. Seed cell suspension in 6-well plates with 4 mL per well at a final density of $1–1.5 \times 10^6$ cells/well. Seed leftover cell suspension into two 100 mm dishes in a final volume of 10 mL/dish (~$1.5–3 \times 10^6$ cells/dish) (*see* **Note 21**). Place 6-well plates and 100 mm dishes in a humidified incubator at 37 °C and 5 % CO_2 and grow until the next day.

3.2.2 Cell Metabolic Labeling with [^{32}P]-Orthophosphate and HA-rD1R Immunoprecipitation

1. On the day of experiment, aspirate medium from 6-well plates, and replenish wells with 1.5 mL of 20 mM HEPES-buffered phosphate-free DMEM containing 10 μg/mL gentamicin and 0.15 mCi/mL [^{32}P]-orthophosphate. Place 6-well plates in a pre-warmed (37 °C) Plexiglas box in an incubator and label cells for 90 min at 37 °C (*see* **Notes 22** and **23**).
2. At the end of labeling period, place dishes on a benchtop covered with absorbent paper or diaper behind a Plexiglas shield, put aside lid of 6-well plates, add 10 μL of 100× ascorbic acid

(final in well 100 μM) and 10 μL of 100× dopamine prepared in ascorbic acid (final in well 10 μM dopamine) to appropriate wells, put lid back on 6-well plates, and incubate cells for 10 min at 37 °C to measure in transfected cells basal and dopamine-induced phosphorylation, respectively.

3. Cover benchtop with protection absorbent paper or diaper, set up a tray filled with ice behind Plexiglas shield, place cells on ice, discard carefully radiolabeled medium in waste plastic bottle using a 10 mL cotton-plugged plastic pipette, discard plastic pipette in Plexiglas container filled with a plastic bag, wash three times with 2 mL of ice-cold PBS using a 10 mL cotton-plugged plastic pipette, and discard plastic pipette in a plastic bag in Plexiglas waste container (*see* **Note 24**).
4. Add 800 μL of RIPA+ buffer, scrape off carefully cells from wells using a Teflon cell lifter, transfer cell lysates to a screw-capped 1.5 mL conical tubes using a P1000 pipette, place tubes on a rotating mixer, and solubilize cell lysates for 60 min at 4 °C (*see* **Note 24**).
5. Centrifuge tubes at 15,000 ×*g* for 15 min at 4 °C to pellet-insoluble cell components.
6. Transfer carefully supernatants (solubilized fraction) to fresh 1.5 mL screw-capped tubes.
7. Pipette two 25 μL aliquots of the supernatants in 1.5 mL Eppendorf tubes for each experimental condition and determine in duplicate the solubilized protein concentration with the Bio-Rad DC protein assay kit and BSA as standard (*see* **Note 25**).
8. Transfer 800 μL of equimolar concentrations of solubilized fractions to 1.5 mL screw-capped conical tubes containing 50 μL of 10 % (v/v) protein A-Sepharose beads prepared in TBS-BSA and preclear lysates on a rotating mixer for 1 h at 4 °C.
9. Centrifuge tubes at 15,000 ×*g* for 10 min at 4 °C to pellet protein A-Sepharose beads and transfer precleared lysates with a P1000 pipette to cold fresh tubes containing 50 μL of rat monoclonal anti-HA affinity matrix and immunoprecipitate HA-rD1R on a rotating mixer overnight at 4 °C. Discard tubes with protein A-Sepharose beads in Plexiglas-shielded radioactive solid waste container (*see* **Note 24**).

*3.2.3 Preparation of Receptor Membranes and [^{3}H]-SCH23390 Binding Assay (See **Note 23**)*

1. Place 100 mm dishes for radioligand binding prepared in Sect. 3.2.1 on ice.
2. Aspirate medium and wash cells once with 5 mL ice-cold PBS.
3. Add 5 mL of lysis buffer, scrape off gently cells from dish surface with Teflon cell lifters, and transfer cell lysates to 15 mL polycarbonate centrifuge tubes on ice.

4. Wash once dishes with 5 mL lysis buffer and transfer washes to cell lysates in centrifuge tubes to get a final volume of 10 mL.
5. Centrifuge cell lysates at 40,000 × *g* for 20 min at 4 °C.
6. Discard supernatant, add 3 mL of lysis buffer, detach pellets by pipetting up and down, and homogenize in centrifuge tubes at a velocity of 17,000 rpm for 15 s with a Brinkmann Polytron.
7. Add 7 mL of lysis buffer to each centrifuge tubes to obtain a membrane wash volume of 10 mL.
8. Centrifuge membrane suspensions at 40,000 × *g* for 20 min at 4 °C.
9. Discard supernatant, add 700 μL of resuspension buffer, and leave on ice while setting up binding assays.
10. Prepare a 10× saturating concentration (50–60 nM) of [^{3}H]-SCH23390 in Milli-Q water.
11. Prepare a 10× *cis*-flupenthixol (100 μM) in Milli-Q water.
12. Set up a three-tier polypropylene rack with 12 × 75 mm polystyrene test tubes to carry total and nonspecific binding reactions as described in Table 1. Take note when using the Brandel semiautomated harvesting system that the rack should be filled with only 48 tubes (Fig. 2).
13. Add 100 μL of membrane preparations to total and nonspecific binding tubes, mix by shaking rack, and incubate for 90–120 min at room temperature.

Table 1
Volume of binding reaction reagents for total and nonspecific binding of [^{3}H]-SCH23390

Binding reaction reagents	Total binding (μL)	Nonspecific binding (μL)
Binding buffer	300	300
Milli-Q water	50	0
10× *cis*-flupenthixol	0	50
10× [^{3}H]-SCH23390	50	50
Membrane preparation	100	100

For each transfection condition, carry total and nonspecific binding reactions in duplicate in a final volume of 500 μL. The maximal binding capacity (B_{max}) of [^{3}H]-SCH23390 is determined using a saturating concentration (5–6 nM) based on the equilibrium dissociation constant (K_D) measured at rD1R (~0.4 nM) in membranes from transfected HEK293 cells [30]. The nonspecific binding dpm value is subtracted from total binding dpm value to calculate B_{max} of [^{3}H]-SCH23390 in pmol/mg membrane proteins using the following equation:

$$B_{max}\left(\frac{\text{pmol}}{\text{mg membrane proteins}}\right)=\frac{\text{Specific Binding dpm Value}}{2,220\ \text{S.A.}\left(\frac{\text{Ci}}{\text{mmol}}\right)\text{Pt}}$$

where 2,220 is a derivative from 1 Ci = 2.22 × 10^{12} dpm to convert the specific binding dpm value in pmol, S.A. is the specific activity of [^{3}H]-SCH23390 in Ci/mmol, and Pt is the total amount of mg proteins in 100 μL of membrane preparations

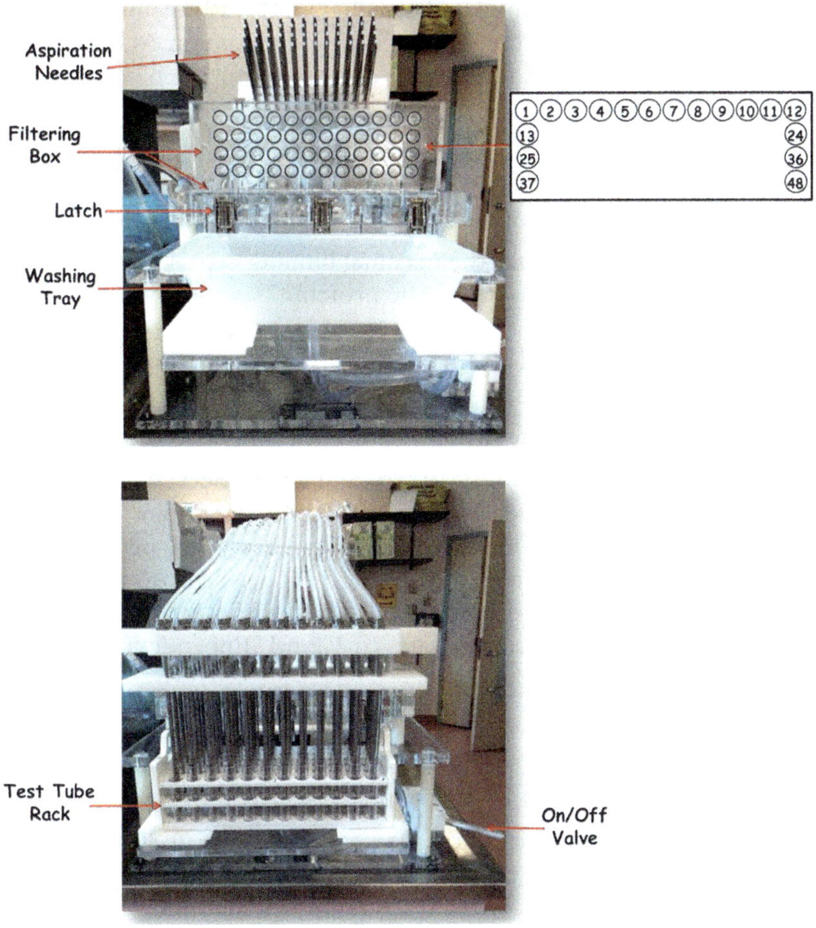

Fig. 2 The Brandel semiautomated harvesting system for rapid filtration and washing of [^{3}H]-SCH23390-radiolabeled-membrane preparations expressing HA-rD1R. The unlatched 48-sample filtering box (4 × 12 matrix) and aspiration needles are shown in the *top panel*. The washing tray is used to clean the apparatus with distilled water prior to filtering binding reactions through glass fiber filters. The diagram on the right depicts the harvest number matching the binding reaction tubes placed in the three-tier polypropylene rack once the filtering box is latched (*bottom panel*). For instance, when the filtering box is latched and on/off valve is set to "harvest" position (*bottom panel*), the harvest number 1–4 corresponds to duplicate total (1–2) and nonspecific (3–4) binding tubes placed in the first front row of the three-tier polypropylene rack (starting at the far *left*)

14. While waiting for incubation of binding reactions to finish, measure protein concentrations of cell lysates with 50 μL aliquots in duplicate using Bio-Rad protein assay kit and BSA as standard.
15. Place a strip of Whatman GF/C glass fiber filter sheet inside the filtering box of Brandel semiautomated harvesting system (Fig. 2), wet filter using a squirt bottle filled with distilled water, latch the filtering box, place the three-tier polypropylene rack with binding reaction tubes on test tube rack holder of harvester, insert aspiration needles in test tubes, filtrate rapidly

through glass fiber filter, and wash membrane bound to filter using three-tube fill-up and aspiration cycles with 5 mL of cold washing buffer per cycle to terminate binding reactions.

16. Remove filter from harvester, put filter circles with [^{3}H]-SCH23390 bound to membranes in plastic scintillation vials, add 5 mL of scintillation liquid, and measure tritium dpm counts in a beta scintillation counter.
17. Calculate the maximal binding capacity (B_{max}) in pmol or fmol per mg membrane proteins.

*3.2.4 Preparation of GRK Immunoblotting Samples (See **Note 23**)*

1. Place 100 mm dishes for GRK immunoblotting prepared in Sect. 3.2.1 on ice.
2. Aspirate medium and wash cells once with 5 mL ice-cold PBS.
3. Add 800 μL of lysis buffer (*see* Sect. **2.4**, item 1) containing protease inhibitors (2 μg/mL aprotinin, 10 μg/mL benzamidine, 10 μg/mL leupeptin, 1 μg/mL pepstatin A, 10 μg/mL soybean trypsin inhibitor, and 10 μg/mL PMSF) (*see* **Note 5**), scrape off gently cells from dishes with Teflon cell lifters, and transfer cell lysates to 1.5 mL Eppendorf tube on ice.
4. Sonicate cell lysates on ice for 25 s.
5. Measure protein concentrations of cell lysates with 25 μL aliquots (*see* **Note 26**) in duplicate using Bio-Rad protein assay kit and BSA as standard.
6. Prepare samples at a concentration of 30 μg protein per 25 μL in 1.5 mL Eppendorf tubes using lysis buffer with protease inhibitors, add 25 μL of 2× SDS sample buffer, and freeze tubes at −20 °C until use for SDS-PAGE and immunoblotting (*see* **Note 27**).

3.2.5 Isolation and Washing of HA-rD1R Immunocomplexes After the Overnight Incubation with Anti-HA Affinity Matrix (Sect. 3.2.2, Step 9)

1. Centrifuge tubes at 15,000 × *g* for 10 min at 4 °C, aspirate carefully supernatant with a P1000 pipette, and discard supernatants into radioactive liquid waste container.
2. Add 1 mL of ice-cold RIPA+ cell lysis buffer to beads, wash beads by inverting tubes six times, pellet beads by centrifugation at 15,000 × *g* for 2 min at 4 °C, remove carefully supernatants with an insulin syringe, and discard supernatants into radioactive liquid waste container. Repeat this step a total of five times.
3. Remove carefully the supernatants (fifth wash), discard supernatants, and remove as much leftover wash buffer from beads as possible using the insulin syringe to dry beads.
4. Add 60 μL of 2× SDS sample buffer to beads and incubate tubes at room temperature for 2 h to dissociate immunocomplexes (*see* **Note 28**).

3.2.6 SDS-PAGE Analysis of P^{32}-Labeled HA-rD1R Immunocomplexes (See **Note 29**)

1. Clean surface of glass plates (18×8 cm) and spacers (1.5 mm thick) with 70 % (v/v) ethanol with Kimwipe to remove any acrylamide waste residue and avoid gel solution to leak.
2. Dry surfaces with Kimwipe.
3. Place spacers along each edge of the inner surface of one glass plate, put the second glass plate onto glass plate with spacers, attach clamps to glass plates, stand the gel sandwich on benchtop to align glass plates and spacers, check for gap between spacers and glass plates, put gel sandwich onto gasket in the casting stand, and secure gel sandwich with cams.
4. Prepare 10 % (v/v) acrylamide gel solution in a 50 mL graduated cylinder. Add to the graduated cylinder 37.5 mL with sterile distilled water, 12.5 mL acrylamide gel solution, 18.75 mL 0.75 M Tris–HCl, pH 8.8, 0.375 mL 10 % (v/v) SDS, 1 mL 10 % (w/v) APS, and 25 μL TEMED (final volume of 50 mL). Seal graduated cylinder with Parafilm, gently mix by inversion twice, fill gel sandwich with ~15 mL, carefully overlay acrylamide gel with 0.5 mL *n*-butanol, and let separating gel polymerize for 10 min (*see* **Note 30**).
5. Pour off *n*-butanol in a container for chemical waste disposal, rinse thoroughly with distilled water, insert a piece of Whatman filter paper in gel sandwich to remove excess of water on glass plates and top of separating gel, and place comb (15 wells, 1.5 mm thick) in gel sandwich about half way through between the top of glass plates and separating gel.
6. Prepare 4 % (v/v) acrylamide gel solution in a 25 mL graduated cylinder. Add to graduated cylinder 5.91 mL sterile distilled water, 1.33 mL acrylamide gel solution, 1.65 mL 0.75 M Tris–HCl, pH 6.5, 100 μL 10 % (v/v) SDS, 1 mL 10 % (w/v) APS, and 10 μL TEMED (final volume of 10 mL). Seal graduated cylinder with Parafilm, gently mix by inversion once, fill gel sandwich with ~4.5 mL, and let stacking gel polymerize for 5–10 min.
7. Remove gently comb from stacking gel, fill wells with 1× running buffer using a beaker or graduated cylinder, load protein markers and samples into individual wells (55 μL), and attach the upper chamber to gel sandwich (*see* **Note 31**).
8. Put a magnetic stir bar in the lower buffer chamber of electrophoresis unit, fill with ~2.5 L of 1× running buffer, remove with a steady hand the gel sandwich and upper chamber from casting stand, and place it carefully in the lower buffer chamber.
9. Add 1× running buffer to upper buffer chamber to submerge the electrode running along the upper chamber ridge (*see* **Note 32**), place the safety lid on the electrophoresis unit, and plug black and red wires into the power supply jacks.

10. Run samples through stacking gels at a voltage of 200 V for ~40–45 min.
11. Once samples migrate into separating gels, increase voltage to 400 V and resolve samples until the loading and lowest protein weight marker run off the gel sandwich (~15–20 min).
12. Turn off the power supply, disconnect the wires, remove safety lid, lift upper buffer chamber with gel sandwich, detach upper buffer chamber from gel sandwich, discard buffer in sink, take off clamps from gel sandwich, slide gently away spacers using a spatula, lift carefully the glass plate attached to gel, remove any stacking gel leftover from separating gel using a spacer, and cut upper left corner of the separating gel with a 45° angle for tracking gel loading orientation.
13. Place separating gel in a Pyrex tray containing the fixing solution, seal with Saran wrap, microwave gel at max power for 1.5 min, transfer gel on precut blot paper, cover with Saran wrap, put blot paper with gel onto gel dryer, and dry gels at 66 °C for 2 h (*see* **Note 33**).
14. Put gel dried on blot paper in phosphor screen cassette, expose for 2 h, scan phosphor screen in phosphorimager, and quantify HA-rD1R levels (*see* **Note 34**).
15. After phosphorimager processing place gel into X-ray film cassette with an intensifying screen, put an autoradiography film onto gel, and expose at −80 °C overnight (*see* **Note 35**).

3.2.7 SDS-PAGE and Immunoblotting of GRK2/3 Input Samples (See Note 36)

1. Prepare 10 % (v/v) gels for SDS-PAGE and load samples (55 μL/well) as described in Sect. 3.2.6, steps 1–12.
2. Place gel in distilled water (*see* **Note 37**).
3. Cut two pieces of blot paper with 1 cm wider margins than gel size and leave in 150 mm polystyrene cell culture dish filled with semidry transfer buffer.
4. Cut PVDF membrane to fit exactly gel size (*see* **Note 38**), place membrane in a 150 mm dish containing methanol, leave for 5 min at room temperature, transfer membrane to a fresh dish, rinse once with water, and place in a dish filled with semidry transfer buffer for 3 min at room temperature.
5. Place first piece of wet blot paper in 150 mm dish onto a flat surface, flip horizontally gel, put onto wet blot paper, lay down PVDF membrane on gel, place the second piece of wet blot paper on the membrane (*see* **Note 39**).
6. Roll gently a plastic or glass pipette over the gel transfer sandwich to remove air bubbles (*see* **Note 40**).
7. Remove safety cover and stainless-cathode plate electrode of Trans-Blot® SD semidry transfer cell, flip horizontally gel

transfer sandwich, and put it onto the platinum-coated titanium anode (*see* **Note 41**).

8. Place stainless-cathode plate electrode on top of the gel transfer sandwich, put the safety cover on the transfer unit, and plug black and red wires into the power supply jacks.
9. Transfer at a voltage of 15 V for 18 min.
10. Remove gel transfer sandwich from apparatus, place membrane in dish with Blotto solution, and incubate on a rocking platform for 1 h at room temperature or overnight at 4 °C (*see* **Note 42**).
11. Discard Blotto solution, rinse quickly three times the PVDF membrane in a 150 mm dish with 1× TBS-T buffer, place membrane in a Parafilm boat (*see* **Note 43**), add 10 mL of primary polyclonal rabbit anti-GRK2/3 antibodies [35] diluted (1:2,500) in 1× TBS-T buffer with 0.01 % (w/v) NaN_3, place on a rocking platform at 4 °C, and incubate overnight (*see* **Note 44**).
12. Save the primary antibody solution in a 15 mL screw-capped tube and store at 4 °C (*see* **Note 45**).
13. Rinse quickly three times the PVDF membrane in a 150 mm dish with 1× TBS-T buffer and then wash three times with 20 mL 1× TBS-T buffer on a rocking platform for 15 min at room temperature.
14. Place PVDF membrane in a Parafilm boat, add 10 mL of secondary HRP-conjugated goat anti-rabbit antibodies diluted (1:5,000) in 1× TBS-T buffer (without NaN_3), and incubate on a rocking platform for 1 h at room temperature (*see* **Note 46**).
15. Discard secondary antibody, rinse quickly three times the PVDF membrane in a 150 mm dish with 1× TBS-T buffer, and then wash three times in 20 mL 1× TBS-T buffer on a rocking platform for 15 min at room temperature.
16. Hold gently PVDF membrane with a forceps, drain the excess of TBS-T buffer by gently gliding the edge on a Kimwipe, and place in Parafilm boat.
17. Mix the two ECL reagents (1:1), add on PVDF membrane in a Parafilm boat, and incubate with shaking for 1 min at room temperature.
18. Grip gently the PVDF membrane with a forceps, drain the excess of ECL detection mixture by gliding gently the edge on a Kimwipe, and put the blotted membrane inside a plastic sheet protector with proteins side up in an X-ray film cassette.
19. Place an autoradiography film onto wrapped membrane in the dark and expose film for 15 s (*see* **Note 47**).

20. Save membrane and store in TBS-T at room temperature if a stripping and reprobing procedure is required (*see* Sect. 3.3.2). Alternatively, the wet PVDF membrane can be wrapped in Saran wrap and store at 4 °C until needed.

3.3 Cross-Linking Studies

3.3.1 Cross-Linking and Coimmuno-precipitation of D1R–GRK Complexes (See Note 48)

1. Place 100 mm dishes in BSC following overnight transfection, aspirate medium, wash cells with 5 mL PBS (pH 7.4, room temperature), remove PBS, add complete MEM (10 mL/dish), and grow cells for an additional 18–24 h in a humidified incubator at 37 °C and 5 % CO_2.
2. Put dishes from the same transfection condition in BSC, aspirate medium, wash cells with 5 mL PBS (pH 7.4, room temperature), remove PBS, trypsinize cells (0.5 mL trypsin/dish), add fresh complete MEM (8 mL/dish), triturate, pool cells, and reseed in three 100 mm dishes (*see* **Note 49**). Repeat for each transfection condition and grow reseeded cells until next day in a humidified incubator at 37 °C and 5 % CO_2.
3. On the day of assay for each transfection condition, pick the two 100 mm dishes assigned to drug treatment conditions (ascorbic acid and dopamine), aspirate cell culture media, add 10 mL of serum-free medium (20 mM HEPES-buffered MEM containing 10 μg/mL gentamicin), and incubate cells for 90 min in a humidified incubator at 37 °C and 5 % CO_2 (*see* **Notes 50** and **51**).
4. Following the 90 min incubation period, add 100 μL of 100× ascorbic acid (0.1 mM final in assay) and 100× dopamine (10 μM final in assay) to separate dishes and incubate with drugs for 10 min at 37 °C.
5. At the end of drug treatment, put dishes on ice, aspirate medium, wash once or twice with 5 mL of ice-cold PBS, remove by aspiration, replenish dishes with 4.5 mL PBS (pH 8.0), add 0.5 mL of 10× DSP per dish (2.5 mM final in dish), and leave cells on ice and incubate for 2 h at 4 °C (*see* **Note 52**).
6. At the end of cross-linking reaction, aspirate DSP–PBS solution, add 800 μL RIPA+ containing protease inhibitors, scrape off carefully cells from dishes with a Teflon cell lifter, transfer cross-linked cell lysates to screw-capped 1.5 mL conical tubes using a P1000 pipette, place tubes on a rotating mixer, and solubilize cross-linked cell lysates for 120 min at 4 °C (*see* **Notes 53** and **54**).
7. Proceed then as described in Sect. 3.2.2, steps 5–9 (*see* **Notes 55** and **56**).
8. Isolate and wash HA-rD1R immunocomplexes as described in Sect. 3.2.4, steps 1–3.

9. Add 2× SDS-DTT sample buffer as follows: 60 μL to dried beads in immunoprecipitation tubes and 40 μL to GRK2/3 input tubes (10 μL) (*see* **Note 55**).
10. Incubate tubes in a water bath at 37 °C for 30 min to cleave cross-linker and afterward freeze samples if not used immediately in SDS-PAGE (*see* **Note 56**).

3.3.2 SDS-PAGE and Immunoblotting of Analyses of HA-rD1R Immunocomplexes and GRK2/3 Input Samples from Cross-Linking Studies

1. Prepare 10 % (v/v) gels for SDS-PAGE and load samples (55 μL/well) as described in Sect. 3.2.5, steps 1–12.
2. Perform immunoblotting procedure as described in Sect. 3.2.7, steps 2–19.
3. Save membrane and store in TBS-T until use for stripping and reprobing to visualize the amount of immunoprecipitated HA-rD1R.
4. Incubate PVDF membrane with ~10–15 mL stripping buffer in a close container on a rocking plate for 30 min at room temperature.
5. Discard stripping buffer in glass container for chemical waste disposal and rinse three times the PVDF membrane with large volume of TBS-T for 5 min on a rocking plate at room temperature.
6. Incubate with Blotto solution on a rocking platform overnight at 4 °C or for 1 h at room temperature (*see* **Note 42**).
7. After overnight incubation in Blotto solution, discard the blocking reagent, rinse quickly three times the PVDF membrane in a 150 mm dish with 1× TBS-T buffer, place membrane in a Parafilm boat (*see* **Note 43**), add 10 mL of primary monoclonal mouse biotinylated anti-HA antibodies diluted (1:1,000) in 1× TBS-T buffer with 0.01 % (w/v) NaN_3, place on a rocking platform at 4 °C, and incubate overnight (*see* **Note 44**).
8. Save the primary antibody solution in a 15 mL screw-capped tube and store at 4 °C (*see* **Note 45**).
9. Rinse quickly three times the PVDF membrane in a 150 mm dish with 1× TBS-T buffer and then wash three times with 20 mL 1× TBS-T buffer on a rocking platform for 15 min at room temperature.
10. Place PVDF membrane in a Parafilm boat, add 10 mL of HRP-conjugated streptavidin diluted (1:5,000) in 1× TBS-T buffer (without NaN_3), and incubate on a rocking platform for 1 h at room temperature (*see* **Note 46**).
11. Discard the HRP-conjugated streptavidin solution, rinse quickly three times the PVDF membrane in a 150 mm dish with 1× TBS-T buffer, and then wash three times in 20 mL 1× TBS-T buffer on a rocking platform for 15 min at room temperature.

12. Hold gently PVDF membrane with a forceps, drain the excess of TBS-T buffer by gently gliding the edge on a Kimwipe, and place in Parafilm boat.
13. Mix the two ECL reagents (1:1), add on PVDF membrane in a Parafilm boat, and incubate with shaking for 1 min at room temperature.
14. Grip gently the PVDF membrane with a forceps, drain the excess of ECL detection mixture by gliding gently the edge on a Kimwipe, and put the blotted membrane inside a plastic sheet protector with proteins side up in an X-ray film cassette.
15. Place an autoradiography film onto wrapped membrane in the dark and expose film for 15 s (*see* **Note 46**).

3.4 Immunofluorescence Confocal Microscopy

1. Following overnight transfection, split transfected HEK293 cells as described in Sect. 3.2.1, step 2.
2. Seed 0.5 mL of cell suspension on etched 12 mm diameter coverslips (*see* **Note 58**) placed in a 24-well plate (60,000 cells/well) and let the cells grow on coverslips 36–48 h at 37 °C in a 5 % CO_2 humidified incubator.
3. On the day of experiment, prepare fresh stock solutions of 50× ascorbic acid (5 mM) and 50× dopamine (0.5 mM prepared in 5 mM ascorbic acid) and leave on ice until used on cells.
4. Place 24-well plates on benchtop, aspirate medium, add 0.5 mL of 20 mM HEPES-buffered MEM in each well, add 10 μL of 50× ascorbic acid (0.1 mM final in assay) and 50× dopamine (10 μM final in assay) to separate wells, and incubate coverslips for 10 min at 37 °C.
5. Aspirate medium and gently wash coverslips with 0.5 mL PBS at room temperature.
6. Aspirate PBS, add 0.3 mL 4 % (v/v) paraformaldehyde in PBS per well with coverslips, and fix cells for 20 min at room temperature in a fume hood.
7. Discard paraformaldehyde in a chemical waste glass bottle and rinse briefly wells once with 0.5 mL PBS at room temperature. Discard PBS washes in a paraformaldehyde waste glass bottle.
8. Incubate twice coverslips in 0.3 mL/well of quenching buffer for 10 min (*see* **Note 59**).
9. Rinse coverslips several times with PBS (3–4 times).
10. Permeabilize cells in 0.3 mL of blocking/permeabilization buffer for 30 min at room temperature.
11. Prepare a humidified chamber to hold coverslips. Place a filter paper (paper towel or Kimwipe can be also used) in the bottom of a 150×20 mm polystyrene cell culture dish and lay a piece of Parafilm on filter paper saturated with distilled water (*see* **Note 60**).

12. Transfer coverslips from the well with a sharp-angled microscopy forceps onto Parafilm in humidified dish cell side up.
13. Decant excess blocking solution by gently wicking coverslips onto a paper towel or Kimwipe.
14. Add on top of coverslips (cell side up) 100 μL of primary antibodies mixture (mouse monoclonal anti-HA antibody, 1:300 of 0.4 μg/μL stock; rabbit polyclonal anti-GRK2/3 antibody, 1:500) diluted in blocking/permeabilization buffer and incubate for 1 h at room temperature.
15. Wash coverslips five times with 100 mL of 1 % (w/v) BSA and 0.4 % (w/v) saponin in PBS for 5 min at room temperature.
16. Add on top of coverslips (cell side up) 100 μL of secondary antibodies mixture (Alexa 488-conjugated goat anti-mouse, 1:500; Texas red-conjugated goat anti-rabbit, 1:500) diluted in blocking/permeabilization buffer and incubate for 30 min at room temperature in the dark.
17. Wash three times with blocking/permeabilization buffer for 10 min at room temperature.
18. Decant gently blocking/permeabilization buffer, add 100 μL of SlowFade® equilibration buffer on coverslips, and pre-equilibrate coverslips for 5–10 min at room temperature.
19. Decant as much equilibration buffer from coverslips as possible (*see* **Note 61**).
20. Add one drop of SlowFade® Gold antifade reagent onto glass slides, mount coverslips cell side down onto SlowFade® Gold antifade reagent on the glass slide with a sharp-angled microscopy forceps, apply nail polish on the edge of coverslips to seal on the glass slide, and incubate at room temperature in the dark for 24 h (*see* **Note 62**).
21. Visualize cells with a confocal laser microscopy and capture images for analysis (*see* **Note 63**).

4 Typical/Anticipated Results

Anticipated results from a typical experiment exploring GRK3-mediated D1R phosphorylation using whole-cell phosphorylation and HA-tagged receptor immunoprecipitation approaches are presented in Fig. 3. In agreement with the presence of putative serine and threonine phosphorylation sites for GRKs in the intracellular regions of D1R [6–10], GRK3 overexpression promotes an increase in the basal and dopamine-induced phosphorylation of rD1R. Truncations of the cytoplasmic tail of rD1R lead to a significant reduction in dopamine-independent (basal) and dependent GRK3-induced receptor phosphorylation (*see* [8] for details).

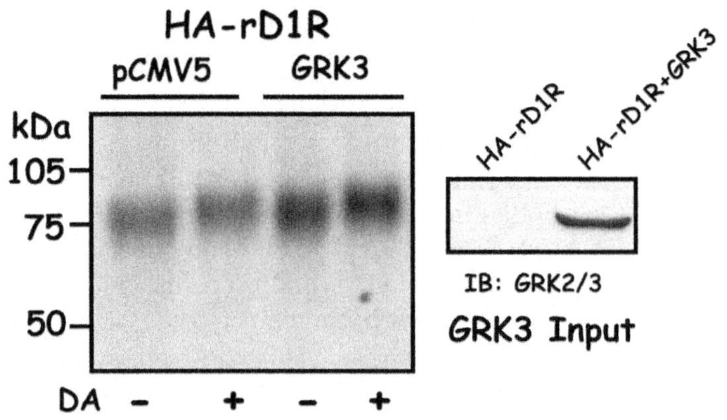

Fig. 3 Demonstration of basal and dopamine-induced phosphorylation of HA-rD1R by GRK3 in HEK293 Cells. The *left panel* shows a representative autoradiogram of phosphorylation of HA-rD1R transfected in HEK293 cells with rat GRK3 or with empty pCMV5 vector. Transfected cells metabolically labeled with [^{32}P]-H_3PO_4 were stimulated in the absence (–) or presence (+) of dopamine (DA) for 10 min. Cells treated with or without DA were solubilized in RIPA+ buffer and subjected to HA-matrix immunoprecipitation and SDS-PAGE using a 10 % (v/v) gel. Receptor phosphorylation depicted in the left panel was visualized following an overnight autoradiography at −80 °C. The relative molecular weight of protein standards is indicated in kilodaltons (kDa). The *right panel* represents a typical blot for immunoblotting (IB) GRK3 input samples using 30 μg of cell lysate proteins resolved by SDS-PAGE with a 10 % (v/v) gel and probed with rabbit polyclonal anti-GRK2/3 antibodies [35]. The electrophoretic mobility of GRK3 is ~80 kDa. The figure is modified from [8]

These findings suggest the presence of GRK3 phosphorylation sites in the cytoplasmic tail of D1R. Interestingly, the extent of GRK3-induced rD1R desensitization is not significantly changed following truncation of the cytoplasmic tail implying that phosphorylation and desensitization of rD1R by GRK3 are dissociable regulatory processes (*see* [8] for details).

A typical experiment using solubilized protein samples obtained from cross-linking studies and the Bio-Rad DC protein assay kit is shown in Fig. 4. In our hands, this assay gives reliable values with a high coefficient of determination (R^2) close to 0.99. Results obtained from a typical protein assay indicate that 10 μL solubilized lysate samples used for GRK3 input samples give on average between 10 and 15 μg total proteins, which is enough to detect expression levels of transfected GRK3 as shown in Fig. 5 (middle panel). Figure 5 also presents anticipated results that can be obtained when studying the formation of complex between D1R and GRK3 or GRK2 (*see* [8] for details) using a DSP cross-linking approach. The specificity of the interaction between GRK3 and HA-rD1R under basal and dopamine exposure is underscored by the lack of GRK3 detection from cells transfected with GRK3 only and subjected to immunoprecipitation with HA affinity matrix (Fig. 5, top panel). These results also emphasized the importance of extensive washes of HA affinity matrix to interpret unequivocally results from cells expressing the receptor and GRK isoforms. In agreement with

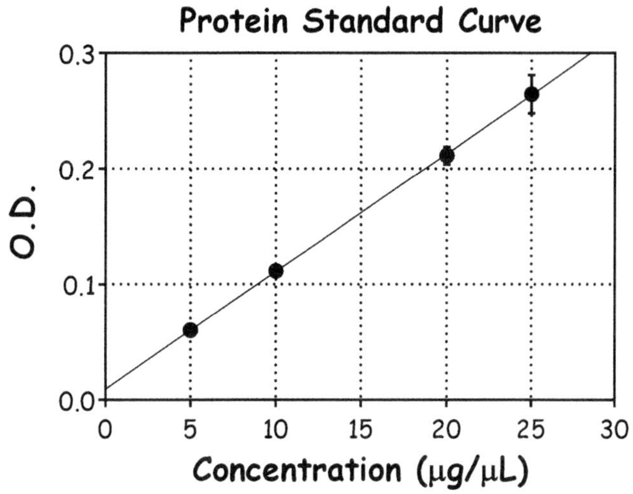

Protein Standards	O.D.	
μg/μL	Y1	Y2
5	0.057	0.063
10	0.106	0.117
20	0.219	0.204
25	0.248	0.281
Protein Samples		
GRK3 (+)	0.274	0.280
HA-rD1R+GRK3 (-)	0.332	0.320
HA-rD1R+GRK3 (+)	0.335	0.308

Results

μg/μL	O.D.
1.052	0.277
1.245	0.326
1.227	0.322

Fig. 4 Representative data obtained with the DC protein assay and solubilized samples from a cross-linking experiment. The *top panel* depicts the protein standard curve generated from four concentrations of BSA (5, 10, 20, and 25 μg/μL). The standard curve is best fitted to a linear relationship with R^2 value of 0.986 ($p < 0.0001$). The *bottom left* panel shows the duplicate values obtained for OD of protein standards and protein samples from solubilized cells transfected with GRK3 alone, HA-rD1R, and GRK3 treated with (+) or without (−) dopamine (DA). The *bottom right panel* reports the average OD values and calculated concentrations of solubilized proteins obtained from the three different experimental conditions. OD, optical density

the classical paradigm of GPCR regulation by GRKs [1, 2] and data presented in Fig. 3, the increase of dopamine-induced rD1R phosphorylation by GRK3 relative to basal can be linked to an augmentation of GRK3 levels associating with the agonist-activated rD1R signaling complex (Fig. 5, top panel). Results obtained with reprobing of the blot shown in the top panel following the stripping procedure clearly indicate that the increase in the extent of dopamine-induced phosphorylation by GRK3 relative to basal is not

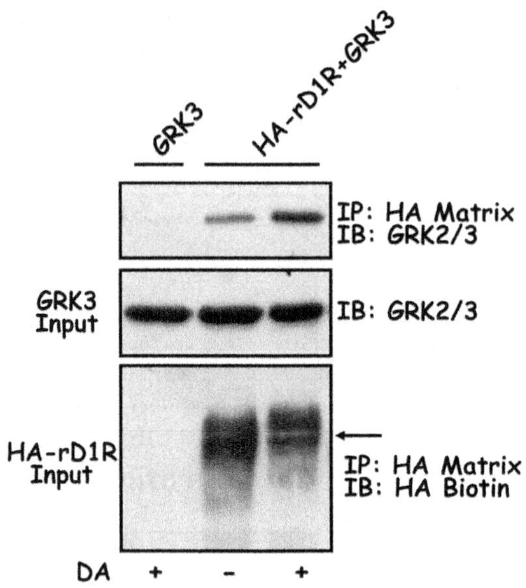

Fig. 5 Detection of basal and dopamine-induced complex formation between HA-rD1R and GRK3 in HEK293 cells using the DSP cross-linking method. The *top* and *middle panels* show representative blots of immunoblotting (IB) of GRK3 from immunoprecipitated (IP) and GRK3 input samples from HEK293 cells transfected with rat GRK3 alone or with GRK3 and HA-rD1R stimulated in the presence (+) or absence (−) of dopamine (DA) for 10 min. Cells treated in the absence (−) or presence (+) of DA were incubated with the cross-linker DSP (Fig. 1), solubilized in RIPA+ buffer, and subjected to HA-matrix immunoprecipitation. Immunocomplexes and GRK3 input samples were treated with DTT to cleave the cross-linker DSP. Proteins were resolved on SDS-PAGE using 10 % (v/v) gels and transferred on a PVDF membrane. Immunodetection of GRK3 was done with the rabbit polyclonal anti-GRK2/3 antibodies [35]. The *bottom panel* displays the IB of HA-rD1R of blot shown in the top panel following stripping and reprobing with a mouse monoclonal biotinylated anti-HA antibody. For clarity, the relative molecular weight of protein standards is not indicated on blots. The electrophoretic mobility of GRK3 is ~80 kDa, while HA-rD1R migrates as a broadband ranging from 75 to 85 kDa. The *arrow* in the *bottom panel* indicates the co-migrating HA-rD1R and GRK3 overlap following transfer on the blot, which partially shields the HA-tagged receptor protein from probing by the biotinylated mouse monoclonal anti-HA antibody. The figure is modified from [8]

explained by the presence of higher levels of immunoprecipitated receptors (Fig. 5, bottom panel). It is also worth mentioning that the reprobed blot demonstrates an upshift in the HA-rD1R band, which is explained by an increase in the receptor phosphorylation level brought by dopamine exposure (Fig. 5, bottom panel). The increase of negatively charged phosphate group on HA-rD1R slows its migration rate on 10 % (v/v) SDS-PAGE.

To some extent similar results with cells co-transfected with HA-rD1R and GRK2 can be anticipated. However, our study suggests that the underlying molecular mechanisms implicated in the regulation of D1R by GRK2 and GRK3 are different (*see* [8] for details). A commonality noted between GRK2 and GRK3 vis-à-vis D1R is that these three proteins remained associated upon receptor

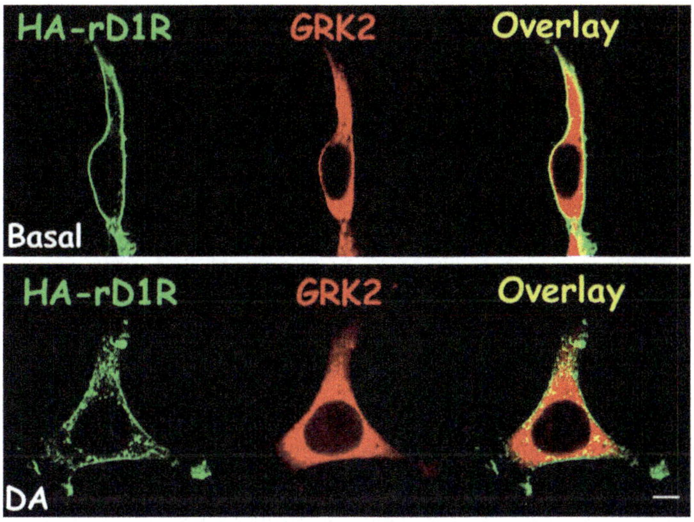

Fig. 6 Co-localization of HA-rD1R and GRK2 in HEK293 cells in the absence and presence of dopamine. The *top* and *bottom panels* show confocal microscopy images of HEK293 cells co-expressing HA-rD1R and rat GRK2 following a 10 min incubation with 100 μM ascorbic acid (Basal) and 10 μM dopamine (DA) at 37 °C, respectively. Immunofluorescence confocal microscopy was done on saponin-permeabilized cells incubated with primary mouse monoclonal anti-HA and rabbit polyclonal anti-GRK2/3 antibodies for immunodetection of HA-rD1R (*green*) and GRK2 (*red*), respectively. Cells were then incubated with secondary Alexa 488 (*green*) and Texas Red-conjugated antibodies. Immunofluorescence images were observed and captured with an Olympus 60× (Plan Apo) oil immersion lens with a Bio-Rad MRC-1024MP confocal laser microscope. An overlay of HA-rD1R and GRK2 images was performed to visualize the co-localization of the two fluorophores. Scale bar, 10 μm

internalization. As indicated by a typical experiment using immunofluorescence and confocal laser microscopy (Fig. 6), one can observe that intracellular vesicles harboring HA-rD1R also display positive labeling for GRK2. Additionally, detection of coincidence in signals from HA-rD1R and GRK2 in the absence of dopamine suggests the constitutive formation of a protein complex between D1R and GRK2 (Fig. 6, top panel), which is in agreement with results obtained with cross-linking studies (*see* [8] for details). Similar findings have been also observed with GRK3 (*see* [8] for details and Fig. 5). Intriguingly, a clathrin-binding motif has been identified in GRK2 (also conserved in GRK3) suggesting a role of GRKs in GPCR internalization [38–42]. Therefore, our immunofluorescence data collected with confocal microscopy may point also to a role of GRK2 and GRK3 in the agonist-induced internalization of D1R.

5 Notes

1. The choice of FBS source should be primarily based on the quality tests performed prior to final purchasing. The end user will request free FBS samples from different manufacturers. Different

lot numbers from one manufacturer can be also obtained if available. In our laboratory, we typically assay FBS lot samples obtained from at least two different manufacturers (e.g., Invitrogen, Wisent, NorthBio). The quality tests are done with HEK293 cells cultured in MEM supplemented with different FBS lots including the one currently in use in the lab (control). HEK293 cells are transfected transiently with 5 μg of plasmid DNA/100 mm dish using empty pCMV5 vector (mock), human D1R-pCMV5, and human D5R-pCMV5. As discussed elsewhere [36, 37], in our hands, this amount of plasmid DNA yields the maximal receptor expression achievable in HEK293 cells. For each FBS tested, transfected cells are seeded in 6-well plates and 100 mm dishes to determine constitutive activation and DA-induced stimulation of AC in intact cells, and maximal binding capacity (B_{max}) of [^{3}H]-SCH23390 in crude membrane preparations using whole-cell cAMP and radioligand binding assays, respectively, as described [36, 37]. Tests are repeated 2–3 times. The measured B_{max} value primarily determines whether the FBS lots are suitable for studies. Additionally, lack or weak DA-induced activation of AC relative to control serum is usually a sign that something is wrong with the FBS lot.

2. The end user prepares 2× HEPES-buffered saline (HBS) by diluting appropriate amounts of premade sterile 1 M HEPES (pH 7.0), 1 M NaCl, and 1 M Na_3PO_4 solutions in Milli-Q water to obtain a final buffer volume of 500 mL or 1 L. Sterilize buffer through a 0.22 μm filter in a BSC, aliquot in 15 mL screw-capped polystyrene conical tubes, and store at −20 °C. Making of premade sterile solutions has been described in detail elsewhere [36, 37].

3. Expression plasmids in pCMV5 for untagged and epitope-tagged human and rat D1R cloned can be obtained from our laboratory upon written request. Requests for GRK2 and GRK3 expression plasmids should be sent to Dr. Robert J. Lefkowitz. Alternatively, the end user can purchase some of these plasmids from Addgene at a reasonable price.

4. Typically, we prepare 500 mL of filter-sterilized RIPA+ buffer without protease inhibitors. We store the RIPA+ cell lysis buffer without protease inhibitors at 4 °C for 3–6 months (and up to 2 years according to Sigma, www.sigmaaldrich.com/content/dam/sigma-aldrich/docs/Sigma/Bulletin/r0278bul.pdf). The end user will need approximately 10 mL of RIPA+ per sample (includes cell solubilization and immunocomplex washing (*see* Sect. 3.2)).

5. We make 1,000× stock solutions of protease inhibitors (2 mg/mL aprotinin, 10 mg/mL benzamidine, 10 mg/mL leupeptin, 1 mg/mL pepstatin A, 10 mg/mL soybean trypsin inhibitor,

and 10 mg/mL PMSF) and store 0.5 mL aliquots in 1.5 mL Eppendorf tubes at −20 °C. All protease inhibitors are prepared in Milli-Q water except for pepstatin A and PMSF, which are dissolved in methanol and DMSO, respectively. Once thawed, protease inhibitors are vortexed to resuspend precipitates prior to adding to RIPA+ cell lysis buffer. This is notably important for PMSF since DMSO crystallizes at low temperature.

6. When using BSA as standard in protein assays, we typically prepare a 1 mg/mL stock solution in Milli-Q water and store 1.0 mL aliquots in 1.5 mL Eppendorf tubes at −20 °C. Tubes can be frozen and thawed several times.

7. The end user must wash protein A-Sepharose beads with TBS-BSA buffer to remove the ethanol, which is used by the manufacturer to store protein A-Sepharose beads. One must first calculate and pipette the total volume of slurry needed for the preclearing step into a 15 mL screw-capped polystyrene tube and centrifuge at 500 × *g* for 5 min at 4 °C. Discard the ethanol supernatant and wash twice with TBS-BSA by centrifugation at 500 × *g* for 5 min at 4 °C. Resuspend with TBS-BSA to obtain a 10 % (w/v) slurry and store up at 4 °C for up to 2 months. Prior to use check for bacterial contamination. Discard if contaminated and prepare fresh batch.

8. We usually prepare 25–50 mL of 2× SDS sample buffer in a 50 mL screw-capped polypropylene conical tube and store it at room temperature when a large number of experiments are performed on regular basis. However, if used sporadically make aliquots and store at −20 °C for up to 6–12 months.

9. DSP (also known as Lomant's reagent or DTSP) is a water-insoluble, thiol-cleavable, homobifunctional, and amine-reactive cross-linking agent. DSP comes as a hygroscopic powder and before opening the bottle, it must be equilibrated at room temperature for 15–20 min. Perform cross-linking experiments using always a fresh 10× stock solution of 25 mM DSP prepared in DMSO. Prepare approximately 30 min prior to incubating cells in the presence or absence of dopamine. Keep stock solution on the benchtop at room temperature and do not put on ice as DMSO freezes at <19 °C.

10. DTT is required to cleave the cross-linker DSP prior to SDS-PAGE and immunoblotting of HA-rD1R immunocomplexes. We typically prepare 1 M (10×) DTT (MW = 154.25) stock solution in Milli-Q water (e.g., 0.31 g dissolved in 2 mL) and store 200 μL aliquots in 1.5 mL Eppendorf tubes at −20 °C for several months.

11. For cross-linking studies, we normally make fresh 1 mL 2× SDS/DTT sample buffer by adding 100 μL of 10× DTT (1 M) to 900 μL premade 2× SDS sample buffer (*see* Sect. 2.2,

item 38 and **Note 8**). Prepare just before making GRK2/3 input samples and washing HA-rD1R immunocomplexes and keep 2× SDS/DTT sample buffer at room temperature.

12. Always wear personal protection equipment when working with acrylamide and *N*, *N'*-methylenebisacrylamide powder or solutions. Acrylamide gels and acrylamide liquid waste should be processed as per the office of risk management, environmental health, and safety guidelines of the end user's institution.

13. The volume of fixing solution used on gels is discarded in a bottle container for chemical waste disposal. We never reuse the solution once gels have been fixed (*see* Sect. 3.2.6).

14. Other anti GRK antibodies have also been generated by Dr. Lefkowitz's group. Alternatively, the end user may purchase mouse monoclonal anti-GRK2 (SC-13143, C-9) and rabbit polyclonal anti-GRK3 (SC-563, C-14) antibodies from Santa Cruz Biotechnology Inc. We have used these antibodies on hippocampal and striatal preparations [8] as well as on untransfected and transfected HEK293 cells (unpublished data).

15. This premade electron microscopy (EM) grade and methanol-free paraformaldehyde and solution is used to make the 4 % w/v paraformaldehyde fixation solution. The use of this premade paraformaldehyde solution has the advantage of preventing hazardous exposure of the end user to paraformaldehyde dust when weighing this irritant chemical. However, whether powder or concentrated solutions (>4 % w/v) are utilized, the end user must always wear gloves and safety glasses (face protection may be required if large quantities are handled). Work with paraformaldehyde powder or concentrated solution must be only performed in a chemical fume hood. Diluted paraformaldehyde solutions (<4 % w/v) can be handled in small quantities on the benchtop. Disposal of paraformaldehyde waste should be done according to the office of risk management, environmental health, and safety guidelines of the end user's institution.

16. For whole-cell phosphorylation and cross-linking studies, HEK293 cells are transfected with wild-type rat expression constructs in pCMV5: HA-rD1R, rGRK2, and rGRK3. Experimental conditions include (1) cells transfected with HA-rD1R alone, (2) cells transfected with the receptor and rGRK2 or rGRK3, and (3) cells transfected with rGRK2 or rGRK3 alone (this transfection condition is employed in cross-linking/coimmunoprecipitation studies as negative control). The end user will need to determine empirically the amount of HA-rD1R expression plasmid to transfect HEK293 cells with receptor alone or with rGRKs to obtain a moderate level of

receptor expression (B_{max}) of 2–4 pmol/mg membrane proteins. The B_{max} value is determined by radioligand binding assay (*see* Sect. 3.2.3) with a saturating concentration of [^{3}H]-SCH23390 (~5–6 nM) as previously described [8, 36, 37]. In our studies, we use 0.03 μg of HA-rD1R plasmid and 1.0 μg of rGRK2/3 plasmids (co-transfection condition) per 100 mm dish. As discussed elsewhere, we have established that a total of 5 μg of plasmid DNA per 100 mm dish is optimal for transfection efficiency and data reproducibility [36, 37]. Therefore, if less than 5 μg of plasmid DNA expression constructs (HA-rD1R ± rGRK2 or rGRK3) per dish is used in transfection, empty expression vector (herein pCMV5) must be added to transfection mix to adjust the total amount of DNA to 5 μg.

17. We recommend gently flicking tubes. Avoid vortexing this solution as this may break plasmid DNAs and reduce transfection efficiency and plasmid expression.

18. The pH of 2× HBS solution at 7.1 is critical and must be within ±0.05 units. If the pH is not properly calibrated, the transfection will not work.

19. Dropwise addition of the solutions in steps 7 and 8 is critical for optimal transfection of cells.

20. A typical whole-cell phosphorylation experiment testing basal and dopamine-induced phosphorylation of HA-rD1R ± GRK2/3 requires three 100 mm dishes for transfection seeded at 2.5×10^6 cells/dish. The number of dishes can be easily increased if additional conditions are tested (e.g., kinase inhibitors or siRNA).

21. The end user will obtain an approximate volume of ~16 mL of cell suspension. Once the number of cells is calculated for each different experimental condition, prepare 17 mL of cell suspension at 300,000–325,000 cells/mL. Leftover of cell suspensions is used to seed dishes to make membrane preparations for radioligand binding and total cell lysates for detection of GRK expression levels by immunoblotting. At this step, the end user can label wells on lid with a marker to indicate cells used for control (basal phosphorylation) and stimulation (dopamine-induced phosphorylation) conditions.

22. The end user will place cells on a flat surface cover with a protection absorbent paper or diaper and label cells behind a Plexiglas shield to protect from P^{32} radiation. If there is no available tissue culture CO_2 incubator for metabolic labeling of cells, a general purpose incubator can be used instead as cells are incubated in HEPES-buffered culture medium. Once cells have been placed in an incubator, the end user monitors

if there is any radiation contamination with a Geiger–Müller detector prior to disposing the protection paper or diaper.

23. For efficiency, we begin processing dishes of cells for radioligand binding assays and GRK immunoblotting samples (*see* Sects. 3.2.3 and 3.2.4) during the 90 min metabolic labeling time.
24. With respect to P^{32} liquid and solid wastes generated in Sect. 3.2.2, step 3, the end user will store the radioactive wastes in a large container shielded with Plexiglas and to be held until they decay to a low activity level (six half-lives) that can be disposed of to landfill and sewer according to municipal regulations. We typically seal screw-capped tubes used in Sect. 3.2.2, step 4, with Parafilm to avoid radioactive leak from tubes.
25. Prior to performing preclearing and immunoprecipitation steps, the end user must measure protein concentrations of solubilized preparations in duplicate using the Bio-Rad DC protein assay and according to the manufacturer protocol. This must be done to verify that experimental samples have similar solubilized protein concentrations. BSA protein standards (0–25 μg/μL) are prepared in RIPA+ cell lysis buffer. In our hands, the solubilized protein concentrations from single wells of a 6-well plate measured in different experiments range from 0.5 to 0.7 μg/μL. If differences in concentrations are larger than 10 %, the end user must normalize the solubilized protein concentration to similar levels using RIPA+ cell lysis buffer.
26. If the OD value of samples is outside the range of the protein standard curve, the end user will need to readjust the volume of cell lysate aliquots.
27. We typically mix immunoblotting samples with 2× SDS sample and freeze samples immediately at −20 °C if used within a week or at −80 °C for long-term storage.
28. We do not boil samples to avoid formation of large molecular weight protein aggregates. Typically, a 2 h incubation of bead matrix in 2× SDS sample buffer is sufficient to dissociate HA-rD1R immunocomplexes. After the incubation with 2× SDS sample buffer, we store samples at 4 °C and perform SDS-PAGE the next day. An overnight incubation in 2× SDS sample buffer at room temperature can also be performed.
29. We use for SDS-PAGE analysis of our samples the Hoefer SE640 wide-mini vertical electrophoresis unit (http://www.hoeferinc.com/media/PDFs/Manuals/SE640-IMB0-English.pdf). However, the procedure described herein can also be performed with other commercial electrophoresis units.
30. *n*-Butanol will remove oxygen and prevent air bubbles allowing the formation of gel with a flat top.

31. Prior to loading onto gels, briefly spin the samples at 15,000 × *g* for 10–15 s to pellet beads and recover volume of samples located on the sides of the tubes.
32. Always pour slowly and away from wells to avoid disturbing the samples. Follow the manufacturer's guidelines with respect to filling the lower or upper chamber with the appropriate buffer volume (http://www.hoeferinc.com/media/PDFs/Manuals/SE640-IMB0-English.pdf).
33. Higher temperature may lead to gel cracking. However, if following the 2 h period the gel is not dried, it may suggest that vacuum and/or temperature is not optimal.
34. Values obtained with the phosphorimager should be normalized for lane background and receptor number.
35. We expose to an X-ray film to obtain quality publishing figures. Phosphorimager gives pixelated or grainy images.
36. For our immunoblotting studies, we routinely used a semidry transfer apparatus from Bio-Rad (http://www.bio-rad.com/webroot/web/pdf/lsr/literature/M1703940.pdf). A wet transfer method can also be used.
37. The end user will always wear gloves to avoid contamination of gel and PVDF membrane.
38. We typically cut the upper left corner of PVDF membrane with a 45° angle (above the stained protein molecular weight markers) for tracking the orientation of loaded and transferred samples.
39. The blot paper is cut with 0.5 cm wider edges relative to the gel and PVDF membrane to ensure that the gel and PVDF membrane are completely sandwiched between blotting papers. However, some manufacturers recommend that the blot papers and membrane are trimmed to the same size of gel for the current to flow evenly through the gel.
40. This is a critical step as air bubble will interfere with optimal transfer of protein bands onto PVDF membrane.
41. Verify that the gel sandwich onto to the platinum-coated titanium anode is properly oriented, i.e., wet blot paper, gel on top of the PVDF membrane, and wet blot paper on the surface of anode.
42. The commercial stained protein molecular weight markers used herein do not vanish during subsequent incubation and washing procedures. If the end user notices that the markers disappear, we recommend marking protein molecular weight marker ladder on PVDF membrane with a pencil prior to incubating with Blotto.
43. Make sure that the size of Parafilm boat is just big enough to contain the membrane in order to utilize the smallest volume of primary antibodies and reduce cost of antibody use.

44. Although most protocols recommend incubating with primary antibodies for 1 h at room temperature, in our hands, we notice that the optimal detection of proteins of interest, notably transmembrane receptors, is obtained with an overnight incubation at 4 °C.

45. The primary antibodies are saved and stored at 4 °C (~4–6 months) to be reused for immunoblotting of other blots. Prior to reusing the primary antibodies, the end user will check for any signs of bacterial contamination, and if so, a fresh solution of primary antibodies should be prepared.

46. NaN_3 is used in the primary antibody solution to delay bacterial growth during storage at 4 °C. Do not add NaN_3 in the secondary antibody solution as it inhibits HRP activity. Washing steps following the incubation with primary antibody solution containing NaN_3 are important to eliminate residual NaN_3 from the blot.

47. Several factors such as protein expression and amount of proteins transferred on the PVDF membrane can determine the exposure time. The end user may have to determine empirically the optimum conditions for exposure time (shorter or longer) if the signal on film is too weak or too strong.

48. For cross-linking and coimmunoprecipitation assays with HA-rD1R and GRK2/3, HEK293 cells are co-transfected using the same plasmid DNA concentrations as in whole-cell phosphorylation studies (*see* Sect. 3.1). However, as per coimmunoprecipitation under cross-linking conditions described in Sect. 3.3, the end user will only need 100 mm dishes seeded with transfected cells to perform these studies.

49. Typically, we used three 100 mm dishes for cross-linking assays. The end user will seed two 100 mm dishes (control and dopamine conditions) with 7.5 mL of cell suspension per dish (~$2.5–3.0\times10^5$ cells/mL) for the cross-linking/coimmunoprecipitation part and 9.0 mL in one 100 mm dish for the radioligand binding assay.

50. Serum-free medium is used to be consistent with the serum-free cell culture condition used during 90 min metabolic labeling with P^{32} in whole-cell phosphorylation assays (*see* Sect. 3.2.2, **step 1**).

51. During the 90 min serum starving of cells, the end user will prepare stock solutions of 25 mM DSP (10×, *see* **Note 9**), 10 mM ascorbic acid (100×), and 1 mM dopamine (100×, made in 100× ascorbic acid). Leave ascorbic acid and dopamine stock solutions on ice.

52. The cross-linking reaction can be done at room temperature for 30 min or on ice for 2 h as per company's guidelines

(www.piercenet.com/instructions/2160544.pdf). In our hands with transfected HEK293 cells, we notice that cells detach when incubation with DSP is performed at room temperature but not on ice. The end user may want establishing the best incubation temperature for cross-linking reaction in her/his hands.

53. The end user will note that the solubilization time of cross-linked cell lysates in RIPA+ buffer is doubled (120 min) relative to whole-cell phosphorylation assays (60 min). Cross-linked cell lysates are harder to solubilize and a longer solubilization is required to obtain higher amount of solubilized proteins for immunoprecipitation.

54. Typically, manufacturers of cross-linking agents recommend quenching the cross-linking reaction by adding to reaction buffer a stop solution of 1 M Tris (pH 7.5) or glycine at a final concentration of 20–50 mM. As we were concerned by cells detaching from dishes upon addition of the quench solution, we opted instead to remove the buffer at the end of the 2 h reaction and add immediately RIPA+ buffer. We reasoned that RIPA+ buffer would not only act as a solubilizing agent but also as an effective quencher since it contains 50 mM Tris. The end user may also need to check if her/his cells tolerate the addition of a quench solution.

55. As discussed in step 7 of Sect. 3.2.2, following centrifugation and prior to preclearing, the end user will pipette two 25 μL aliquots from each solubilized extracts of cross-linked cell lysates to determine protein concentration. In addition, for cross-linking studies, the end user will save two 10 μL aliquots from each solubilized extract in 1.5 mL Eppendorf tubes to measure GRK2/3 input using SDS-PAGE and immunoblotting. The end user will usually need only one 10 μL GRK2/3 input sample which is sufficient; the second aliquot serves as backup. The tubes are kept on ice until the addition of 2× SDS-DTT sample buffer (*see* **Note 11**).

56. In contrast to the 1 h preclearing period for solubilized lysates from non-cross-linked samples (e.g., whole-cell phosphorylation), we recommend preclearing solubilized lysates from cross-linked proteins be performed for 2–3 h. In our hands, a longer preclearing time significantly reduced the nonspecific background resulting from immunoprecipitation and immunoblotting procedures.

57. If not used immediately we typically keep our samples at −20 °C (up to a month) or −70 °C (>1 month).

58. Prior to using glass coverslips, they must be etched to help adherence of cells to the glass coverslips. Typically, we put coverslips into a 150×20 mm glass petri dish, cover them with

concentrated sulfuric acid (H_2SO_4), and incubate on rocking platform for 1 h in a fume hood at room temperature. Decant H_2SO_4 into a bottle to be reused or disposed as a chemical waste and rinse the glass coverslips with running distilled water for 1 h. On a rocking platform place the petri dish, wash glass coverslips with distilled water for several hours using multiple washes, and perform a final wash with distilled water overnight. Place individually glass coverslips on a Whatman filter paper in a fresh glass petri dish and sterilize by autoclaving. Avoid tipping the dish before autoclaving as it will make coverslips stick together.

59. Paraformaldehyde causes strong autofluorescence, which can be blocked by a reduction of aldehyde groups using incubation with the quenching buffer.
60. The humidified chamber allows incubating the coverslips with a small volume of antibody and prevents the coverslips to dry off.
61. Removing as much equilibrium buffer as possible is important to avoid diluting the antifade reagent.
62. Incubation in the dark is critical to reduce photobleaching.
63. The end user should minimize the exposure of fluorescently labeled samples to light with neutral density filters and expose coverslips only when imaging and capturing a signal to diminish photobleaching.

Acknowledgments

This research was supported by an operating grant from the Canadian Health Research Institutes (MOP-81341) to MT. Boyang Zhang holds an Ontario Graduate Scholarship (OGS). Caroline Lefebvre is the recipient of scholarships from OGS and Fonds de recherche du Québec-Santé.

References

1. Gainetdinov RR, Premont RT, Bohn LM, Lefkowitz RJ, Caron MG (2004) Desensitization of G protein-coupled receptors and neuronal functions. Annu Rev Neurosci 27:107–144
2. Kelly E, Bailey CP, Henderson G (2008) Agonist-selective mechanisms of GPCR desensitization. Br J Pharmacol 153(Suppl 1):S379–S388
3. Luttrell LM, Lefkowitz RJ (2002) The role of beta-arrestins in the termination and transduction of G-protein-coupled receptor signals. J Cell Sci 115(Pt 3):455–465
4. Missale C, Nash SR, Robinson SW, Jaber M, Caron MG (1998) Dopamine receptors: from structure to function. Physiol Rev 78(1):189–225
5. Neve KA, Neve RL (1997) molecular biology of dopamine receptors. In: Neve KA, Neve RL (eds) The dopamine receptors. Humana Press, Totowa, NJ, pp 27–76
6. Tiberi M, Nash SR, Bertrand L, Lefkowitz RJ, Caron MG (1996) Differential regulation of

dopamine D1A receptor responsiveness by various G protein-coupled receptor kinases. J Biol Chem 271(7):3771–3778
7. Jackson A, Iwasiow RM, Chaar ZY, Nantel MF, Tiberi M (2002) Homologous regulation of the heptahelical D1A receptor responsiveness: specific cytoplasmic tail regions mediate dopamine-induced phosphorylation, desensitization and endocytosis. J Neurochem 82(3):683–697
8. Sedaghat K, Tiberi M (2011) Cytoplasmic tail of D1 dopaminergic receptor differentially regulates desensitization and phosphorylation by G protein-coupled receptor kinase 2 and 3. Cell Signal 23(1):180–192. doi:10.1016/j.cellsig.2010.09.002
9. Lamey M, Thompson M, Varghese G, Chi H, Sawzdargo M, George SR, O'Dowd BF (2002) Distinct residues in the carboxyl tail mediate agonist-induced desensitization and internalization of the human dopamine D1 receptor. J Biol Chem 277(11):9415–9421
10. Gardner B, Liu ZF, Jiang D, Sibley DR (2001) The role of phosphorylation/dephosphorylation in agonist-induced desensitization of D1 dopamine receptor function: evidence for a novel pathway for receptor dephosphorylation. Mol Pharmacol 59(2):310–321
11. Rankin ML, Marinec PS, Cabrera DM, Wang Z, Jose PA, Sibley DR (2006) The D1 dopamine receptor is constitutively phosphorylated by G protein-coupled receptor kinase 4. Mol Pharmacol 69(3):759–769
12. Gainetdinov RR, Bohn LM, Walker JK, Laporte SA, Macrae AD, Caron MG, Lefkowitz RJ, Premont RT (1999) Muscarinic supersensitivity and impaired receptor desensitization in G protein-coupled receptor kinase 5-deficient mice. Neuron 24(4):1029–1036
13. Gainetdinov RR, Bohn LM, Sotnikova TD, Cyr M, Laakso A, Macrae AD, Torres GE, Kim KM, Lefkowitz RJ, Caron MG, Premont RT (2003) Dopaminergic supersensitivity in G protein-coupled receptor kinase 6-deficient mice. Neuron 38(2):291–303
14. Daigle TL, Caron MG (2012) Elimination of GRK2 from cholinergic neurons reduces behavioral sensitivity to muscarinic receptor activation. J Neurosci 32(33):11461–11466. doi:10.1523/JNEUROSCI.2234-12.2012
15. Daigle TL, Ferris MJ, Gainetdinov RR, Sotnikova TD, Urs NM, Jones SR, Caron MG (2014) Selective deletion of GRK2 alters psychostimulant-induced behaviors and dopamine neurotransmission. Neuropsychopharmacology 39(10):2450–2462. doi:10.1038/npp.2014.97
16. Kinoshita S, Sidhu A, Felder RA (1989) Defective dopamine-1 receptor adenylate cyclase coupling in the proximal convoluted tubule from the spontaneously hypertensive rat. J Clin Invest 84(6):1849–1856. doi:10.1172/JCI114371
17. Ohbu K, Kaskel FJ, Kinoshita S, Felder RA (1995) Dopamine-1 receptors in the proximal convoluted tubule of Dahl rats: defective coupling to adenylate cyclase. Am J Physiol 268(1 Pt 2):R231–R235
18. Albrecht FE, Drago J, Felder RA, Printz MP, Eisner GM, Robillard JE, Sibley DR, Westphal HJ, Jose PA (1996) Role of the D1A dopamine receptor in the pathogenesis of genetic hypertension. J Clin Invest 97(10):2283–2288
19. Sanada H, Jose PA, Hazen-Martin D, Yu PY, Xu J, Bruns DE, Phipps J, Carey RM, Felder RA (1999) Dopamine-1 receptor coupling defect in renal proximal tubule cells in hypertension. Hypertension 33(4):1036–1042
20. Watanabe H, Xu J, Bengra C, Jose PA, Felder RA (2002) Desensitization of human renal D1 dopamine receptors by G protein-coupled receptor kinase 4. Kidney Int 62(3):790–798
21. Felder RA, Sanada H, Xu J, Yu PY, Wang Z, Watanabe H, Asico LD, Wang W, Zheng S, Yamaguchi I, Williams SM, Gainer J, Brown NJ, Hazen-Martin D, Wong LJ, Robillard JE, Carey RM, Eisner GM, Jose PA (2002) G protein-coupled receptor kinase 4 gene variants in human essential hypertension. Proc Natl Acad Sci U S A 99(6):3872–3877
22. Zeng C, Sanada H, Watanabe H, Eisner GM, Felder RA, Jose PA (2004) Functional genomics of the dopaminergic system in hypertension. Physiol Genomics 19(3):233–246
23. Bates MD, Caron MG, Raymond JR (1991) Desensitization of DA1 dopamine receptors coupled to adenylyl cyclase in opossum kidney cells. Am J Physiol 260(6 Pt 2):F937–F945
24. Zhou X, Sidhu A, Fishman PH (1991) Desensitization of the human D1 dopamine receptor: evidence for involvement of both cyclic AMP-dependent and receptor-specific protein kinases. Mol Cell Neurosci 2:464–472
25. Black LE, Smyk-Randall EM, Sibley DR (1994) Cyclic AMP-mediated desensitization of D1 dopamine receptor-coupled adenylyl cyclase in NS20Y neuroblastoma cells. Mol Cell Neurosci 5(6):567–575
26. Jiang D, Sibley DR (1999) Regulation of D(1) dopamine receptors with mutations of protein kinase phosphorylation sites: attenuation of the rate of agonist-induced desensitization. Mol Pharmacol 56(4):675–683

27. Ventura AL, Sibley DR (2000) Altered regulation of the D(1) dopamine receptor in mutant Chinese hamster ovary cells deficient in cyclic AMP-dependent protein kinase activity. J Pharmacol Exp Ther 293(2):426–434
28. Bates MD, Olsen CL, Becker BN, Albers FJ, Middleton JP, Mulheron JG, Jin SL, Conti M, Raymond JR (1993) Elevation of cAMP is required for down-regulation, but not agonist-induced desensitization, of endogenous dopamine D1 receptors in opossum kidney cells. Studies in cells that stably express a rat cAMP phosphodiesterase (rPDE3) cDNA. J Biol Chem 268(20):14757–14763
29. Lewis MM, Watts VJ, Lawler CP, Nichols DE, Mailman RB (1998) Homologous desensitization of the D1A dopamine receptor: efficacy in causing desensitization dissociates from both receptor occupancy and functional potency. J Pharmacol Exp Ther 286(1):345–353
30. Chaar ZY, Jackson A, Tiberi M (2001) The cytoplasmic tail of the D1A receptor subtype: identification of specific domains controlling dopamine cellular responsiveness. J Neurochem 79(5):1047–1058
31. Zamanillo D, Casanova E, Alonso-Llamazares A, Ovalle S, Chinchetru MA, Calvo P (1995) Identification of a cyclic adenosine 3′,5′-monophosphate -dependent protein kinase phosphorylation site in the carboxy terminal tail of human D1 dopamine receptor Neurosci Lett 188(3):183–186
32. Mason JN, Kozell LB, Neve KA (2002) Regulation of dopamine D(1) receptor trafficking by protein kinase A-dependent phosphorylation. Mol Pharmacol 61(4):806–816
33. Dicker F, Quitterer U, Winstel R, Honold K, Lohse MJ (1999) Phosphorylation-independent inhibition of parathyroid hormone receptor signaling by G protein-coupled receptor kinases. Proc Natl Acad Sci U S A 96(10):5476–5481
34. Violin JD, Ren XR, Lefkowitz RJ (2006) G-protein-coupled receptor kinase specificity for beta-arrestin recruitment to the beta2-adrenergic receptor revealed by fluorescence resonance energy transfer. J Biol Chem 281(29):20577–20588. doi:10.1074/jbc.M513605200
35. Arriza JL, Dawson TM, Simerly RB, Martin LJ, Caron MG, Snyder SH, Lefkowitz RJ (1992) The G-protein-coupled receptor kinases beta ARK1 and beta ARK2 are widely distributed at synapses in rat brain. J Neurosci 12(10):4045–4055
36. Plouffe B, D'Aoust JP, Laquerre V, Liang B, Tiberi M (2010) Probing the constitutive activity among dopamine D1 and D5 receptors and their mutants. Methods Enzymol 484:295–328. doi:10.1016/B978-0- 12-381298-8.00016-2
37. Plouffe B, Tiberi M (2013) Functional analysis of human D1 and D5 dopaminergic G protein-coupled receptors: lessons from mutagenesis of a conserved serine residue in the cytosolic end of transmembrane region 6. Methods Mol Biol 964:141–180. doi:10.1007/978-1-62703-251-3_10
38. Shiina T, Arai K, Tanabe S, Yoshida N, Haga T, Nagao T, Kurose H (2001) Clathrin box in G protein-coupled receptor kinase 2. J Biol Chem 276(35):33019–33026
39. Baig AH, Swords FM, Szaszak M, King PJ, Hunyady L, Clark AJ (2002) Agonist activated adrenocorticotropin receptor internalizes via a clathrin-mediated G protein receptor kinase dependent mechanism. Endocr Res 28(4):281–289
40. Chen Z, Gaudreau R, Le Gouill C, Rola-Pleszczynski M, Stankova J (2004) Agonist-induced internalization of leukotriene B_4 receptor 1 requires G-protein-coupled receptor kinase 2 but not arrestins. Mol Pharmacol 66(3):377–386. doi:10.1124/mol.66.3
41. Mangmool S, Haga T, Kobayashi H, Kim KM, Nakata H, Nishida M, Kurose H (2006) Clathrin required for phosphorylation and internalization of beta2-adrenergic receptor by G protein-coupled receptor kinase 2 (GRK2). J Biol Chem 281(42):31940–31949. doi:10.1074/jbc.M602832200
42. Ribeiro FM, Ferreira LT, Paquet M, Cregan T, Ding Q, Gros R, Ferguson SS (2009) Phosphorylation-independent regulation of metabotropic glutamate receptor 5 desensitization and internalization by G protein-coupled receptor kinase 2 in neurons. J Biol Chem 284(35):23444–23453. doi:10.1074/jbc.M109.000778

Chapter 8

Ubiquitination of Dopamine Receptor Studied by Sequential Double Immunoprecipitation

Kamila Skieterska, Pieter Rondou, and Kathleen Van Craenenbroeck

Abstract

G protein-coupled receptors are integrated in a complicated network of interacting proteins and are often regulated by posttranslational modifications. Ubiquitination is a posttranslational modification in which a 76-amino acid polypeptide, ubiquitin, is covalently attached to the substrate protein. Ubiquitination targets proteins for degradation but can also regulate other cellular processes such as endocytosis, trafficking, and DNA repair. Determination of the ubiquitination status of receptors can bring new insights into the understanding of the molecular basis of G protein-coupled receptors (GPCR) signaling and regulation, and as GPCRs are the main target of pharmaceutical drugs, this research may help in developing new tools for medical treatment.

Here, we introduce a double sequential immunoprecipitation technique which is successfully used for studying the ubiquitination profile of the dopamine D_4 receptor. In a first step, lysates containing ectopically expressed D_4 receptor are subjected to a first immunoprecipitation round. After denaturation of the eluate, the receptor is subjected to a second immunoprecipitation step to eliminate receptor-interacting proteins that could also be ubiquitinated. After final sodium dodecyl sulfate-polyacrylamide gel electrophoresis and Western blot analysis, ubiquitinated receptor species can be detected by specific antibodies.

Key words Ubiquitination, GPCR, Dopamine receptor, Immunoprecipitation

1 Introduction

1.1 Enzymatic Reactions Involved in protein Ubiquitination

Ubiquitination is a posttranslational modification in which a 76-amino acid polypeptide, ubiquitin, is covalently linked, predominantly via an isopeptide bond to the lysine residue of a substrate protein. This reaction is performed by the sequential action of three enzymes. In the first, ATP-depended step, ubiquitin is activated by the ubiquitin-activating enzyme E1. Next, ubiquitin is transferred from E1 to the ubiquitin-conjugating enzyme E2. Finally, ubiquitin is covalently attached to the substrate by E3 ubiquitin ligases, which are multiprotein complexes that provide specificity to the ubiquitination process by recognition of the substrate [1]. Ubiquitination is reversible, and the removal of

Mario Tiberi (ed.), *Dopamine Receptor Technologies*, Neuromethods, vol. 96,
DOI 10.1007/978-1-4939-2196-6_8,

ubiquitin is mediated by deubiquitinating enzymes (DUBs) that hydrolyze Ub-protein isopeptide bonds.

Distinct types of ubiquitination are generated by attachment of a single ubiquitin moiety to one or more lysine residues of the substrate protein, called monoubiquitination and multi-monoubiquitination, respectively. Moreover, ubiquitin itself contains seven lysine residues (K6, K11, K27, K29, K33, K48, and K63), and all of them can act as acceptors for other ubiquitin molecules giving rise to polyubiquitin chains of different configurations.

Different types of ubiquitination (mono-, multi-mono-, poly-) lead to different functional effects [2]. The first identified and best characterized type is lysine 48-linked polyubiquitination which functions as a signal for protein degradation by the 26S proteasome. Today it is known that ubiquitination can also regulate other cellular processes such as endocytosis, trafficking, and DNA repair [3]. Because ubiquitination is involved in the regulation of so many processes, it is not surprising that impairments in the ubiquitination system are associated with a variety of human diseases ranging from cancer, neurodegenerative disorders, viral infection to diabetes, muscle wasting, and inflammation [4].

1.2 Ubiquitination of G protein-Coupled Receptors (GPCRs)

Although it is widely accepted that ubiquitination can play a significant regulatory role in many cellular processes, there is only a limited knowledge about the importance of this modification on signaling and regulation of dopamine receptors and GPCRs in general, which warrants future research.

The best characterized role of ubiquitin in GPCR regulation in mammalian cells is functioning as a sorting signal for lysosomal degradation. In this case, agonist-activated GPCRs are targeted to lysosomes for degradation via the highly conserved endosomal-sorting complex required for transport (ESCRT) pathway, which has been best described for the protease-activated receptor 2 (PAR2) [5] and chemokine CXCR4 receptor [6].

Many GPCRs have been reported to exist in a basal ubiquitinated state. Such receptors are mainly present both in the endoplasmic reticulum (ER) and at the plasma membrane. Ubiquitination of GPCRs during synthesis in the ER functions mainly as a quality control system since misfolded receptors will be ubiquitinated and targeted for degradation via the ER-associated degradation (ERAD) pathway [7, 8]. However, some properly folded GPCRs are directed from the ER to the proteasome and upon deubiquitination are released towards the cell surface [9]. Additionally, constitutive ubiquitination of plasma membrane receptors was also reported. In these cases, ubiquitination seems to regulate, via different mechanisms, cell surface expression of the receptors. For example, the CXCR7 receptor is basally ubiquitinated and upon agonist stimulation undergoes deubiquitination in a process that requires receptor phosphorylation and β-arrestin recruitment.

When the ligand is removed, the receptor is again ubiquitinated and recycled back to the cell membrane [10].

Recent studies have also shown that signaling of one GPCR can influence ubiquitination of another GPCR. In this case, GPCRs at the cell surface can be transiently ubiquitinated after activation of another receptor and this can remarkably change cell responsiveness. Such a phenomenon was described for the angiotensin II type 1 receptor (AT_1R) in response to dopamine D_5 receptor (D_5R) activation [11]. In these studies it was shown that disruption of the D_5R gene in mice caused increased blood pressure and AT_1R expression. Additionally, activation of D_5R induced ubiquitination of AT_1R at the cell surface and its degradation via the proteasome. All these data suggest that D_5R negatively regulates AT_1R expression.

Examples of a differential regulatory role of ubiquitination can be found in another area of GPCR research, which is currently attracting more and more attention, namely, biased agonism. Biased agonism is the phenomenon in which activation of the same GPCR by different ligands leads to distinct responses [12]. For the μ-opioid receptor (MOR), it was shown that the agonists DAMGO and morphine differentially regulate its internalization and recycling [13]. Recently, it was reported that morphine and DAMGO cause differential β-arrestin recruitment and ubiquitination of MOR. Activation of MOR with DAMGO promotes β-arrestin1-mediated ubiquitination and dephosphorylation of the receptor, whereas stimulation of MOR with morphine leads to β-arrestin2-dependent internalization, prolonged phosphorylation, and slower resensitization [14].

1.3 Methods for Studying Ubiquitination

Different methods for studying ubiquitination exist. In in vitro ubiquitination assays, the purified protein of interest, ubiquitin, and the proper E1, E2, and E3 enzymes are mixed in the presence of ATP and the ubiquitination reaction is catalyzed in vitro. This assay can be used, for example, to detect the specificity between E2 and E3 or E3 and a substrate protein [15, 16]. It may also be applied for the verification whether a specific lysine residue undergoes ubiquitination, but in this case the variants of the protein of interest with mutated distinct possible ubiquitination sites need to be available. Next, mass spectrometry-based methods provide the possibility to obtain direct evidence for ubiquitination on a particular lysine of the substrate. This is possible due to the di-glycine remnant that remains attached to the ubiquitinated residue after tryptic digestion. The remnant is derived from the C-terminus of ubiquitin and results in a mass shift of 114.04 Da that can be detected by MS/MS [17]. The possible disadvantage of this method is the fact that the same remnant on target substrates can also originate from other ubiquitin-like proteins such as NEDD8 and ISG15, making it impossible to identify the original modification [18].

Also the bioluminescence resonance energy transfer (BRET) [10, 19] and biomolecular fluorescence complementation (BIFC) assays [20] were successfully used to study ubiquitination in living cells. Both techniques allow studying dynamic processes of ubiquitination which occur in a living cell. This allows the researcher to investigate, for example, the effect of receptor activation or blocking on its ubiquitination status in real time. BIFC additionally can provide information in which cell compartment the ubiquitinated protein is localized. The drawback of these methods is the need for overexpression of the proteins of interest fused to tag:Rluc (*Renilla* luciferase) and one of GFP (green fluorescent protein) variants in BRET or split YFP (yellow fluorescent protein) in BIFC, what makes them unsuitable for studying ubiquitination in native conditions.

Finally, the most common method for studying ubiquitination is immunoprecipitation (IP), which is also described in this chapter. IP is the technique in which the protein of interest (e.g., a GPCR) is isolated from the solution (typically a cell lysate) by using an antibody that specifically binds to this protein through its variable domain. Next, the constant domain of the antibody is coupled to a solid resin which is created by agarose/polyacrylamide/Sepharose or magnetic beads to which most often recombinant Protein A or G has been immobilized. The high affinity of Protein A and G for binding antibodies leads to the isolation of a protein complex containing the antibody, the protein of interest, and all its interacting partners.

For ubiquitination studies, special variants of IP were developed in which at different stages of the procedure the protein-protein interactions are destroyed to allow the detection of ubiquitination signal originating from the protein of interest and not from other interacting partners that can also be ubiquitinated. For example, protein-protein interactions can be broken prior to the IP procedure, by boiling lysates in a buffer containing high concentration of the detergent sodium dodecyl sulfate (SDS). In this case, all proteins become denatured before IP, and therefore, the antibody used for IP should be able to detect the protein of interest independently of its conformation. Additionally, the concentration of SDS needs to be decreased before IP as in such denaturing condition the antibody used for IP will also be denatured. On the other hand, protein-protein interactions can also be broken after IP, by eluting (denaturing) samples in buffer containing SDS and reducing agents like dithiothreitol (DTT) or β-mercaptoethanol. After IP and elution, the isolated proteins are separated on the polyacrylamide gel, and it is possible to detect both ubiquitinated and non-ubiquitinated forms of the protein of interest due to the difference in molecular weight of those forms and thus their differential migration during polyacrylamide gel electrophoresis (PAGE). When a specific protein is monoubiquitinated, the modified form is 8 kDa heavier than the non-ubiquitinated species of the protein. Multi-monoubiquitinated proteins will have their original molecular weight increased by 8 kDa multiplied by the number of attached ubiquitin molecules. In case of polyubiquitination, immunodetection often reveals a smeary, high molecular weight pattern, representing multiple species of the protein of interest with polyubiquitin chains of different length. In this situation, however, it is impossible to distinguish which band(s) represent(s) ubiquitinated protein of interest or to exclude that other ubiquitinated proteins contribute to the detected signal. In order to circumvent latter issues, the sequential double IP was developed to specifically assess ubiquitination levels for the protein of interest.

Such procedure was successfully used to study the ubiquitination pattern of the dopamine D_4 receptor (D_4R) [21, 22] and is discussed in detail in Sect. 3 of this chapter. In this case, the protein of interest (e.g., D_4R) is first overexpressed in eukaryotic cells, such as human embryonic kidney cells (HEK293 or HEK293T), which are frequently used for ubiquitination studies [6, 9–11, 19–22]. Next, cells are lysed in radioimmunoprecipitation (RIPA) buffer which contains (next to general protease inhibitors) also an inhibitor of deubiquitinating enzymes, *N*-ethylmaleimide (NEM). After the first IP, proteins are eluted in a sample buffer containing high concentration of SDS and DTT. In this way all proteins interacting with the receptor are removed. The ubiquitin stays covalently attached to the receptor through the isopeptide bond which is formed between the C-terminal glycine of ubiquitin and the lysine

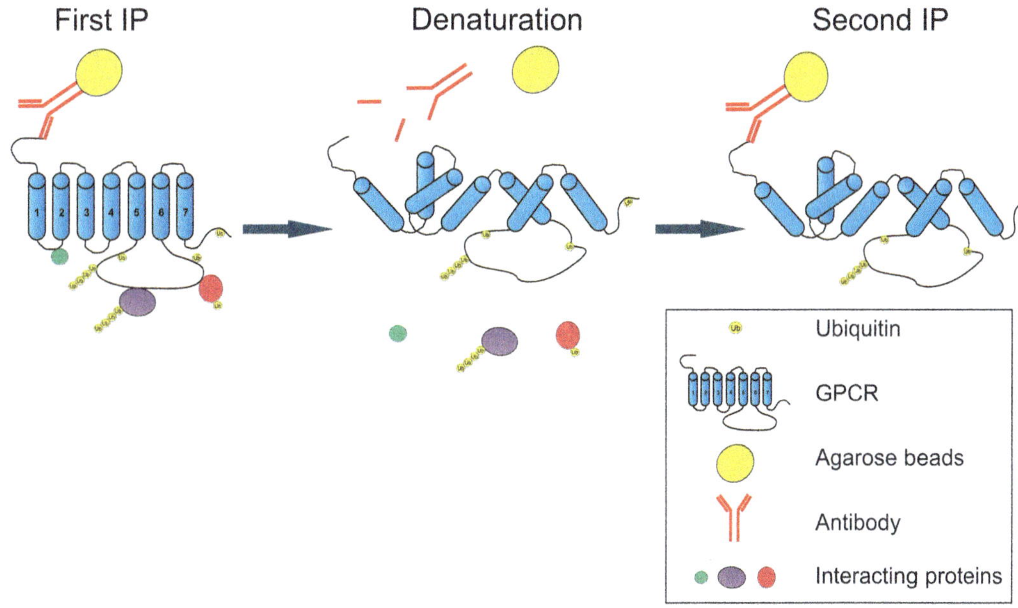

Fig. 1 Schematic representation of the sequential double IP procedure. Cell lysate contains all proteins present in the cell. After the first IP, GPCRs are isolated together with their interacting proteins. During denaturation, protein-protein interactions are disrupted, but ubiquitin stays attached to the receptor via a covalent isopeptide bond. The ubiquitination signal detected after the second IP only represents the modified receptor

residue of the receptor (Fig. 1). The obtained eluate is then diluted with RIPA buffer to decrease the level of SDS and DTT, in order to allow permissive conditions for the second IP round. After this second IP, the receptor is again eluted from the beads, followed by SDS-PAGE and Western blotting procedure. Finally, the receptor ubiquitination signal can be detected with specific antibodies (against Ub or against a Ub-linked tag in case of prior ectopic expression). Additionally, using antibodies specifically recognizing different ubiquitin chains (e.g., lysine 48-linked, lysine 63-linked), also the type of ubiquitination can be determined. This kind of information can also be obtained when tagged ubiquitin molecules with mutations of specific lysines (e.g., lysine 48, 63) are ectopically expressed. Determination of the type of ubiquitination can already give an indication about the function of this specific modification.

2 Materials

2.1 Culturing Cells and Transient Transfection

1. Culture medium for human embryonic kidney (HEK 293T) cells: Dulbecco's modified Eagle's medium (DMEM; Invitrogen) supplemented with 10 % (v/v) fetal calf serum (FCS), 0.03 % (v/v) L-glutamine, 100 units/ml penicillin, and 100 μg/ml streptomycin. Store at 4 °C, before adding to the cells pre-warm to 37 °C.
2. DMEM with 2 % (v/v) FCS and DMEM without FCS (serum free medium—SFM) and concentrations of L-glutamine, penicillin, and streptomycin as in Sect. 2.1, item 1 for transient transfection. Store at 4 °C, before adding to the cells pre-warm to 37 °C.
3. Polyethylenimine—PEI (high molecular weight, water-free (25 kDa branched), Sigma-Aldrich: 40.872-7) dissolve in Milli-Q water to final concentration of 1 mg/ml and adjust pH to 7 with HCl. Filter sterilize and aliquot. Can be stored at −20 °C for more than 1 year, for frequent use store at 4 °C.
4. Expression plasmids based on pcDNA3 carrying genes (encoding proteins of interest): for example, plasmids encoding HA D_4R, Flag-ubiquitin. Store at −20 °C.
5. Phosphate-buffered saline (PBS). Store at 4 °C to have cold buffer.
6. Incubator in which cells will be grown in a controlled environment (37 °C, 98 % humidity, 5 % CO_2).
7. Laminar flow cabinet for culturing cells and performing transfections.
8. Sterile tissue culture-treated flasks, 10-cm dishes, and pipets.

2.2 Cell Lysis

1. Radioimmunoprecipitation buffer (RIPA): 150 mM NaCl; 50 mM Tris/HCl pH 7.5; 1 % NP-40; 0.1 % SDS; and 0.5 % deoxycholic acid sodium salt (Acros Organics: 218591000). Store at 4 °C (stable for approximately 1 month).

   At the day of the experiment, add protease inhibitors: aprotinin (2.5 μg/ml, Sigma-Aldrich: A6279), pefablock (1 mM, Sigma-Aldrich: 76307), leupeptin (10 μg/ml, Acros Organics: 328350050), and phosphatase inhibitor β-glycerol phosphate disodium salt pentahydrate (10 mM, Fluka BioChemika: 50020). Keep lysis buffer with inhibitors on ice.
2. *N*-ethylmaleimide (NEM, Sigma-Aldrich: E3876), the inhibitor of deubiquitinating enzymes, is dissolved at the day of experiment in RIPA buffer to a final concentration of 10 mM. When NEM is added, it is indicated in the Methods section. Store at 4 °C.

2.3 Immunoprecipitation and Elution

1. Primary antibody recognizing the epitope of the protein of which the ubiquitination status will be investigated. This can be a protein-specific epitope or a tag-epitope in case of exogenous expression of tagged protein of interest. Store at 4 or at −20 °C (depends on the antibody).
2. Protein A UltraLink Resin (Thermo Scientific: 53139). Store at 4 °C.
3. 4× Laemmli sample buffer: 4 % SDS; 50 % glycerol; 0.2 % bromophenol blue; 65 mM Tris/HCl pH 6.8, supplemented with freshly added DTT (final concentration 50 mM). Store at room temperature.
4. 5× Laemmli sample buffer: 5 % SDS; 50 % glycerol; 0.2 % bromophenol blue; 65 mM Tris/HCl, pH 6.8 supplemented with freshly added DTT (final concentration 50 mM). Store at room temperature.

2.4 Sodium Dodecyl Sulfate-Polyacrylamide Gel Electrophoresis (SDS-PAGE)

1. Acrylamide/bis solution (37.5:1, Serva: 10681.01). Store at 4 °C.
2. 10 % (v/v) SDS dissolved in Milli-Q water. Store at room temperature.
3. 10 % (v/v) ammonium persulfate (APS, Sigma-Aldrich: A3678) dissolved in Milli-Q water. Store at 4 °C.
4. *N*,*N*,*N*′,*N*′-Tetramethylethylenediamine (TEMED, Bio-Rad: 101-0801). Store at 4 °C.
5. 1.5 M Tris/HCl pH 8.8. Store at room temperature.
6. 0.5 M Tris/HCl pH 6.8. Store at room temperature.
7. Electrophoresis buffer: 49 mM Tris; 0.38 M glycine; 0.1 % SDS. Store at room temperature.

8. PageRuler ™ Prestained Protein Ladder (Thermo Scientific: 26616). Store at –20 °C.
9. Biotinylated protein ladder (Cell Signaling: 7727). Store at –20 °C.
10. SDS-PAGE equipment: this instruction assumes the use of a Mini-PROTEAN Tetra Cell from Bio-Rad. The setup can be readily adapted to other formats.

2.5 Western Blotting

1. Transfer buffer: 47 mM Tris; 37 mM glycine; 0.03 % SDS; 20 % methanol. Store at room temperature.
2. Nitrocellulose membrane: Protran BA85 (GE Healthcare Life Sciences Whatman™: 10401196). Cut the membrane before blotting to obtain fragments of the same size as the gel (5 cm × 8.5 cm).
3. Chromatography paper, 3 mm Chr (Whatman™: 3030917). Cut before blotting to obtain fragments of the same size as gel and membrane (5 cm × 8.5 cm).
4. Electroblotting equipment: the instruction assumes the use of a Mini Trans-Blot Module from Bio-Rad. The setup can be readily adapted to other formats but not to a (semi-)dry one.

2.6 Immuno-detection

1. TBS buffer: 50 mM Tris/HCl pH 7.5; 150 mM NaCl. Store at room temperature.
2. TBS-T buffer: TBS buffer with addition of 0.1 % Tween 20. Store at room temperature.
3. Blocking solution: 5 % (v/v) nonfat dry milk or 5 % bovine serum albumin (BSA) or commercial blocking buffer (e.g., Odyssey blocking buffer from LI-COR Biosciences: 927-40000) diluted in a TBS buffer. Store at 4 °C.
4. Specific primary antibodies diluted in blocking buffer with addition of 0.1 % Tween 20. Store at 4 °C.
5. Specific secondary antibodies coupled to infrared dye or horseradish peroxidase (HRP) diluted in blocking buffer with addition of 0.1 % Tween 20. Store at 4 °C.
6. Substrate for HRP: Western Lightning Plus-ECL enhanced chemiluminescence substrate (Perkin Elmer: NEL105001EA). Mixed substrate and enhancing solutions should be stored at 4 °C and are stable for 1 day.
7. Stripping buffer: 62.5 mM Tris/HCl pH 6.8; 2 % SDS (store at room temperature). Directly before stripping procedure, add β-mercaptoethanol to a final concentration of 100 mM.
8. Detection equipment: Kodak Image Station 440CF from Kodak for HRP-coupled antibodies and Odyssey from LI-COR Biosciences for infrared dye-coupled antibodies.

3 Methods

3.1 Cell Culture and Transient Transfection

We investigate ubiquitination of the dopamine D_4 receptor (D_4R) in HEK 293T cells which are commonly used for ubiquitination studies. As these cells do not express endogenous D_4R, transient transfection with plasmid encoding D_4R is therefore necessary. Often endogenous ubiquitination signals detected after Western blot immunodetection are very weak, and in this case also overexpression of ubiquitin can be recommended (*see* Fig. 2 IB: anti-ubiquitin).

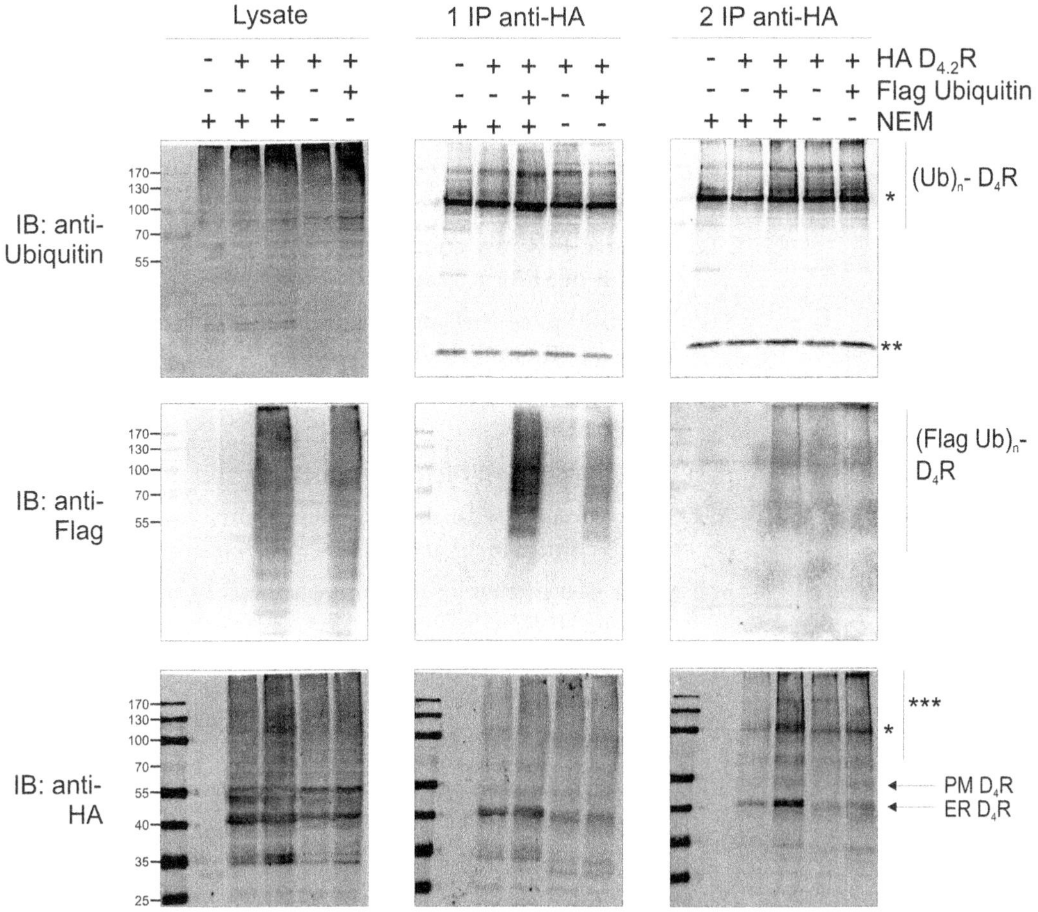

Fig. 2 Comparison of results from ubiquitination assays performed with or without addition of NEM in the RIPA lysis buffer. HEK 293T cells overexpressing HA $D_{4.2}R$ or HA $D_{4.2}R$ with Flag ubiquitin were lysed in RIPA buffer with or without addition of 10 mM NEM. Next, sequential double IP was performed with anti-HA (16B12). Ubiquitination signal was detected using anti-ubiquitin (P4D1) and anti-mouse IRDye 800 or HRP-coupled anti-Flag M2. Receptor was visualized with rat anti-HA and anti-rat IRDye 680LT. *Association of two heavy chains (100 kDa), **light chain (25 kDa) of the mouse anti-HA antibody used for IP. ***High MW HA $D_{4.2}R$-specific signals. PM - plasma membrane, mature $D_{4.2}R$; ER - ER-retained, immature $D_{4.2}R$

1. Grow HEK 293T cells in DMEM supplemented with 10 % FCS (*see* Sect. 2.1, item 1) in a controlled environment (37 °C, 98 % humidity, 5 % CO_2).
2. Twenty-four hours before transfection, seed cells on 10-cm dish at a density of 2.5×10^6 cells/dish. To increase the amount of starting material for the ubiquitination assay, it is recommended to prepare two 10-cm dishes for each experimental sample.
3. On the day of transfection:
   - (a) Change medium on the cells to DMEM supplemented with 2 % FCS (*see* Sect. 2.1, item 2). Add 9 ml of fresh medium per 10-cm dish.
   - (b) Prepare DNA mix containing 10 μg of total DNA per 10-cm dish and 490 μl of SFM. If you use several plasmid constructs for transfection keep the same amount of specific plasmid in every sample. If necessary use an empty vector plasmid to equalize the total amount of DNA between the samples.
   - (c) Prepare PEI mix containing 25 μl of PEI (1 mg/ml) and 475 μl of SFM per one 10-cm dish.
   - (d) Mix 500 μl of PEI mix with 500 μl of DNA mix (amount per one 10-cm dish).
   - (e) Vortex immediately and incubate for 10 min at room temperature.
   - (f) Add transfection mixture to the cells and mix gently.
   - (g) Incubate cells with the transfection mixture for 4–6 h in a tissue culture incubator (37 °C, 98 % humidity, 5 % CO_2).
   - (h) After 4–6 h, change medium on the cells to DMEM containing 10 % FCS.
   - (i) Grow cells for the next forty-eight h to obtain the highest protein expression levels.
4. Forty-eight hours after transfection, collect the cells:
   - (a) Remove dishes with cells from the incubator and wash them twice with ice cold PBS.
   - (b) Scrape cells with a spatula and transfer to 15-ml tubes.
   - (c) Spin down the cells at $110 \times g$ for 10 min at 4 °C.
   - (d) Aspirate PBS. Freeze cell pellets at −70 °C. The cell pellets can be kept for several weeks before you start the IP or you can proceed after 1 h of freezing to the cell lysis step. We advise to keep cells at −70 °C for at least 1 h as this will improve isolation of GPCRs from the membrane environment during the lysis procedure.

3.2 Cell Lysis in RIPA Buffer

1. Lyse cells in RIPA buffer supplemented with protease/phosphatase inhibitors and NEM (*see* Sect. 2.2, items 1 and 2). Presence of NEM in the lysis buffer allows preservation of the bonds between proteins and ubiquitin, and this results in isolation of an increased amount of the modified protein of interest (*see* Fig. 2 IB: anti-Flag, 1 and 2 IP). Depending on the amount of the cells collected use, 300–500 μl of RIPA buffer. Typically, 500 μl of RIPA buffer is sufficient to lyse about 10×10^6 cells. Transfer the cell lysate to a 1.5 ml microcentrifuge tube.
2. Incubate the cell lysate for 1 h at 4 °C applying end-to-end rotation.
3. Centrifuge the samples at 8,000 × *g* for 10 min at 4 °C.
4. Transfer the supernatant (cleared lysate) to a new 1.5 ml tube.
5. Transfer 40 μl of supernatant to another 1.5 ml tube. Add to this supernatant 10 μl of 5× Laemmli buffer (*see* Sect. 2.3, item 4) supplemented with 50 mM DTT and incubate at 37 °C for 10 min. This supernatant (input control) will be used to check

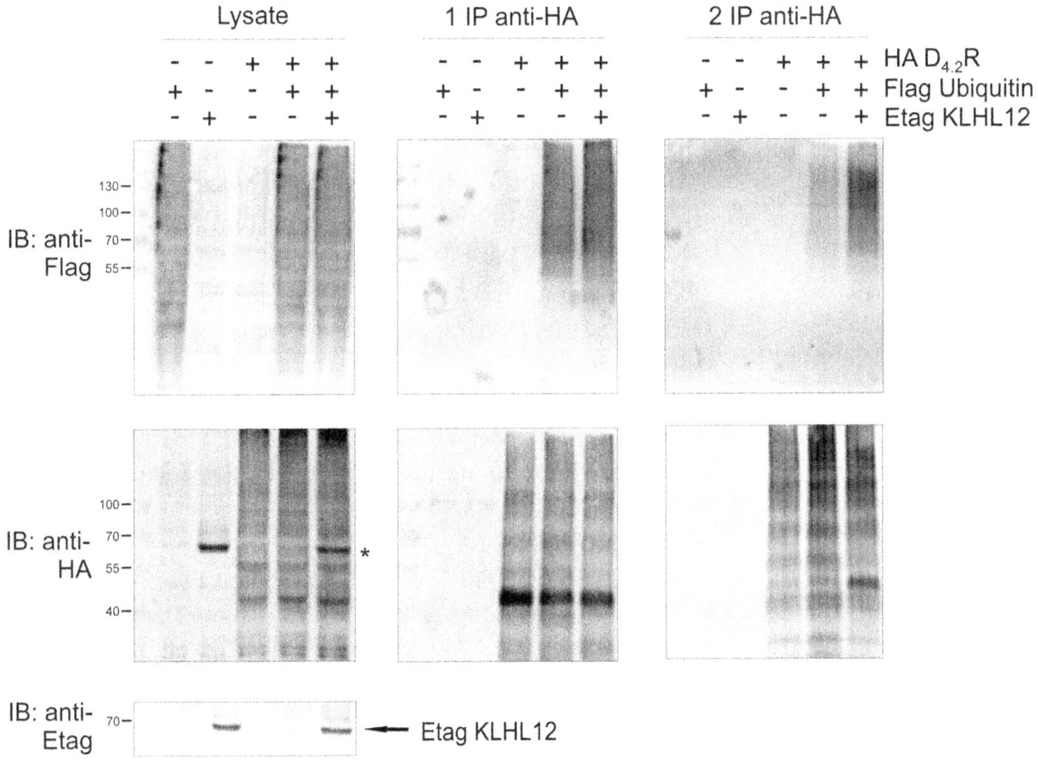

Fig. 3 Ubiquitination of the dopamine D_4 receptor (D_4R) detected via sequential double IP. HEK 293T cells were transiently transfected as indicated. Forty-eight hours posttransfection cells were lysed and sequential double IP was performed with anti-HA (16B12). After Western blot analysis, proteins were visualized with HRP-coupled anti-Flag M2, rat anti-HA, and anti-rat IRDye 680LT or rabbit anti-Etag and anti-rabbit IRDye 800. KLHL12 is the protein that interacts with D_4R and enhances its ubiquitination [21]. *Band representing Etag KLHL12

if all proteins of interest are well expressed at equal level (*see* Fig. 3—Lysate).

6. The rest of the lysate will be used for double sequential IP.

3.3 Double Sequential Immunoprecipitation

1. Add 2 μg of primary antibody directed against the epitope of the protein of interest (or its tag in case of tagged receptor overexpression) to the cleared lysates from Sect. 3.2, step 6 (*see* **Note 1**).
2. Incubate cell lysates with the antibody for 2–4 h at 4 °C using end-to-end rotation.
3. Prepare agarose beads conjugated with Protein A or G (*see* **Note 2**). Per sample you need to prepare 20 μl of pure beads (delivered as 50 % slurry) (*see* **Note 3**). Wash the beads three times with 1 ml of RIPA buffer from Sect. 3.2, step 1 (centrifugation after each washing step: 110×*g*, 4 °C, 1 min). Beads for multiple samples are prepared at once in one tube.
4. Add the washed beads to the lysates containing antibody and incubate overnight at 4 °C using end-to-end rotation.
5. The next day, wash the beads three times with 1 ml of RIPA buffer with freshly added phosphatase and protease inhibitors. After each washing step spin down your samples at 110×*g* for 1 min at 4 °C. In this step all proteins present in the lysate, which are not bound to the immunoprecipitated protein of interest, are removed (*see* **Notes 4** and **5**).
6. After the final washing step, aspirate all RIPA buffer from the beads and add 50 μl of 4× Laemmli buffer (*see* Sect. 2.3, item 3) supplemented with 50 mM DTT to each sample.
7. Denature the samples at 37 °C for 10 min (*see* **Note 6**) under constant shaking.
8. Transfer the whole eluate (50 μl, without beads) to a new 1.5 ml tube. Transfer 12.5 μl from this eluate (1/4 of the first elution fraction) to another 1.5 ml tube, and keep it for further analysis on Western blot to check whether the first IP was successful (*see* Fig. 3—1 IP).
9. Add again 50 μl of 4× Laemmli buffer with DTT to the beads and denature the samples at 37 °C for 10 min (to ensure complete elution from the beads).
10. Transfer second elution fraction to the tube from step 8 (mix fraction one and two).
11. Repeat steps 9 and 10 to have all three elution fractions in one tube.
12. Dilute eluate (from all three fractions) with 850 μl of RIPA buffer (with protease/phosphatase inhibitors and NEM) to decrease the concentration of SDS and DTT from the Laemmli buffer, before proceeding to the second round of IP.

13. Repeat steps from 1 to 7, but add only 1.5 μg of antibody for the second IP and 15 μl of pure beads (after the first IP the samples contain less proteins). After elution you will obtain your final sample (*see* Fig. 3—2 IP).

3.4 SDS-PAGE

1. Prepare discontinuous polyacrylamide gels (separating and stacking gel) with the proper concentration of acrylamide in the separating gel (depending on the molecular weight of your proteins of interest). In Table 1 amounts of different products that need to be mixed to prepare two mini gels are listed. APS and TEMED initiate the reaction of polymerization of the gel and should be added to your mixture at the end.
2. Load 20 μl (for a gel with 15 wells) or 25 μl (for a gel with 10 wells) of your samples and protein ladders (prestained and biotinylated) and run the gel at 100 V and 250 mA in the electrophoresis buffer (*see* Sect. 2.4, item 7) until the bromophenol blue dye, which is included in the 4× and 5× Laemmli buffer, reaches the bottom of the separating gel (90–120 min).

3.5 Western Blotting: Wet Transfer Setup

1. Prepare a transfer sandwich: on the site of the transfer cassette which will be placed closer to the cathode (in the Mini Trans-Blot Module, this is the black part of the cassette), place elements in the following order: a sponge, three chromatography papers, separating gel (stacking gel should be removed), nitrocellulose membrane, three chromatography papers, sponge. All elements should be put for 20 s in the transfer buffer before adding them to the sandwich. Try to eliminate air bubbles between all parts of the sandwich, especially between the gel and the membrane.

Table 1
Products and amounts required for preparation of two mini gels for Mini-PROTEAN Tetra Cell system (Bio-Rad)

Separating gel	8.5 % 40–100 kDa	10 % 20–70 kDa	12 % 20–60 kDa	15 % 10–40 kDa	Stacking gel	4 %
Water	5.09 ml	4.7 ml	4.1 ml	3.3 ml	Water	3 ml
37.5 % acrylamide/ bis solution	2.21 ml	2.6 ml	3.2 ml	4 ml	37.5 % acrylamide/ bis solution	625 μl
1.5 M Tris/ HCl pH 8.8	2.5 ml	2.5 ml	2.5 ml	2.5 ml	0.5 M Tris/ HCl pH 6.8	1.26 ml
10 % SDS	100 μl	100 μl	100 μl	100 μl	10 % SDS	50 μl
10 % APS	100 μl	100 μl	100 μl	100 μl	10 % APS	50 μl
TEMED	10 μl	10 μl	10 μl	10 μl	TEMED	5 μl

Additionally, information about recommended concentration of acrylamide for best separation of proteins based on their molecular weight is included

2. Place the cassette in the transfer chamber, put a small container with ice next to it, and fill the chamber with the transfer buffer (*see* Sect. 2.5, item 1). Perform the transfer for 1 h at 100 V and 250 mA.

3.6 Detection of Ubiquitinated proteins

1. When the transfer is finished, remove the membrane from the transfer sandwich and place it in the box with blocking solution (*see* Sect. 2.6, item 3). Incubate for minimum 30 min at room temperature upon shaking.
2. Add primary antibody against epitope of protein of interest, for example, anti-HA antibody (*see* Fig. 3) diluted in the blocking buffer with addition of 0.1 % Tween 20. The best dilution should be determined for each antibody. In general, for the antibodies against exogenous tags, a good starting dilution is 1:1,000 and for antibodies against endogenous epitope 1:100–1:500 (*see* **Note 7**). Incubate for 1 h at room temperature or overnight at 4 °C upon shaking. In Table 2 examples of dilutions of frequently used antibodies in our laboratory are presented.
3. Wash the membranes three times 5 min in TBS-T buffer.
4. Add secondary antibody conjugated with fluorescent dye (e.g., infrared dyes—IRDye) or enzyme (e.g., horseradish peroxidase—HRP) which will recognize primary antibody (*see* **Note 8**). Dilute secondary antibody in blocking buffer supplemented with 0.1 % Tween 20. Typical dilutions of secondary antibodies are listed in Table 3.
5. Wash the membranes three times 5 min in TBS-T buffer.
6. Detect proteins on the membrane. Different detection systems are available. In our laboratory we use Kodak Image Station

Table 2
Examples of optimized dilutions of primary antibodies used for immunoblotting

Antibody	Company	Catalog number	Dilution for IB
Anti-ubiquitin (P4D1)	Cell Signaling	3936	1:1,000
Anti-D_4R (D16)	Santa Cruz Biotechnology	Sc-31481	1:250
Anti-Flag HRP (M2)	Sigma-Aldrich	A8592	1:2,000
Anti-Flag	Sigma-Aldrich	F3165	1:4,000
Anti-HA (16B12) mouse	Covance	MMS-101P-1000	1:2,000
Anti-HA rabbit	GeneTex	GTX29110	1:2,000
Anti-HA rat (3F10)	Roche	11867423001	1:2,000
Anti-Etag	Abcam	ab3397	1:2,000
Anti-c-myc (4A6)	Millipore	05-724	1:1,000

Table 3
Examples of optimized dilutions of secondary antibodies used for immunoblotting

Antibody	Company	Catalog number	Dilution for IB
Anti-mouse IRDye 800	LI-COR Biosciences	926-32210	1:15,000
Anti-mouse IRDye 680 LT	LI-COR Biosciences	926-68020	1:20,000
Anti-rabbit IRDye 800	LI-COR Biosciences	926-32211	1: 15,000
Anti-rabbit IRDye 680 LT	LI-COR Biosciences	926-68021	1:20,000
Anti-goat IRDye 800	LI-COR Biosciences	926-32214	1:15,000
Anti-rat IRDye 680 LT	LI-COR Biosciences	926-68029	1:20,000
Anti-mouse HRP	Cell Signaling	7076	1:2,000
Anti-rabbit HRP	Cell Signaling	7074	1:2,000
Anti-biotin HRP	Cell Signaling	7075	1:2,000

440CF for HRP-coupled antibodies and Odyssey from LI-COR Biosciences for infrared dye-coupled antibodies.

Optional (*see* **Note 9**):

7. Add stripping buffer onto the membrane. Incubate for 30 min at 65 °C.
8. Wash the membranes three times 5 min in TBS-T buffer.
9. Add blocking buffer from step 1 and incubate for 30 min at room temperature.
10. To detect ubiquitin or another protein, repeat steps 2–6 with other primary and secondary antibodies.

4 Notes

1. The antibody used for IP can recognize endogenous proteins or exogenous tags (e.g., Flag, His, HA), added to the protein of interest with the help of genetic engineering methods. Both approaches have advantages and disadvantages. The main advantage of the antibodies directed against endogenous proteins is the possibility to isolate the protein under native, endogenous conditions, without the need of overexpression. Unfortunately, there is a lack of good (specific and sensitive) antibodies for a lot of proteins, especially GPCRs. Antibodies against a tag are generally more sensitive and specific. Moreover, the amount of protein of interest that can be immunoprecipitated increases upon its prior overexpression, further enhancing the chances of detecting ubiquitin signals afterwards. Finally, also prior co-expression of tagged ubiquitin molecules can enhance ubiquitin detection levels using tag-directed antibodies

as compared to anti-ubiquitin antibodies (*see* Fig. 2—compare ubiquitination signal detected with anti-ubiquitin and anti-Flag antibodies).

2. The type of beads which is the most suitable for IP is chosen based on the origin of the antibody used in the procedure (species of animal in which IgG was produced). In general, Protein G beads are used to bind and purify IgG from human, mouse, and rat, whereas Protein A beads are recommended for isolation of IgG from human and rabbit but most of the time detect also very well mouse antibodies. There are also immobilized Protein A/G beads available, which combine the IgG-binding domain of both Protein A and protein G, enabling purification of numerous IgG subclasses and species.

   If the protein that undergoes IP has a common tag (e.g., HA, Flag), the use of beads directly coupled to the specific antibody recognizing the tag (Flag beads, HA beads) can be considered. This will help to save time (4 h incubation with such beads is sufficient to obtain high immunoprecipitation efficiency). However, immobilization can reduce the antibody's affinity to the antigen which will reduce IP efficiency.
3. Per sample 20 μl of pure beads needs to be added to the lysate. We always prepare a mixture of 20 μl of beads and 30 μl of RIPA (the same buffer which was used for washing the beads). From this 50 μl mixture, we transfer around 40 μl (using a pipet tip from which the ending is cut, to obtain larger diameter) to the tube containing lysate with antibody.
4. If nonspecific signals are observed after immunodetection with specific antibodies, extra washing steps can be introduced. Also, increasing the salt concentration in the RIPA washing buffer can help to remove proteins nonspecifically bound to the beads.
5. When proteins that bind nonspecifically to the beads interfere with visualization of the proteins of interest, a preclearing of the lysate can be recommended. In this additional step, lysates are incubated with the beads (without addition of antibody) for 30–60 min at 4 °C with rotation. Then, samples are centrifuged for 1 min (3,000 × *g*, 4 °C); cleared lysates are transferred to clean microcentrifuge tube and can be used for IP.
6. When ubiquitination of dopamine receptors is studied, we recommend using a respectively lower temperature for elution procedure (37 or 50 °C instead of 95 °C in many elution protocols). Dopamine receptors, as well as many other class A GPCRs, have a molecular weight of around 50 kDa. If the elution of proteins is performed at 37 °C, the IgG used for IP will appear at the height of 100 kDa (association of two heavy chains of the antibody) and of 25 kDa (light chain). On the other hand, when 95 °C is used, the antibody band will be observed at 50 kDa, and this immunoglobulin band will mask the band of a lot of

GPCRs. Additionally, denaturation of the receptor performed at 95 °C will result in the formation of receptor aggregates which migrate as very-high molecular weight smears. Another procedure to overcome the problem of detection of the antibody band at the height of the protein of interest will be the use of a secondary antibody (1) that is conformation specific and can recognize only the non-denatured form of the antibody (antibody used for IP is denatured and will not be recognized) or (2) that recognizes only the light chain of the primary antibody. Also, usage of primary antibody directly coupled to the enzyme (e.g., HRP) or infrared dye or produced in another animal species than the antibody used for IP can be a good alternative (remember that for closely related species, cross recognition may occur). One more possibility is to cross-link antibody used for IP to the beads (e.g., by usage of the Thermo Scientific Pierce Crosslink IP kit). This procedure results in the formation of the antibody-bead complex resistant to denaturation.

7. As sometimes the amount of antibody is limited, it is possible to incubate two membranes with the same antibody in one box. Membranes should be placed in the box "back to back," meaning that the proteins on the membranes are exposed to the antibody. Antibody diluted in the blocking buffer with addition of Tween 20 can also be reused several times for incubation with different membranes. Antibody can be stored for 1 month at 4 °C (if antibody is diluted in a blocking buffer containing milk or BSA, then addition of NaN_3 is recommended), for longer storage freezing of the antibody is advised.
8. It is possible to use primary antibody which is immediately coupled to the enzyme or fluorescent dye. This will eliminate the need to use secondary antibody and will save time. If you plan to use antibody recognizing a common tag most of the time, there are primary antibodies available conjugated with the enzyme (e.g., HRP).
9. In case of ubiquitination assays, you will need to detect at least two proteins on your membranes: the protein of which ubiquitination status you want to investigate (dopamine receptor) and ubiquitin. Often it is better to incubate the membrane with antibody recognizing ubiquitin after the stripping procedure. This extra denaturation procedure helps to expose ubiquitin epitopes that can be detected by the antibody.

Acknowledgments

This work was supported by Research Foundation—Flanders (FWO), project nr. G.0109.09N. KS has a pre-doctoral FWO fellowship; PR and KVC have post-doctoral FWO fellowships.

References

1. Shenoy SK (2007) Seven-transmembrane receptors and ubiquitination. Circ Res 100: 1142–1154
2. Komander D (2009) The emerging complexity of protein ubiquitination. Biochem Soc Trans 37:937–953
3. Husnjak K, Dikic I (2012) Ubiquitin-binding proteins: decoders of ubiquitin-mediated cellular functions. Annu Rev Biochem 81:291–322
4. Petroski MD (2008) The ubiquitin system, disease, and drug discovery. BMC Biochem 9:S7
5. Hasdemir B, Bunnett NW, Cottrell GS (2007) Hepatocyte growth factor-regulated tyrosine kinase substrate (HRS) mediates post-endocytic trafficking of protease-activated receptor 2 and calcitonin receptor-like receptor. J Biol Chem 282:29646–29657
6. Marchese A, Raiborg C, Santini F, Keen JH, Stenmark H, Benovic JL (2003) The E3 ubiquitin ligase AIP4 mediates ubiquitination and sorting of the G protein-coupled receptor CXCR4. Dev Cell 5:709–722
7. Petaja-Repo UE, Hogue M, Laperriere A, Bhalla S, Walker P, Bouvier M (2001) Newly synthesized human δ opioid receptors retained in the endoplasmic reticulum are retrotranslocated to the cytosol, deglycosylated, ubiquitinated, and degraded by the proteasome. J Biol Chem 276:4416–4423
8. Dores MR, Trejo JA (2012) Ubiquitination of G protein-coupled receptors: functional implications and drug discovery. Mol Pharmacol 82:563–570
9. Milojevic T, Reiterer V, Stefan E, Korkhov VM, Dorostkar MM, Ducza E, Ogris E, Boehm S, Freissmuth M, Nanoff C (2006) The ubiquitin-specific protease Usp4 regulates the cell surface level of the A2A receptor. Mol Pharmacol 69:1083–1094
10. Canals M, Scholten DJ, de Munnik S, Han MKL, Smit MJ, Leurs R (2012) Ubiquitination of CXCR7 controls receptor trafficking. PLoS One 7:e34192
11. Li H, Armando I, Yu P, Escano C, Mueller SC, Asico L, Pascua A, Lu Q, Wang X, Villar VAM (2008) Dopamine 5 receptor mediates Ang II type 1 receptor degradation via a ubiquitin-proteasome pathway in mice and human cells. J Clin Invest 118:2180–2189
12. Urban JD, Clarke WP, Von Zastrow M, Nichols DE, Kobilka B, Weinstein H, Javitch JA, Roth BL, Christopoulos A, Sexton PM (2007) Functional selectivity and classical concepts of quantitative pharmacology. J Pharmacol Exp Ther 320:1–13
13. Raehal KM, Schmid CL, Groer CE, Bohn LM (2011) Functional selectivity at the μ-opioid receptor: implications for understanding opioid analgesia and tolerance. Pharmacol Rev 63:1001–1019
14. Groer CE, Schmid CL, Jaeger AM, Bohn LM (2011) Agonist-directed interactions with specific β-arrestins determine μ-opioid receptor trafficking, ubiquitination, and dephosphorylation. J Biol Chem 286(36):31731–31741
15. Wertz IE, O'Rourke KM, Zhou H, Eby M, Aravind L, Seshagiri S, Wu P, Wiesmann C, Baker R, Boone DL (2004) De-ubiquitination and ubiquitin ligase domains of A20 downregulate NF-κB signalling. Nature 430(7000): 694–699
16. Zhao Q, Liu L, Xie Q (2012) In vitro protein ubiquitination assay. Methods Mol Biol 876:163–172
17. Anania VG, Pham VC, Huang X, Masselot A, Lill JR, Kirkpatrick DS (2013) Peptide level immunoaffinity enrichment enhances ubiquitination site identification on individual proteins. Mol Cell Proteomics 13:145–156, mcp. M113. 031062
18. Sylvestersen KB, Young C, Nielsen ML (2013) Advances in characterizing ubiquitylation sites by mass spectrometry. Curr Opin Chem Biol 17:49–58
19. Perroy J, Pontier S, Charest PG, Aubry M, Bouvier M (2004) Real-time monitoring of ubiquitination in living cells by BRET. Nat Methods 1:203–208
20. Fang D, Kerppola TK (2004) Ubiquitin-mediated fluorescence complementation reveals that Jun ubiquitinated by Itch/AIP4 is localized to lysosomes. Proc Natl Acad Sci U S A 101:14782–14787
21. Rondou P, Haegeman G, Vanhoenacker P, Van Craenenbroeck K (2008) BTB protein KLHL12 targets the dopamine D4 receptor for ubiquitination by a Cul3-based E3 ligase. J Biol Chem 283:11083–11096
22. Rondou P, Skieterska K, Packeu A, Lintermans B, Vanhoenacker P, Vauquelin G, Haegeman G, Van Craenenbroeck K (2010) KLHL12-mediated ubiquitination of the dopamine D4 receptor does not target the receptor for degradation. Cell Signal 22:900–913

Part III

Visualization and Imaging of Dopamine Receptors and Ligands

Chapter 9

Dopamine Receptors in the Subthalamic Nucleus: Identification and Localization of D5 Receptors

Lionel Froux, Diana Suarez-Boomgaard, Jerome Baufreton, Alicia Rivera, Maurice Garret, and Anne Taupignon

Abstract

Herein we present methodological approaches for the identification and characterization of dopamine receptors in the subthalamic nucleus, a component nucleus of the basal ganglia, at pre-and postsynaptic locations and of their roles with an emphasis given to the dopamine D5 receptor subtype. This chapter focuses on the possible sources of divergence between electrophysiological studies and describes the pharmacological tools available for functional studies of this receptor. The procedures for single-cell reverse transcription PCR (polymerase chain reaction) identification of dopamine D5 receptor mRNA and the immunochemical detection of the receptor at cellular and subcellular levels are presented.

Key words D5 dopamine receptor, Subcellular localization, Single-cell RT-PCR, Immunohistochemistry, Electron microscopy

1 Introduction

The basal ganglia (BG) are a network of several interconnected subcortical nuclei. They form somatotopically organized cortico-cortical loops controlling voluntary movement, as well as cognitive and motivational aspects of motor behavior [1]. The subthalamic nucleus (STN) plays a prominent role in the BG for several reasons: (1) It is an input structure of the network, in the "hyperdirect" pathway; (2) it is the sole excitatory (glutamatergic) nucleus in the BG; (3) together with these properties, the pacemaker activity of subthalamic neurons may confer on the STN the ability to drive the whole BG network; and finally, (4) high-frequency electrical stimulation of the STN is used as a symptomatic treatment for Parkinson's disease in humans.

Dopamine (DA) is critical for the functions supported by the BG. It originates from midbrain DA neurons in the *substantia nigra pars compacta* and the ventral tegmental area. Axons from

Mario Tiberi (ed.), *Dopamine Receptor Technologies*, Neuromethods, vol. 96,
DOI 10.1007/978-1-4939-2196-6_9, © Springer Science+Business Media New York 2015

these neurons extend through the medial forebrain bundle to provide rich DA innervation of the striatum and more limited innervation of other basal ganglia regions, including the STN [2, 3]. In the STN, DA is released from *en passant* tyrosine hydroxylase immunoreactive axons that form synapses. DA fibers and release sites are sparse, and local electrical stimulation evokes low extracellular DA concentrations [4]. Thus, a 1 s train of stimulations at 50Hz evoked the action potential-dependent release of DA in the STN, with a peak concentration estimated to be in the 100–500 nM range, tenfold lower than that measured upon a single stimulation in any striatal territory using the same voltammetric method [2, 4–6]. It is unknown whether DA release also occurs at nonsynaptic sites.

1.1 Dopamine Receptor (DAR) Repertoire in the STN

Anatomical and functional studies, using techniques such as radioligand binding, immunohistochemistry for light and electron microscopy, electrophysiology, and single-cell RT-PCR, have established that D1 (D1R, D5R) and D2 (D2R, D3R, D4R) receptor families are present and functional in the STN (for review, *see* [7, 8]). D1R are expressed on afferent (presynaptic) terminals in the primate STN, whereas D5R are expressed postsynaptically in subthalamic neurons of rats and monkeys [9, 10, 7, 11]. Activation of D1 receptor family increases the firing frequency of STN neurons in rat brain slices [12, 13]. D5R activation enhances evoked and spontaneous burst-firing duration, acting on the Ca_V1 channel in vitro and in vivo, through the adenylyl cyclase-cAMP-PKA cascade in burst-competent STN neurons [10, 12, 31].

The D2 receptor family is present at pre- and postsynaptic sites within the STN. Activation of postsynaptic D2R and D3R depolarizes subthalamic neurons, increases their firing frequency and excitability, but reduces their bursting potency. Thus, postsynaptic D2R have the ability to turn burst firing into single-spike firing [14, 15, 12]. Presynaptic D2R or D3R activation reduces the initial release probability of GABA and glutamate by afferent terminals [16, 17]. D4R may be expressed presynaptically, possibly indicating an involvement in the control of GABA afferent transmission to the STN [18].

1.2 Demonstration of DAR Action

The identity of DARs acting in the STN was revealed by patch-clamp experiments in whole-cell, cell-attached, and perforated-patch configurations in rodent brain slices (*see* **Note 1**). We will review the procedures and comment on the potential source of discrepancies between the studies. Our main technical focus is the identification of D5R action.

In the cell-attached configuration, the membrane, the neuron metabolism, and, thus, the ionic buffering, channel activity, and transduction pathways of DARs remain intact. The perforated patch, using the pore-forming molecule gramicidin to obtain electrical access to neurons, is thought to have a similar protective action. In any case, whole-cell patch clamp is disadvantageous and may emphasize presynaptic actions of DARs to the detriment of postsynaptic actions that may necessitate the preservation of neuron cytosol (*see* **Note 2**).

1.2.1 DAR Agonists

The commonly used agonists of D1 and D2 receptor families (SKF81297 and SKF82958, and quinpirole), at concentrations of 5–10 μM, did not induce co-activation of receptors in these 2 families. These concentrations, used in perfusion on brain slices, seem a good compromise between specificity and power of action [12] (*see* **Note 3**). SKF81297, SKF82958, and SKF38393, at concentrations of 2–5 μM in Krebs medium perfusion, acted on burst firing. There was no qualitative difference in the action of these drugs, which all induced a significant increase in burst duration, accompanied by an increase in action potentials per burst [10] (*see* **Note 4**). This phenomenon was caused by activation of the postsynaptic membrane receptors and protein kinase A followed by Ca_v1 channel phosphorylation. It was marked in approximately 90 % burst-competent neurons [10] (*see* **Note 5**). The burst-firing capacity was linked to differential coupling of apamin-sensitive SK channels to Ca^{2+} channels [19]. Furthermore, co-application of SKF81297, SKF82958, or SKF38393 together with the antagonist SCH23390 had no effect (*see* Sect. 2.1, item 2, and **Note 6**). Similarly, SKF81297 (5 μM) did not change the amplitude of inhibitory postsynaptic currents or the short-term dynamics of the pallido-subthalamic GABAergic synapses, controlled by D2R and/or D3R receptors [17]. To determine which subtype of D2 receptor family was acting on Ca_V 2.2 channels expressed at the subthalamic neuron membrane, PD168077 was applied to acutely dissociated subthalamic neurons. The affinity of PD168077 for D4R is about 10 nM, whereas its affinity for D2 and D3 receptors is approximately 3 μM. In contrast to quinpirole (10 μM), PD168077 (1 μM) did not alter the voltage-dependent calcium current amplitude [14] ruling out a role of D4R.

1.2.2 DAR Antagonists

DAR antagonists may be used to probe the action of physiologically released DA or to examine the action of exogenously applied DAR agonists. Given the location of the STN in the rodent brain, brain slices containing the STN do not usually retain the whole repertoire of afferent cell bodies [20, 21]. The experimental results of most groups studying the STN slices have revealed that presynaptic fibers, although presumably severed from cell bodies, release GABA or glutamate onto subthalamic neurons, producing spontaneous and miniature postsynaptic potentials and currents. Dopaminergic fibers are unlikely to release DA in the same mode as glutamatergic fibers release glutamate. Rather, DA release is thought to be generally designed for volume transmission, and its diffusion may cause spillover onto extrasynaptic receptors [2]. No evidence of DAR antagonist action and, thus, tonic action of DA, has been detected in the STN in brain slices. This may simply be due to the fact that either DA is washed out of slices continuously bathed in Krebs-derived solutions or severed DA afferents do not retain the properties of intact dopamine neurons. As noted earlier, local electrical stimulation induced transient increases in extracellular DA concentration. Stimulation consisted of 1 s trains at 50 Hz, designed to reproduce burst firing in dopaminergic axons. The DA increase lasted approximately 5 s and peaked in the 100–500 nM range [4]. The features of such a phasic DA release have not yet been reproduced experimentally.

A number of DAR antagonists discriminate between D1 and D2 receptor families, since their affinity for receptors in these two families differs 100- to 1,000-fold. Thus, the Ki values of D2-class antagonists, such as sulpiride and raclopride, are below 50 nM for D2 receptor family and above 5 μM for D1 receptor family. SCH23390 is even more discriminant, with a ratio of 1000 (Ki of 0.5 nM for D1/D5 receptors *vs* 500 nM or more for D2/D3/D4 receptors) [22]. SCH23390 and raclopride can facilitate discrimination between D1 and D2 receptor families when used in brain slices at concentrations of 10 and 5 μM, respectively. Thus, in the STN, raclopride did not prevent the action of any of the 3 agonists (SKF81297, SKF82958, or SKF38393, 3–5 μM), and the co-application of SCH23390 with any of the 3 agonists had no effect [10].

1.2.3 Dopamine

Exogenous DA has only been tested so far using bath application or superfusion of brain slices. Its effect at high concentrations (30–100 μM), on the spontaneous firing rate of subthalamic neurons in brain slices, was investigated in conjunction with a variety of dopaminergic agonists and antagonists. Results obtained with these dopaminergic drugs have led to the suggestion of an unconventional pharmacology of DAR in the STN [23]. However, studies using different approaches suggest that this unconventional pharmacological regulation of STN firing could be linked to the distinct modulation of firing rate parameters (intrinsic firing, Ca^{2+} or K^{+} conductances, and inhibitory or excitatory synaptic potentials/currents) by several DAR subtypes.

DA at concentrations detected in the STN on local electrical stimulation using fast voltammetry has not been tested in brain slices from normal animals. The only evidence of DA action in the 100–500 nM range was obtained in a 6–OHDA experimental model of Parkinsonism: GABAergic but not glutamatergic transmission was inhibited by DA at 100 and 300 nM [24].

1.2.4 No Specific Tools to Distinguish Between D1 Receptor Subtypes

No subtype-selective ligands of receptors in the D1 family have been experimentally tested on neurons. Thus, there are no validated pharmacological tools for distinguishing between D1 and D5 action. For the moment, agonists and antagonists for D1 receptor family can be used to activate DARs, as described above, in brain slices obtained from transgenic animals bearing a null or truncated DAR [10]. A mouse line bearing a null mutation for the D1R was generated by Drago and coll. [25]. Mutant mice have a neomycin phosphotransferase gene inserted in the region of the D1R gene leading to the excision of the third intracellular loop and downstream encoded sequence. Their use is freely granted by Dr Drago on request. They can also be purchased from the Jackson Laboratory (Bar Harbor, Maine, USA). D1R mutant mice exhibit normal coordination and open field activity but fewer rearing events than their nonmutant littermates (*see* **Note 7**). Sibley et al. generated D5R-deficient mice by including a neomycin gene in the D5R gene to disrupt the reading frame downstream from the fourth intracellular loop [26]. MTAs (material transfer agreement) may be obtained from Dr Sibley. Mice homozygous for the mutant D5R show discrete deficits (*see* **Note 8**).

1.2.5 Cis-(Z)-Flupenthixol

Constitutive activity was recognized from the earliest studies as a typical feature of recombinant D5R [27, 28]. Given that burst-competent neurons in the STN express D5R but not D1R, the constitutive activity of D5R may be involved in the motor deficits observed in experimental Parkinsonism. Interestingly, the burst-potentiating action resulting from the constitutive activity of D5R may be maintained in the dopamine-depleted state, i.e., when the synergistic action of the other DARs disappears in the absence of their ligand [12].

A number of psychoactive drugs have been shown to behave as inverse agonists of recombinant D5R, the most effective and specific being fluphenazine, (+)butaclamol, and *cis*-(Z)-flupenthixol [29, 30]. The 3 drugs acted on evoked bursts from subthalamic neurons in brain slices [31]. *Cis*-(Z)-flupenthixol alone produced the changes expected from an inverse D5R agonist (*see* **Note 9**). Then, *cis*-(Z)-flupenthixol was used in behavioral experiments, in vivo and in vitro electrophysiological recordings of STN neurons, and ex vivo functional neuroanatomy studies. Taken together, the data indicated that subthalamic D5R, acting on Ca_V1 channels, were involved in the pathophysiology of Parkinson's disease and that administering an inverse agonist of these receptors was likely to lessen motor symptoms [31].

The combined use of agonists, antagonists, inverse antagonists, and KO mice revealed a cellular function of D5R involved in the physiology as well as physiopathology of STN. The identity of the receptor (D5R *vs* D1R) has been unequivocally demonstrated by single-cell RT-PCR. Cellular and subcellular localizations of D5R have been defined by immunohistochemistry. The two methods are discussed in the following section.

1.3 Single-Cell Reverse Polymerase Chain Reaction and Immunohisto chemistry

Single-cell reverse polymerase chain reaction (scRT-PCR) offers a unique advantage for exploring the mRNA repertoire encoding DARs in single neurons with a known functional profile. Thus, whole-cell recordings of subthalamic neurons in brain slices and acutely dissociated subthalamic neurons are followed by scRT-PCR analysis of mRNA encoding D1 or D2 receptors [14, 10]. We describe in this chapter the procedure used in [10] to detect mRNA of receptors in the D1 receptor family in burst-competent subthalamic neurons.

D5R localization at the cellular and subcellular levels contributed to deciphering the role of this DAR in the STN. It is particularly interesting to demonstrate not only the presence of D5R in the STN but also its pre- and/or postsynaptic localization and, therefore, its relevance to the BG network. Typically, antibodies are generated against a peptide sequence of 10–12 amino acids exclusive to the dopamine receptor under study. In the case of D5R localization, this procedure requires a highly specific antibody that recognizes the antigen without cross-reactivity to other DARs, e.g., D1R. In the experiment below, a polyclonal previously prepared and extensively characterized D5R antibody developed in rabbit was used [32], although commercial alternatives are also available (e.g., D5R antibody #BP124S from Acris).

2 Materials

2.1 Single-Cell RT-PCR

All solutions and reagents must be RNase-free.

1. Sterilized 1.5-ml microcentrifuge tubes.
2. Automatic pipettes capable of dispensing 1–20 μl and 20–200 μl.
3. Sterilized, RNase-free disposable tips for automatic pipettes.
4. Disposable latex gloves.
5. RNase-free DNase I.
6. RNase-free water.
7. Thin-walled PCR tubes.
8. Hexamer random primers.
9. The four deoxyribonucleotide triphosphates.
10. Dithiothreitol.
11. Ribonuclease inhibitor.
12. Moloney murine leukemia virus reverse transcriptase.
13. Taq DNA polymerase.
14. Tris buffered at pH 8.
15. 100 mM $MgCl_2$.

2.2 Immunohistochemistry for Light Microscopy

1. Sodium pentobarbital.
2. Needles (25 gauge) and 1.5 ml syringes.
3. Perfusion pump.
4. 0.1 mM phosphate-buffered saline, pH 7.4 (PBS).
5. Paraformaldehyde.
6. 10 ml glass collecting vials.
7. Sucrose.
8. Dry ice.
9. Freezing microtome.
10. 6-Well plates.
11. Sodium azide.
12. Hydrogen peroxide solution 30 % (w/w).
13. Polyclonal D5R antibody (*see* Sects. 1 and 1.3).
14. Triton X-100.
15. Biotinylated goat anti-rabbit IgG antibody.
16. Horseradish peroxidase-conjugated streptavidin.
17. 3-3′-diaminobenzidine tetrahydrochloride hydrate (DAB).
18. Nickel ammonium sulfate.
19. Silane-coated microscope slides.
20. 50 %, 70 %, 96 %, and 100 % ethanol.
21. Xylene.
22. Coverslip.
23. Histology slide mounting medium.

2.3 Immunohistochemistry for Electron Microscopy

1. Sodium pentobarbital.
2. Needles (25 gauge) and 1.5 ml syringes.
3. Perfusion pump.
4. 0.1 mM phosphate-buffered saline, pH 7.4 (PBS).
5. Paraformaldehyde.
6. Glutaraldehyde.
7. 10 ml glass collecting vials.
8. Vibratome (Leica).
9. 6-Well plates.
10. Sodium azide.
11. Hydrogen peroxide solution 30 % (w/w).
12. Polyclonal D5R antibody.
13. Biotinylated goat anti-rabbit IgG antibody.
14. Horseradish peroxidase-conjugated streptavidin.

15. 3-3′-diaminobenzidine tetrahydrochloride hydrate (DAB).
16. Osmium tetroxide ($OsSO_4$).
17. Glass petri dishes (44 × 12 mm).
18. 50 %, 70 %, 96 %, and 100 % ethanol.
19. Uranyl acetate dehydrate.
20. Acetone.
21. Durcupan ACM (araldite base embedding agent for electron microscopy).
22. Oven incubator.
23. Microscope slides.
24. Plastic sheets.
25. Dissecting scalpel blade.
26. Embedding capsule.
27. Ultramicrotome.
28. Diamond knife.
29. Microdissecting forceps.
30. Nickel grids.
31. Grid storage box.
32. Lead citrate.
33. Sodium hydroxide.

3 Methods

3.1 Single-Cell RT-PCR

3.1.1 Harvesting Cell Content

1. Patch a neuron and then hold it into a whole-cell configuration for 5–20 min. Assess its electrical properties (e.g., burst competency, defined by spontaneous burst firing, plateau potentials, or post-inhibitory rebounds; *see* **Note 10**). Apply negative pressure through the pipette holder and maintain for at least 2 min while the seal and access resistances are monitored. Stop aspiration immediately if the seal breaks (*see* **Note 11**).
2. Withdraw quickly the pipette from the slice, expel its contents into an RNase-free microcentrifuge tube, and immediately place it in a −80 °C freezer (*see* **Note 12**).

3.1.2 Single-Neuron RT-PCR Procedure

1. Perform reverse transcription using protocols similar to those previously described [33, 34]. Like Karagiannis and colleagues, we find that overnight incubation at 37 °C with reverse transcriptase is more efficient than 1-h incubation. The resulting reverse transcription product is kept at −80 °C until used for PCR amplification (*see* **Note 13**).
2. Design the primers based on GenBank rat D1 and D5 receptor sequences (accession numbers M35077 and NM012768,

respectively). Sense primers are D1R-for1: AAGCAGCCTT CATCCTGATTAGC, D1Rfor2: GCATGGACTCTGTCTGT CCTTATA, D5Rfor1: CCATCCTCATCTCCTTCATCCCG, and D5R-for2: ACTCAATTGGCACAGAGACAAGG. Antis ense primers are D1R-rev2: ACAGAAGGGCACCATACAG TTCG, D1R-rev3: GGAGCCAGCAGCACACAAACACC, D5R-rev1: CAGGATGAAGAAAGGCAACCAGC, and D5R-rev3: TGCAGAAAGGAACCATACAGTTC (*see* **Note 14**). A two-stage amplification strategy is designed to detect the low-abundance dopamine receptor mRNAs (*see* **Note 15**).

3. In the first step, amplify 10 μl of a single-cell template cDNA in a 50 μl PCR reaction mixture, using 0.25 μM D1R and D5R-for1 and D5R-rev2 primers. Perform 20 cycles: 45 s at 94 °C, 45 s at 60 °C, and 60 s at 72 °C.
4. Then, use a 1 μl aliquot of this PCR product as a template for a second round of PCR amplification, with each pair of specific nested primers (for2 and rev3). Perform PCR amplification as described above, but using 35 cycles.
5. Sequence and analyze the PCR products on a 2 % agarose gel stained with ethidium bromide.

3.1.3 Controls

1. After harvesting the cell content of 3 neurons, fill a pipette as described in **Note 10**. Lower the pipette into the slice and withdraw it without attempting to seal a neuron. Process the pipette medium as described in Sect. 3.1.1 and use it as a control for possible contamination. If amplification products are detected, the preceding three harvests are discarded.
2. In addition, harvest and treat a 4th neuron in the same way as the others, omitting the reverse transcriptase. This serves as a control for possible nuclear genomic contamination. However, note that the dimension of a patch pipette tip makes harvesting of the nucleus very unlikely.
3. Also run a PCR control experiment, using water instead of cDNA.

ScRT-PCR is very demanding, since all steps in the harvesting are crucial. With the blind-patch technique, approximately 1 in 2 neurons does not yield any amplification products, presumably as no significant cell content is harvested. We expect harvesting with a visualized setup to be more productive. However, when scRT-PCR was used to define the expression profile of subthalamic neurons and harvesting performed on a visualized setup, detectable levels of D2 or D3 receptor mRNA were found in 37 % and 20 % neurons, respectively, but not in all neurons, whereas quinpirole, a broad-spectrum D2 receptor agonist, acted on virtually all neurons in the STN [14, 15]. This finding was attributed to the low abundance of mRNA encoding DARs in subthalamic neurons [14].

3.2 Cellular and Subcellular D5R Identification in the STN by Immunohistochemistry

3.2.1 Immunohistochemistry for Light Microscopy

3.2.1.1 Tissue Preparation

1. Deeply anesthetize the rat with sodium pentobarbital (60 mg/kg, i.p.) and perfuse it transcardially with 0.1 M phosphate-buffered saline, pH 7.4 (PBS), followed by 4 % paraformaldehyde (w/v) in PBS.
2. Remove the brain and postfix by immersion in the same fixative at 4 °C for 2 h.
3. Cryoprotect the brain by incubating it in a 30 % sucrose solution (w/v) in PBS until it sinks to the bottom of the recipient, usually after 72 h. Then, freeze it rapidly in dry ice.
4. Cut the brain with a freezing microtome to obtain coronal free-floating sections 30 μm thick. Store the sections in 0.02 % sodium azide (w/v) in PBS at 4 °C until required for immunohistochemistry. Under these conditions, it may be stored for several months.

3.2.1.2 Immunohistochemistry (*See* **Note 16**)

1. Quench the endogenous peroxidase activity of the tissue by incubating it with 3 % hydrogen peroxide in PBS for 15 min (*see* **Note 17**).
2. Incubate with the D5R antibody diluted in 0.2 % Triton X-100 and 0.1 % sodium azide in PBS at 4 °C for 72 h (*see* **Note 18**).
3. Sequentially incubate for 1 h with biotinylated goat anti-rabbit IgG followed by horseradish peroxidase-conjugated streptavidin, both diluted in 0.2 % Triton X-100 in PBS (*see* **Note 19**).
4. Develop the peroxidase reaction with a fresh dilution of 0.05 % 3-3′-diaminobenzidine (DAB) tetrahydrochloride, 0.03 % nickel ammonium sulfate, and 0.01 % H_2O_2 in PBS (*see* **Note 20**).
5. Stop the peroxidase reaction by transferring immunostained sections into fresh PBS.
6. Mount the immunostained sections on silanized glass slides and air-dry.
7. Dehydrate the sections by passing through a series of increasing ethanol concentrations (50 %, 70 %, 96 %, and 100 %) and clear with xylene.
8. Coverslip with a mounting medium.

3.2.2 Immunohistochemistry for Electron Microscopy

3.2.2.1 Tissue Preparation

1. Deeply anesthetize the rat with sodium pentobarbital (60 mg/kg, i.p.) and perfuse it transcardially with 0.1 M phosphate-buffered saline, pH 7.4 (PBS), followed by 4 % paraformaldehyde (w/v) and 0.1 % glutaraldehyde in PBS.
2. Remove the brain and postfix by immersion in the same fixative at 4 °C for 2 h.
3. Cut the brain with a vibratome to obtain coronal free-floating sections 40–50 μm thick. Collect the sections in PBS and process immediately for immunohistochemistry.

3.2.2.2 Immunochemistry

Perform immunohistochemistry as described in Section "Immunohistochemistry" but with the following modifications:

1. Do not include Triton X-100 in any buffer in order to preserve membrane structures.
2. Develop the peroxidase reaction using only 0.05 % DAB and 0.01 % H_2O_2 in PBS.
3. After the peroxidase reaction, wash the sections with PBS and postfix with 1 % $OsSO_4$ in PBS for 1 h (*see* **Note 21**).

3.2.2.3 Resin Embedding

1. Dehydrate the immunostained sections in ascending series of ethanol dilutions (50 %, 70 %, 96 %, and 100 %) for 40 min each and then with 1 % uranyl acetate in 70 % ethanol and, finally, with acetone (100 %).
2. Embed the sections with 1:1 acetone-Durcupan ACM resin for 1 h and then overnight in Durcupan ACM resin. Alternatively, other araldite embedding agents may be used.
3. Mount sections on glass slides with additional Durcupan ACM resin and cover with plastic sheets to flatten them. Polymerize at 60 °C for 48 h. After polymerization, remove the plastic sheets.
4. Select STN areas, cut out from the slides, and glue onto previously Durcupan-polymerized blocks.
5. Obtain ultrathin sections (50 nm) with an ultramicrotome (*see* **Note 22**). Collect the sections on nickel grids and contrast with 0.25 % lead citrate in 0.1 M NaOH.
6. Examine with an electron microscope.

4 Conclusion

Patch-clamp experiments together with scRT-PCR identification and immunohistochemistry have framed DA control of electrical activity of subthalamic neurons. However, there are a number of unresolved questions regarding DARs in the STN. The absence of functional D1Rs at post-synaptic sites in pacemaker and burst-incompetent neurons within the STN remains ambiguous. Evidence of the action of DARs on identified glutamatergic afferents of subthalamic neurons is also missing. It is not known whether DARs control the cortico-subthalamic, thalamo-subthalamic, or glutamatergic afferents from midbrain structures. Finally, a crucial question is that of possible changes in DAR expression and properties in persistent DA-depleted states. In any case, caution should be exercised in inferring conclusions on the basis of equating DA absence in normal animals and persistent DA-depleted states. As constitutive D5R activity appears to contribute to motor symptoms, determining whether the expression of D5R changes in experimental models of Parkinson's disease and in humans is likely to result in a better understanding of the physiopathology of this disease.

There are also a number of unsolved issues but virtually no tools to investigate them. Numerous pharmacological tools are currently used to investigate the transduction pathways of recombinant G protein-coupled receptors expressed in heterologous systems, but few have been used and validated in neurons. Accordingly, knowledge of the DAR transduction pathways in subthalamic neurons is limited. It would be useful to have tools that do not cross the cell membrane and can be dissolved in pipette medium and are thus capable of acting at the single-neuron level to differentiate unambiguously between pre- and postsynaptic action sites. Such tools are also likely to be valuable for distinguishing functionally different subthalamic neuron subpopulations. New tools should be designed with this capacity. "Uncoupling peptides," modeled on the G protein-binding site(s) of D1R and D5R, are most urgently needed.

Wholly unresolved questions include possible synergistic actions of the various DARs and their heteromerization. Evidence from native cells and heterologous expression systems support the hypothesis that G protein-coupled receptors traffic and signal within higher-order complexes [35]. Since D2R and D3R mRNAs were found in the same neurons [14], these receptors may assemble in heteromers at the membrane of the subthalamic neuron subpopulation co-expressing these two subtypes. Heteromerization of D5R and D2R may also occur, since D5R-D2R hetero-oligomers were found in HEK293T cells stably expressing tagged human D5 and D2 receptors [36]. Furthermore, agonist-dependent, molecular cross talk between D5R and $GABA_A$ receptors was reported to underlie a codirectional decrease in $GABA_A$ and D5R function [37]. In addition, the disparity between D1R and D5R subcellular localization, as revealed by recombinant receptor studies [38], and the demonstration that D1R-NMDA complexes may be a source of NMDA receptors for lateral diffusion to the synapses [39] raise the intriguing question of the capacity of D5R to contribute to complexes in a similar way.

The lack of tools to investigate these questions is not only true for the STN but also for the other brain regions expressing more than one receptor subtype (e.g., the *substantia nigra reticulata*, the entorhinal cortex, the hippocampus [40–42].

Finally, a question critically relevant to physiology and physiopathology must be tackled in the future. To date, constant, high DA concentrations have been used in brain slices to investigate its action. It is highly unlikely that these concentrations mimic the tone of ambient DA in the STN. It is even more unlikely that they reflect the acute changes in DA that may occur when dopaminergic neurons fire in bursts. Fortunately, new tools may soon help to shed light on the possible roles of tonic and phasic DA in the STN. DAT-cre and TH-cre mouse lines are already being used to express channelrhodopsins in *substantia nigra pars compacta* neurons using the Cre-Lox strategy [43–45]. This should make it possible to answer the remaining questions regarding the affinity states of DARs (high affinity for D2-like receptors *vs* low affinity for D1-like receptors [2]) and their possible relative changes in occupancy during tonic and phasic firing of DA neurons.

5 Notes

1. Frontal, horizontal, sagittal, and parasagittal slices obtained from rats or mice, varying in age from 3 weeks to 2 months, are routinely used.
2. Note that comparing the action of 10–100 μM DA, using extracellular, single-unit recordings, on the one hand, and whole-cell patch clamp, on the other hand, have led to contrasting results: virtually all the subthalamic neurons increased their firing rate, whereas only 1 in 3 was responsive, respectively [23].
3. It is often requested that the action of an agonist of receptors in one family is tested in the presence of an antagonist of receptors in the other family. We suggest conducting this experiment in 2 steps: preincubating the slices in an antagonist solution for a few minutes followed by co-application of the antagonist together with the agonist.
4. This experiment used the blind-patch method, thus obtaining recordings from subthalamic neurons deep in the slice, which required a patch pipette with a resistance of 10–12 MOhms. We recently recorded spontaneous burst firing and evoked bursts using the visualized patch method. We observed that using pipettes with a resistance below 10 MOhms, a usual condition in the visualized patch method, often led to run down of burst competency, i.e., the attenuation, or even disappearance, of plateau potentials and rebound bursts in about 30 min, whereas all parameters (temperature, perfusion rate, composition of intrapipette, and perfusion solutions) remained unchanged in the blind and visualized experiments. In the same way, dialysis of one or more intracellular components, necessary to maintain the transduction pathways of D1 receptor family, may be accentuated at low access resistances.
5. To avoid possible drug-induced priming effects, a slice is perfused once with the agonist under study.
6. In our experiments, successive applications of an agonist and antagonist to a brain slice did not always fully inhibit the action

of the agonist. However, preincubation or pre-perfusion of an antagonist alone plus co-application of antagonist and agonist was effective.

7. D1R mutant mice die on weaning unless appropriate zootechnical measures are implemented to help them access sufficient food.
8. No significant qualitative difference between mice and rats regarding the molecular parameters specifically targeted by DARs has yet been reported, but the type of animal model should be taken into account when comparing global responses, such as firing patterns or behaviors.
9. *Cis*-(Z)-flupenthixol is a neuroleptic. In order to inhibit any possible action via D2R, it was tested at 1 μM in the continuous presence of raclopride (5 μM) in vitro and its action compared to that of raclopride in vivo.
10. Capillaries for pipettes are from a dedicated stock. Always wear gloves to handle them. Only pull 2–4 pipettes at a time; store them in a dedicated container. Fill pipettes with 10 μl standard pipette medium from a freshly opened (every day), dedicated stock using a sterile plastic tuberculin (1 ml) syringe pulled to an appropriate diameter over a flame. Pass the silver filament in the pipette holder through a flame before and after each harvest. Apply a positive pressure to the pipette holder before lowering the pipette into the bath and do not remove until contact with a neuron is detected (as usual for standard patch clamping).
11. Use a negative pressure provided by depressing a tuberculin connected to the pipette holder suction tube by 0.1–0.2 ml.
12. To empty the pipette, gently break the tip on the side of an RNase-free microcentrifuge tube. The pipette medium is then forced into the microcentrifuge tube by applying a positive pressure to a tuberculin tightly connected to the pipette by a dedicated tubing connection. Perform reverse transcription as soon as possible after harvesting.
13. All the steps of the reverse transcription enzymatic reaction must be performed in complete RNase-free conditions.
14. "Primer-Blast" software may be used to find appropriate primers for any mRNA of interest (http://www.ncbi.nlm.nih.gov/tools/primer-blast/index.cgi?LINK_LOC=BlastHome). To avoid amplifying genomic DNA, it is necessary to design intron over spanning primers. However, this is not possible for the intronless D1R and D5R [46]. In these cases, a control without reverse transcription is suggested (see controls in Sect. 3.1.3).

15. This strategy is of great interest for detecting mRNAs with low expression from single neurons. In addition, the use of 4 different specific primers greatly enhances the specificity of the amplification process. However, this strategy is not suited to obtaining quantitative data.
16. All the steps were performed with gentle shaking.
17. Wash with PBS between all the steps (3 times for 10 min each).
18. Proper D5R antibody dilution is previously determined by testing various concentrations. Choose the highest antibody concentration which produces the most intense immunohistochemical staining without background. In the case of the polyclonal D5R antibody, this was 1:500. We routinely incubate D5R antibody previously pre-absorbed with the peptide used in its production as a negative control.
19. Avoid using sodium azide in the horseradish peroxidase-conjugated streptavidin dilution buffer as it inhibits enzyme activity.
20. DAB is carcinogenic to humans. When handling this chemical, avoid all contact and work with appropriate protection equipment. We prepare a 10× DAB stock solution in 0.05 M Tris Buffer (pH 7.6), which is stored at −20 °C in an appropriate volume for one reaction (aliquots of 500 μl). Nickel ammonium may be used to intensify the immunohistochemical signal.
21. We prepare 4× stock solution by dissolving OsO_4 in distilled water, which takes a long time. Stock solution may be stored at −20 °C for several months. OSO_4 is highly toxic.
22. A pale gold to silver color indicates that sections have the correct thickness.

Acknowledgments

We wish to thank the Regional Council of Aquitaine and Fondation de France who partly supported this work by grants 12006005 (FCAN scheme), and 2005013850 and 00016810, respectively. Financial support was also provided by CNRS and University of Bordeaux. D. Sibley (NINDS, Bethesda) and J. Drago (Monash University) allowed us to use the D5R and D1R mutant mice they engineered. J. Waddington (Royal College of Surgeons, Dublin) donated the D5R mutant mice we used to generate a colony. L.F received a PhD fellowship from MRT.

References

1. Haber SN (2003) The primate basal ganglia: parallel and integrative networks. J Chem Neuroanat 26(4):317–330
2. Rice ME, Patel JC, Cragg SJ (2011) Dopamine release in the basal ganglia. Neuroscience 198:112–137. doi:10.1016/j.neuroscience.2011.08.066
3. Smith Y, Kieval JZ (2000) Anatomy of the dopamine system in the basal ganglia. Trends Neurosci 23(10 Suppl):S28–S33
4. Cragg SJ, Baufreton J, Xue Y, Bolam JP, Bevan MD (2004) Synaptic release of dopamine in the subthalamic nucleus. Eur J Neurosci 20(7):1788–1802. doi:10.1111/j.1460-9568.2004. 03629.x
5. Benoit-Marand M, Borrelli E, Gonon F (2001) Inhibition of dopamine release via presynaptic D2 receptors: time course and functional characteristics in vivo. J Neurosci 21(23):9134–9141
6. Gonon FG (1988) Nonlinear relationship between impulse flow and dopamine released by rat midbrain dopaminergic neurons as studied by in vivo electrochemistry. Neuroscience 24(1):19–28
7. Rommelfanger KS, Wichmann T (2010) Extrastriatal dopaminergic circuits of the Basal Ganglia. Front Neuroanat 4:139. doi:10.3389/fnana.2010.00139
8. Wilson CJ, Bevan MD (2011) Intrinsic dynamics and synaptic inputs control the activity patterns of subthalamic nucleus neurons in health and in Parkinson's disease. Neuroscience 198:54–68. doi:10.1016/j.neuroscience.2011.06.049
9. Flores G, Liang JJ, Sierra A, Martinez-Fong D, Quirion R, Aceves J, Srivastava LK (1999) Expression of dopamine receptors in the subthalamic nucleus of the rat: characterization using reverse transcriptase-polymerase chain reaction and autoradiography. Neuroscience 91(2):549–556
10. Baufreton J, Garret M, Rivera A, de la Calle A, Gonon F, Dufy B, Bioulac B, Taupignon A (2003) D5 (not D1) dopamine receptors potentiate burst-firing in neurons of the subthalamic nucleus by modulating an L-type calcium conductance. J Neurosci 23(3):816–825
11. Svenningsson P, Le Moine C (2002) Dopamine D1/5 receptor stimulation induces c-fos expression in the subthalamic nucleus: possible involvement of local D5 receptors. Eur J Neurosci 15(1):133–142
12. Baufreton J, Zhu ZT, Garret M, Bioulac B, Johnson SW, Taupignon AI (2005) Dopamine receptors set the pattern of activity generated in subthalamic neurons. FASEB J 19(13):1771–1777. doi:10.1096/fj.04-3401hyp
13. Loucif AJ, Woodhall GL, Sehirli US, Stanford IM (2008) Depolarisation and suppression of burst firing activity in the mouse subthalamic nucleus by dopamine D1/D5 receptor activation of a cyclic-nucleotide gated non-specific cation conductance. Neuropharmacology 55(1):94–105. doi:10.1016/j.neuropharm.2008.04.025
14. Ramanathan S, Tkatch T, Atherton JF, Wilson CJ, Bevan MD (2008) D2-like dopamine receptors modulate SKCa channel function in subthalamic nucleus neurons through inhibition of Cav2.2 channels. J Neurophysiol 99(2):442–459. doi:10.1152/jn.00998.2007
15. Zhu ZT, Shen KZ, Johnson SW (2002) Pharmacological identification of inward current evoked by dopamine in rat subthalamic neurons in vitro. Neuropharmacology 42(6): 772–781
16. Shen KZ, Johnson SW (2000) Presynaptic dopamine D2 and muscarine M3 receptors inhibit excitatory and inhibitory transmission to rat subthalamic neurones in vitro. J Physiol 525(Pt 2):331–341
17. Baufreton J, Bevan MD (2008) D2-like dopamine receptor-mediated modulation of activity-dependent plasticity at GABAergic synapses in the subthalamic nucleus. J Physiol 586(8):2121–2142. doi:10.1113/jphysiol.2008. 151118
18. Floran B, Floran L, Erlij D, Aceves J (2004) Activation of dopamine D4 receptors modulates [3H]GABA release in slices of the rat thalamic reticular nucleus. Neuropharmacology 46(4):497–503. doi:10.1016/j.neuropharm.2003.10.004
19. Hallworth NE, Wilson CJ, Bevan MD (2003) Apamin-sensitive small conductance calcium-activated potassium channels, through their selective coupling to voltage-gated calcium channels, are critical determinants of the precision, pace, and pattern of action potential generation in rat subthalamic nucleus neurons in vitro. J Neurosci 23(20):7525–7542
20. Beurrier C, Ben-Ari Y, Hammond C (2006) Preservation of the direct and indirect pathways in an in vitro preparation of the mouse basal ganglia. Neuroscience 140(1):77–86. doi:10.1016/j.neuroscience.2006.02.029
21. Bosch C, Mailly P, Degos B, Deniau JM, Venance L (2012) Preservation of the hyperdirect pathway of basal ganglia in a rodent brain slice. Neuroscience 215:31–41. doi:10.1016/j.neuroscience.2012.04.033

22. Missale C, Nash SR, Robinson SW, Jaber M, Caron MG (1998) Dopamine receptors: from structure to function. Physiol Rev 78(1): 189–225
23. Tofighy A, Abbott A, Centonze D, Cooper AJ, Noor E, Pearce SM, Puntis M, Stanford IM, Wigmore MA, Lacey MG (2003) Excitation by dopamine of rat subthalamic nucleus neurones in vitro-a direct action with unconventional pharmacology. Neuroscience 116(1):157–166
24. Shen KZ, Zhu ZT, Munhall A, Johnson SW (2003) Dopamine receptor supersensitivity in rat subthalamus after 6-hydroxydopamine lesions. Eur J Neurosci 18(11):2967–2974
25. Drago J, Gerfen CR, Lachowicz JE, Steiner H, Hollon TR, Love PE, Ooi GT, Grinberg A, Lee EJ, Huang SP et al (1994) Altered striatal function in a mutant mouse lacking D1A dopamine receptors. Proc Natl Acad Sci U S A 91(26):12564–12568
26. Hollon TR, Bek MJ, Lachowicz JE, Ariano MA, Mezey E, Ramachandran R, Wersinger SR, Soares-da-Silva P, Liu ZF, Grinberg A, Drago J, Young WS 3rd, Westphal H, Jose PA, Sibley DR (2002) Mice lacking D5 dopamine receptors have increased sympathetic tone and are hypertensive. J Neurosci 22(24):10801–10810
27. Tiberi M, Caron MG (1994) High agonist-independent activity is a distinguishing feature of the dopamine D1B receptor subtype. J Biol Chem 269(45):27925–27931
28. Demchyshyn LL, McConkey F, Niznik HB (2000) Dopamine D5 receptor agonist high affinity and constitutive activity profile conferred by carboxyl-terminal tail sequence. J Biol Chem 275(31):23446–23455. doi:10.1074/jbc.M000157200
29. D'Aoust JP, Tiberi M (2010) Role of the extracellular amino terminus and first membrane-spanning helix of dopamine D1 and D5 receptors in shaping ligand selectivity and efficacy. Cell Signal 22(1):106–116. doi:10.1016/j.cellsig.2009.09.020
30. Martin MW, Scott AW, Johnston DE Jr, Griffin S, Luedtke RR (2001) Typical antipsychotics exhibit inverse agonist activity at rat dopamine D1-like receptors expressed in Sf9 cells. Eur J Pharmacol 420(2–3):73–82
31. Chetrit J, Taupignon A, Froux L, Morin S, Bouali-Benazzouz R, Naudet F, Kadiri N, Gross CE, Bioulac B, Benazzouz A (2013) Inhibiting subthalamic D5 receptor constitutive activity alleviates abnormal electrical activity and reverses motor impairment in a rat model of Parkinson's disease. J Neurosci 33(37):14840–14849. doi:10.1523/JNEUROSCI.0453-13.2013
32. Khan ZU, Gutierrez A, Martin R, Penafiel A, Rivera A, de la Calle A (2000) Dopamine D5 receptors of rat and human brain. Neuroscience 100(4):689–699
33. Christophe E, Roebuck A, Staiger JF, Lavery DJ, Charpak S, Audinat E (2002) Two types of nicotinic receptors mediate an excitation of neocortical layer I interneurons. J Neurophysiol 88(3):1318–1327
34. Karagiannis A, Gallopin T, David C, Battaglia D, Geoffroy H, Rossier J, Hillman EM, Staiger JF, Cauli B (2009) Classification of NPY-expressing neocortical interneurons. J Neurosci 29(11):3642–3659. doi:10.1523/JNEUROSCI.0058-09.2009
35. Smith NJ, Milligan G (2010) Allostery at G protein-coupled receptor homo- and heteromers: uncharted pharmacological landscapes. Pharmacol Rev 62(4):701–725. doi:10.1124/pr.110.002667
36. So CH, Verma V, Alijaniaram M, Cheng R, Rashid AJ, O'Dowd BF, George SR (2009) Calcium signaling by dopamine D5 receptor and D5-D2 receptor hetero-oligomers occurs by a mechanism distinct from that for dopamine D1-D2 receptor hetero-oligomers. Mol Pharmacol 75(4):843–854. doi:10.1124/mol.108.051805
37. Liu F, Wan Q, Pristupa ZB, Yu XM, Wang YT, Niznik HB (2000) Direct protein-protein coupling enables cross-talk between dopamine D5 and gamma-aminobutyric acid A receptors. Nature 403(6767):274–280. doi:10.1038/35002014
38. Kruusmagi M, Kumar S, Zelenin S, Brismar H, Aperia A, Scott L (2009) Functional differences between D(1) and D(5) revealed by high resolution imaging on live neurons. Neuroscience 164(2):463–469. doi:10.1016/j.neuroscience.2009.08.052
39. Ladepeche L, Dupuis JP, Bouchet D, Doudnikoff E, Yang L, Campagne Y, Bezard E, Hosy E, Groc L (2013) Single-molecule imaging of the functional crosstalk between surface NMDA and dopamine D1 receptors. Proc Natl Acad Sci U S A 110(44):18005–18010. doi:10.1073/pnas.1310145110
40. Zhou FW, Jin Y, Matta SG, Xu M, Zhou FM (2009) An ultra-short dopamine pathway regulates basal ganglia output. J Neurosci 29(33):10424–10435. doi:10.1523/JNEUROSCI.4402-08.2009
41. Duan TT, Tan JW, Yuan Q, Cao J, Zhou QX, Xu L (2013) Acute ketamine induces hippocampal synaptic depression and spatial memory impairment through dopamine D1/D5 receptors. Psychopharmacology 228(3):451–461. doi:10.1007/s00213-013-3048-2

42. Medin T, Rinholm JE, Owe SG, Sagvolden T, Gjedde A, Storm-Mathisen J, Bergersen LH (2013) Low dopamine D5 receptor density in hippocampus in an animal model of attention-deficit/hyperactivity disorder (ADHD). Neuroscience 242:11–20. doi:10.1016/j.neuroscience.2013.03.036
43. Tritsch NX, Ding JB, Sabatini BL (2012) Dopaminergic neurons inhibit striatal output through non-canonical release of GABA. Nature 490(7419):262–266. doi:10.1038/nature11466
44. Brown MT, Bellone C, Mameli M, Labouebe G, Bocklisch C, Balland B, Dahan L, Lujan R, Deisseroth K, Luscher C (2010) Drug-driven AMPA receptor redistribution mimicked by selective dopamine neuron stimulation. PLoS One 5(12):e15870. doi:10.1371/journal.pone.0015870
45. Rossi MA, Sukharnikova T, Hayrapetyan VY, Yang L, Yin HH (2013) Operant self-stimulation of dopamine neurons in the substantia nigra. PLoS One 8(6):e65799. doi:10.1371/journal.pone.0065799
46. Beaulieu JM, Gainetdinov RR (2011) The physiology, signaling, and pharmacology of dopamine receptors. Pharmacol Rev 63(1):182–217. doi:10.1124/pr.110.002642

Chapter 10

MALDI Mass Spectrometry Imaging of Dopamine and PET D1 and D2 Receptor Ligands in Rodent Brain Tissues

Richard J.A. Goodwin, Mohammadreza Shariatgorji, and Per E. Andren

Abstract

Both pharmaceutical and neurobiological research require a molecular understanding of the complex biochemistry occurring in the brain at a molecular level. To date, this has relied on indirect measurement of labelled compounds or by sample homogenization and subsequent analysis. However, recent advancements in the field of mass spectrometry imaging (MSI) now enabled the direct analysis of molecules from tissue sections. Drugs and endogenous compounds can be simultaneously desorbed/ionized and their abundance measured and mapped across a tissue section, in a multiplexed way. The technologies allow near cellular spatial resolution analysis and quantitative data to be collected. Sample preparation is a crucial step for successful target analyte detection. The use of standard solvent based and novel solvent-free MALDI matrix application methods have been reported as effective for label-free detection of D1 and D2 dopamine receptor antagonists. Furthermore, recently published protocols describe how neurotransmitters previously undetectable directly by MSI can be successfully analyzed following on-tissue derivatization. Mass spectrometry imaging is becoming established as a significant tool for neuroscience and pharmaceutical research and development.

Key words Mass spectrometry imaging, MALDI, Neurotransmitter, Dopamine, Dry matrix, Derivatization, Raclopride, PET ligand, Imaging, Label-free

1 Introduction

The ability to map the spatial distribution and abundance of analytes in a tissue section is vital if a full biomolecular understanding of tissue physiology, disease mechanism, or pharmaceutical efficacy is to be achieved. This is particularly pertinent for neurobiology where subtle changes or imbalances in brain chemistry can cause physiological or neurodegenerative disorders. There is a demand from researchers to be able to detect and measure endogenous biomolecules such as proteins, peptides, lipids, neurotransmitters and their receptors, as well as exogenous compounds such as pharmaceutical therapeutics and positron emission tomography (PET) ligands, directly from brain tissue sections. For exogenous compounds indirect analysis is possible using a reporter or label-based targets. However, the drawback is the

Mario Tiberi (ed.), *Dopamine Receptor Technologies*, Neuromethods, vol. 96,
DOI 10.1007/978-1-4939-2196-6_10, © Springer Science+Business Media New York 2015

inability to distinguish between the intact parent compound and any subsequent metabolites which can result in tracking the abundance and distribution of the label, not the target. For example, such a situation can occur when performing quantitative whole body autoradiography (QWBA), a technique accepted for regulatory submissions to authorities worldwide [1]. There are numerous other radiolabelled techniques that can be used to infer information indirectly. For example, dopamine receptors cannot be directly analyzed but can be studied by use of positron emission tomography (PET) imaging using C-11 radiolabelled dopamine ligands such as D2 dopamine receptor antagonist 3,5-dichloro-*N*-{[(2S)-1-ethylpyrrolidin-2-yl]methyl}-2-hydroxy-6-methoxybenzamide (raclopride) and the D1 dopamine receptor antagonist 7-chloro-3-methyl-1-phenyl-1,2,4,5-tetrahydro-3-benzazepin-8-ol (SCH 23390) [2, 3]. While such noninvasive techniques enable multiple, longitudinal, experiments to be performed, there does remain the issue that once again it is the label that is measured or tracked. A further significant obstacle is the complexity in generating, developing, and validating any such labelled probes. For endogenous targets such as the small-molecule neurotransmitter dopamine (DA), a biogenic amine, quantitative analysis has been limited to methods such as electrochemical analysis following tissue homogenization and separation by liquid chromatography (LC) and as a result there is a loss of all spatial information. Alternatively, researchers have relied on indirect histochemical, immunohistochemical (IHC), and ligand-based assays to detect these small-molecule transmitter substances [4–6]. For IHC the antibodies can often be limited in their ability to distinguish between different transmitters [7]. While there are drawbacks to label-based techniques, including those alluded to above and others (such as assays being a singleplex assessment per probe), once developed they do provide a valuable resource of neurological study and have revolutionized modern neuroscience [8–11].

For a better understanding of neurological disorders, as well as to enable the rapid and effective development of new therapies, the development of assays that allow untargeted and unlabelled assessment of endogenous and exogenous compounds is required. A technology that offers such untargeted and label-free analysis is mass spectrometry imaging (MSI) [12–17]. Traditional mass spectrometric analysis has enabled the study of a wide class of compounds and allows the differentiation between a target analyte, related metabolites, and endogenous molecular constituents of a sample. However, it has relied on the homogenization of the tissue prior to analysis, and hence all spatial distribution information is lost. Mass spectrometry imaging however performs discrete analysis directly from the sample surface and so spatial information is retained. A full mass spectrum can be collected with limited prior information so the analysis can be untargeted. Once the data set has been collected, it can then be interrogated and the distribution and abundance of any chosen molecular mass be mapped.

With a wide range of ionization sources coupled to an equally wide range of mass analyzers, there are multiple modalities capable of performing MSI. However, the most commonly used ionization source is matrix assisted laser desorption ionization (MALDI) [12, 18, 19]. Other effective but less commonly utilized sources include desorption electrospray ionization (DESI) [20–22], secondary ionization mass spectrometry (SIMS) [23, 24], and liquid extraction surface analysis (LESA) [25, 26]. MALDI analysis involves coating the tissue section with a matrix that co-crystallizes with the analytes in the sample and aids energy transfers from the ionizing laser to the sample. The spatial resolution of any experiment, irrespective of the ionization method, is determined by the dimensions of the sampling area and the distance between the sampling positions. For MALDI this is limited by the laser spot diameter and the distance that spot rasters (moves) between subsequent firings. MALDI mass spectrometry imaging (MALDI-MSI) has been successfully used for the detection and quantification of proteins, peptides, lipids, endogenous metabolites, and exogenous small molecules, often simultaneously, from a wide range of tissue samples [27–31]. Numerous reviews on the use of MSI have been published, with some focused on neuroscience [16, 32]. A broad overview of the applicability of MSI analysis for both exogenous compounds and endogenous metabolites is presented in a special issue of the Journal of Proteomics (Vol. 75, Issue 16, 2013) [33]. Of particular interest is the review by Prideaux and Stoeckli that summarizes drug and metabolite studies using MSI [34].

While MALDI data is not directly quantifiable, methods for determining tissue quantification have been developed. Approaches include the use off-line approaches such as adjacent tissue sections laser microdissection [35], or simultaneous on-line methods such as quantitation spotting of standards onto vehicle control tissue sections [36]. More complex strategies use homogeneously applied deuterated standards [37]. Reviews covering sample preparation have been published and explain in detail how to optimize sample methodologies to improve robustness and validity of data generated [38]. While only a simple protocol is required to perform basic MSI analysis, there has been extensive modifications and adaptations required to enable successful detection of certain targets. For MALDI-MSI detection of D1 and D2 receptor PET ligands raclopride and SCH23390 MALDI matrix selection and application method are crucial [35]. For certain endogenous compounds, optimization of sample preparation is insufficient to enable detection, and the use of on-tissue derivatization is required [39, 40]. The direct MALDI analysis of dopamine in rodent brain tissue sections required the development of an on-tissue derivatization strategy [41]. This chapter provides an overview of how researchers can perform MALDI MSI analysis of tissue sections with a comprehensive description of the methods associated with optimal sample collection and preparation protocols. The focus is on analysis of rodent brain tissue sections but the methods are adaptable to all tissue sections and we describe how researchers can obtain high quality validated MALDI MSI data. The protocol describes sample preparation for the analysis of unlabelled exogenous dopamine receptor ligands using both the standard solvent-based MALDI matrix application methods (for compounds such as olanzapine [42, 43] and solvent-free matrix application required for analysis of receptor ligands such as raclopride and SCH22390. The protocol also describes methods for direct MALDI analysis of endogenous dopamine in brain tissue sections.

2 Materials

The selection of animal model (species, strain, age, sex), drug treatment, or surgery should be determined by the researcher and no specific pre-experiment husbandry requirements are necessary unless required for experimental reasons. There are no constraints on method of drug administration and all routes are routinely used [intraperitoneal (i.p.), intravenous (i.v.), peroral (p.o.), subcutaneous (s.c.)] prior to MSI analysis. Animals must be housed, handled, acclimatized, and sacrificed as directed by national and local institutional ethical committees.

The MALDI mass spectrometer used for MSI analysis is typically determined by those locally available to the researcher but most commonly have a MALDI source coupled to a time-of-flight (TOF) mass analyzer. Increasingly, MALDI sources are however being fitted with a wide range of available mass analyzers (FTICR, Orbitrap, etc) to perform MSI with high mass accuracy and spectral resolution. The breadth and combination of laser ionization and mass analyzers systems is continuing to expand, with new and modified ionization systems enabling higher resolution or greater efficiency readily becoming available. Therefore, researchers should undertake specific research on the mass spectrometer to be employed for their research, and no variety, model, or manufacture specific detail is provided within this chapter. References to specific research contain details on the mass spectrometer systems employed.

2.1 Reagents

All water, solvents, and chemicals used should be of highest analytical grade available due to the sensitivity of the mass spectrometric analysis.

MALDI matrix α-cyano-4-hydroxycinnamic acid (CHCA): 10 mg/mL in 50 % acetonitrile 50 % water and 0.1 % trifluoroacetic acid (TFA).

Derivatization reagent 2,4-diphenyl-pyranylium tetrafluoroborate (DPP-TFB): 1 mg/mL in 100 % methanol.

Derivatization solution 150 μL of the derivatization reagent (DPP-TFB) 1.5 mL of 50 % methanol 50 % water, buffered by 1 μL triethylamine.

3 Methods

3.1 Sample Dissection

The method by which the animal is euthanized will affect subsequent results and cervical dislocation or anaesthetization with isoflurane prior to decapitation or dissection are the preferred methods for allowing rapid collection of brains [44]. In order to perform neurological studies it will often be necessary to dose and sacrifice over a short time period, often only minutes after i.v administration. As the time from sacrifice to dissection can significantly affect subsequent measured target abundance, particularly neuropeptides [45–47], care should be taken to allow a consistent period between each sacrifice and dissection. It is better to take slightly longer than necessary for all animals than have some procedures performed at different rates.

Following the animal sacrifice and the removal of the skull, the brain can be removed, with care taken while removing the brain from the meninges. This process requires experience and is not described extensively here. Detailed descriptions can be obtained from other Springer published methods [44]. Once removed, the brain should be placed on to a clean cutting plate. If sagittal sections are required for analysis, then the brain should be dissected down the midline using a medium size scalpel while gently supporting the brain with flat forceps or tweezers. If coronal sections are needed, then no further dissection is required. All tissues should then be immediately snap-frozen.

3.2 Sample Snap-Freezing

If no pre-freezing processing is required (*see* **Note 1**) there are three options for snap-freezing rodent brains. The first is free floating in liquid nitrogen. This should be avoided for whole rat brains as they are likely to fracture. Mouse brains however, due to their smaller size, fracture less often. The most reliable way to freeze without damage is to use isopentane (2-methylbutane) chilled on dry ice. This should be performed in a fume hood while using all appropriate personal protective equipment (including gloves and safety glasses). Sufficient isopentane to allow brain to free-float should be put in a disposable weigh boat or suitable container and placed on a bed of dry ice. Cubes of dry ice can be placed carefully into the liquid to aid chilling. Once chilled adequately the isopentane will stop bubbling when new dry ice flakes are added. This should be prepared prior to animal sacrifice and dissection. Whole brains can be placed in the solvent directly using forceps, while dissected hemispheres should first be placed onto a small folded piece of aluminum foil for support as this will help the brain maintain its shape and subsequently enable good quality and correctly orientated tissue sections to be cut (*see* **Note 2**). The third option is to snap-freeze using isopropanol (isopropyl alcohol). This is less toxic and volatile than isopentane and but does not reach such a low temperature as rapidly. More significantly, however, isopropanol leaves a solvent residue on the sample that needs to be evaporated prior to long-term storage or subsequent processing. Chilled isopropanol is an acceptable method if isopentane cannot be used. Once frozen, the solvent residue can be removed by blotting with lint-free tissue or by briefly holding under a stream of nitrogen.

Samples should then be placed in a labelled container, either masking tape labelled aluminum foil, plastic bag or sufficiently large plastic tube. These should be prechilled to prevent warming of the brain tissue. Samples should be transferred on dry ice and stored in a −80 °C freezer. It is not sufficient to store at −20 °C even for short periods of time [48]. Only fresh-frozen tissues should be used for MS imaging analysis and no formalin fixation or paraffin embedding fixation should be used (*see* **Note 3**).

3.3 Tissue Sectioning

The quality of any MSI data collected will be determined by the quality of the tissue sections and therefore every care should be taken to collect highest quality samples. Sections are cut using a cryostat microtome fitted with a roll plate and both the stage and chamber temperature should be set at approximately −16 to −18 °C. Lower temperatures can cause either tissue sections to crack or roll but individual optimization will be required. The cryostat should be extensively cleaned prior to use if previously used for traditional histology as the contamination of embedding media can cause ion suppression from the sample. Particular care should be taken to clean all surfaces with which the tissue samples come into contact.

Prior to commencing sample sectioning all slides should be labelled and prechilled to chamber temperature. For some MALDI mass spectrometers the use of metal or metal-coated microscope slides will be required and researchers should check with instrument manufacture guidelines prior to use. In addition to slides for MALDI analysis, all standard glass microscope slides should be prechilled if adjacent tissue sections are required for traditional histology. Slides should be in containers suitable for −80 °C storage. Typically microscope mailer boxes are used as they enable samples to be removed from storage without risking repeated thawing of other samples.

A brain to be sectioned should be transferred from freezer to cryostat on dry ice and placed into the chilled chamber using chilled forceps and allowed to equilibrate for a minimum of 5 min.

The orientation of sections is determined by experimental requirements but for sagittal sections it is usual to have the hemisphere orientated with the midline facing outward towards the operator. Again using forceps or tweezers, the brain should then be picked up in the required orientation, while distilled water is pipetted onto the surface of the interchangeable metal sample disks used to hold sample within the cryostat (Fig. 1). Approximately 300 μL of water should be used to mount a rat brain hemisphere and 150 μL is required for a mouse brain. The water should form a 1 cm diameter dome. The brain should then be quickly but gently placed into the water which will freeze in a matter of seconds. Additional water can be pipetted around the side of the brain to add additional support. Care should be taken to try and have no

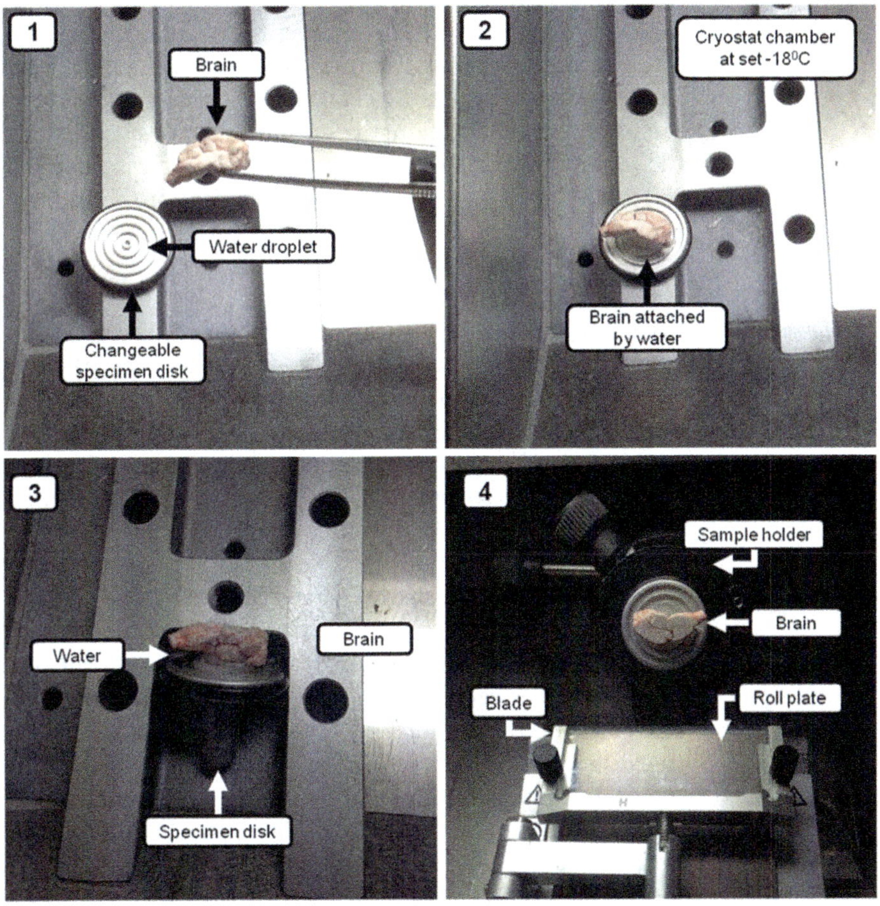

Fig. 1 Images of a sagittal rat brain being mounted using water as support media in a cryostat microtome. Multiple brains can be held in cryostat prior to sectioning. Labelled are major components of a standard cryostat. (1) Brain held in required orientation using tweezers before being placed into a droplet of water. (2) Brain placed onto water and left to freeze. (3) Water support holding brain in correct orientation. (4) Brain held in sample holder prior to sectioning

water above the plane that sections are to be cut. Optimal cutting temperature media (OCT) should not be used to mount samples as this, like the paraffin, causes contamination of both the cutting blade and sample surface. Such material is readily ionized in the mass spectrometer and causes extensive suppression of other analytes (*see* **Note 4**).

Different models of cryostat microtome are fitted with a range of cutting blade (Fig. 1.4), either disposable or non-disposable. Whenever possible, disposable low-profile blades should be used to the reduced risk of contamination. In addition, starting an experiment with a sharper blade makes it easier to cut better quality tissue sections. If using non-disposable blades then they should be cleaned prior to use. All cutting blades should be prechilled prior to starting the experiment. A roll guard should be used if fitted to cryostat as it prevents tissue rolling over itself as it is cut. It should be positioned in line with the cutting blade and the optimal position determined by small corrective movement forward and backwards.

The water mounted sample on the metal sample stage should then fixed into the cutting clamp (moving part of the cryostat) and orientated so that sections are cut from narrowest to widest part of the tissue. This means that as the tissue is cut it is forced under the roll guard by an increasingly wider section of tissue and reduces the chance of the tissue catching or creasing on the cutting plate (behind the blade), the roll guard or the blade itself.

3.3.1 Tissue Thickness

Tissue section thickness is crucial for optimal MSI analysis. Researchers have used sections ranging from 5 to 50 μm thick but the current consensus is that sections between 10 and 14 μm are optimal [49, 50]. This provides sufficient material for ionization while also enabling easy sectioning. Thinner tissue sections (<5 μm) can be difficult to produce when samples are not supported by embedding media. Ideally, tissue sections should be approximately one cell layer deep, and therefore thicker sections (>30 μm) can be unrepresentative of a single cell layer being mapped. It is not ideal to have a tissue depth greater than the width of the laser spot diameter. Trimming of the brain to the correct depth can be done by cutting thicker sections (>50 μm) and then reducing to thinner sections required for analysis. Once a section is cut it needs to be transferred to the sample slide as quickly as possible as if left in the cryostat chamber it will start to curl. The sections can be moved by gentle touching with a prechilled fine paint brush, a pencil, or any static-free tool. The prechilled microscope slide or MALDI targets should be positioned next to the cutting blade (with care) and the section gently pushed onto the slide. Once in the correct position a finger should be placed underneath the slide to warm and thaw-mount the brain section in place. The slide should then be returned to the cold surface of the cryostat chamber to refreeze the section in place. This process of positioning a section and thaw-mounting by warming from underside can be repeated until the required number of sections are in place. Care should be taken to try and not warm adjacent sections when placing others onto the slide. The wearing of a disposable paper mask during sectioning processes lets operator lean towards sample without having to hold breath and can reduce the chance of accidentally breathing into chamber and affecting section temperature.

It is worthwhile cutting sections for traditional histology, immunohistochemistry (IHC), or laser microdissection [51] directly preceding or immediately following those cut for MSI analysis to ease inter-assay comparison. Furthermore, as sections are often required from different depths of the brain the intervening sections can be collected for traditional homogenization and extraction analysis using pre-weighed and prechilled collection tubes. Multiple brains can be sectioned for the same experiment simply by mounting each sample individually and interchanging the brains within the microtome as required. Once all sections have been collected from a brain, it can be released from the metal sample stage by warming from underside and once the supporting ice has melted the sample can be returned, on dry ice, to the −80 °C freezer along with the tissue sections on slides that are not to be immediately analyzed. Remember that any control tissue sections will also need to be cut and thaw-mounted, especially as control tissue is required for quantitation experiments.

Once ready to perform analysis the samples should be taken from the −80 °C freezer, or directly from cryostat, and transferred on dry ice and then dried under a gentle stream of room temperature nitrogen for approximately 3 min. The sample can also be warmed by hand from underside during the drying process. Care should be taken to minimize condensation formation onto the slide surface and if this does occur, blowing water droplets over the tissue sections should be avoided, as this can cause analytes to be delocalized and contaminate other parts of the sample. Some practitioners prefer to desiccate from frozen for approximately 20 min. Once dried the sample can be stored in a desiccator for up to 48 h prior to any further sample processing.

3.4 Sample Preparation

There are a number of variations in sample preparation needed depending on analysis to be performed. The first is to decide if quantitation is required (Sect. 3.4.1) or if just relative abundance measurements are sufficient. To analyze dopamine ligands a solvent-free MALDI matrix may need to be applied (Sect. 3.4.2) but for other compounds a standard solvent-based matrix is required (Sect. 3.4.4). For dopamine to be analyzed directly on-tissue derivatization is required (Sect. 3.4.3) with subsequent standard solvent-based wet MALDI matrix applied (Sect. 3.4.4). It is always worth scanning samples prior to processing or matrix application to obtain an optical record of the samples prior to analysis.

This is quickly performed using a mid-range desktop scanner, though higher grade histology scanners can be used. Samples should be kept off the scanner bed surface by placing supporting microscope slide at each edge of the slide. Scanning should be performed after all drying of desiccation steps. Repeat scanning is also recommended following quantitation and matrix application as an aid to final presentation of the results obtained during the MSI experiment. It is possible to image after acquiring MSI data by removal of MALDI matrix (*see* **Note 6**).

3.4.1 Quantitation

A quantitative measurement of a target analyte requires the application of a calibration curve, spotted onto control tissue, and analyzed during the same experiment. Simultaneous analysis is required to mitigate any inter-analysis variation caused by factors such as fluctuations in matrix application or mass spectrometer performance. Typically a minimum of 5 calibration points, ranging over the anticipated abundance of the target in tissue, are spotted in a solvent and water solution (typically 50/50). The volume spotted (typically 0.2–0.5 μL per calibration point) should be allowed to dry prior to subsequent processing. Once calibration spots have been applied and allowed to dry the sample is best returned to the freezer for a minimum of 1 h. The sample should then be transferred on dry ice to a source of nitrogen and dried as previously described. This is to mitigate any effects that the wetting by the calibration solution has on matrix adhesion and crystallization. More complex methods of obtaining more accurate quantitation are possible but require the use of a deuterated standard [37, 43].

3.4.2 Solvent-Free Dry Matrix Application

The usual method of applying the matrix in a solvent based solution (Sect. 3.4.4) [14, 18, 38] does not produce sufficient analyte ionization and detection of some PET ligands, such as raclopride [35]. Therefore, the use of a solvent-free dry matrix is required. This method, initially used for detection of lipids [52] has also been successfully applied to the detection of small molecules [51]. The α-cyano-4-hydroxycinnamic acid (CHCA) matrix needs to be manually ground in a mortar and pestle until fine powder. The matrix should change color and become a light pale yellow. Several grams of the MALDI matrix can be ground at one time and stored until required (*see* **Note 5**). The matrix should be lightly re-ground immediately prior to use or reuse. The solvent-free dry matrix is applied using a small sieve (Fig. 2). Attempts to use fine micron pore sieves to apply only the finest particles prove ineffective due to constant blocking. The method works satisfactorily using a small domestic sieve, typically used for tea, purchased from a local store.

When ready to apply matrix, the slide should be removed from storage and dried under a gentle stream of nitrogen. This should be for approximately 30 s, with the sample warmed by hand from the underside. Excessive drying or desiccation is not required at

Fig. 2 MALDI matrix manually ground using a mortar and pestle prior to application of the fine powdered, using a sieve, over brain tissue sections

this stage. The slide should be placed on a piece of paper on a flat surface, ideally in a fume hood. Using a large spatula the ground CHCA matrix should be evenly dusted over the brain sections by holding the sieve 10 cm above the sections. The whole slide can be covered to excess and this should only take 30 s. The excess matrix should then be tipped back into the stock and the slide with the brain sections tapped gently with the spatula to dislodge any matrix not adhering to the tissue section. The slide and sections should then have nitrogen gently blown over to remove excess matrix. The samples should again be placed on paper on a flat surface and the process of matrix application and excess removal repeated a further three times. The sample is now ready for MALDI mass spectrometric analysis (Fig. 3).

3.4.3 On-Tissue Derivatization and Quantitation of Dopamine

For the direct detection of dopamine in situ, chemical derivatization is required to enable sufficient ionization. Pyrylium salts, e.g., 2,4-diphenyl-pyranylium tetrafluoroborate (DPP-TFB) are reacted with primary amines to produce *N*-alkyl or *N*-aryl-pyridinium derivatives [41]. The DPP-TFB derivatization reagent is dissolved in 100 % methanol to prepare a 1 mg/mL stock. The derivatization solution is then prepared using 150 μL of the derivatization stock in 1.5 mL of 50 % methanol 50 % water, buffered by 1 μL triethylamine. Brain tissue sections to be derivatized need to be taken from −80 °C storage and be desiccated for 15 min prior to further processing. The derivatization solution needs to be homogeneously applied to the brain tissue sections. Where available, this should be by an automated sprayer to reduce risk of variation in

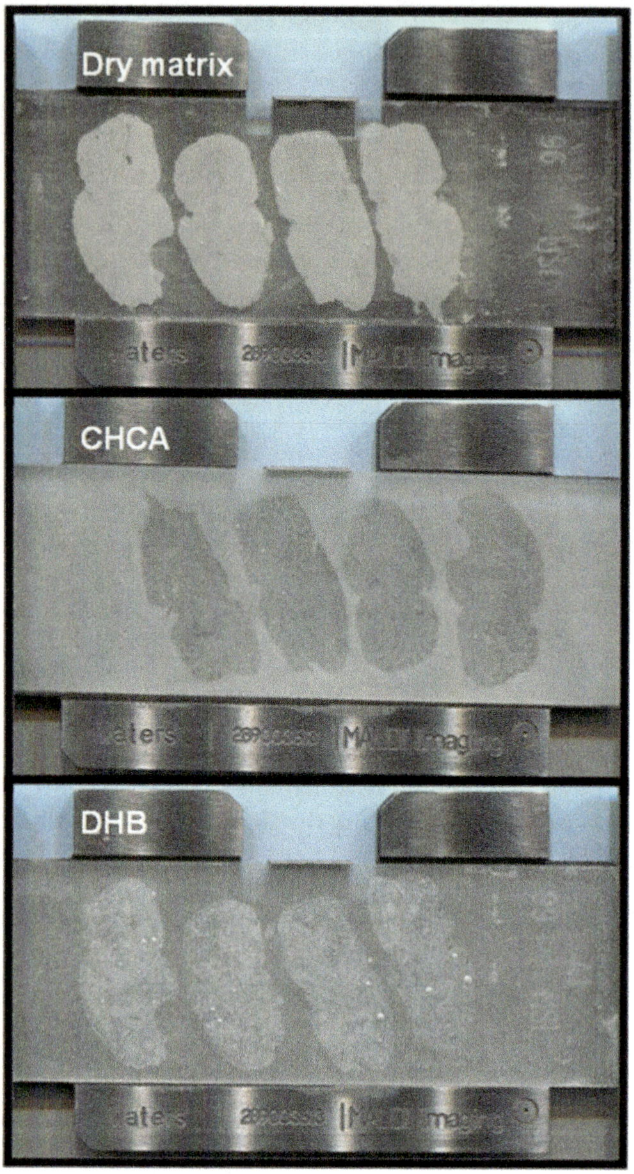

Fig. 3 Rat brain tissues sections prior to and post matrix application. (*Top*) Sagittal rat brain tissue sections (14 μm thick) thaw-mounted onto MALDI compatible glass slides with a solvent-free dry CHCA MALDI matrix application. Excess matrix removed by gentle tapping and a stream of nitrogen. Matrix is seen to adhere only to the tissue sections. (*Middle*) Sections following standard solvent-based wet CHCA MALDI matrix. The matrix is homogeneously applied over the entire slide though appears less thick on tissues. This is due to crystal size and formation on tissue compared to slide. (*Bottom*) Sections following DHB MALDI matrix applications. Homogeneous coating is of a lighter white color and appears thicker on tissue sections

application of the sample over surface. If not available, a manual airbrush or thin layer chromatography sprayer can be used. In all cases, a fine even mist needs to be generated that does not wet the surface and risk delocalization of analytes. In case of using an automatic sprayer (ImagePrep, Bruker Daltonics), the derivatization solution is applied in three stages, namely, a spraying cycle (2.5 s), an incubation period (15 s), and a drying period (50 s). The prepared slides are then incubated for 60 min in a chamber whose atmosphere is saturated with the vapor arising from a 50 % methanol solution. To acidify the tissue sections, 1 mL of a solution containing acetic acid, water, and methanol (10:45:45) can be sprayed over the tissue using the same set up as above. The treated tissue sections can then be incubated once more in the methanol vapor-saturated chamber for 30 min. For quantitative imaging deuterated dopamine should be spotted on a control tissue section to generate a calibration curve as previously described (Sect. 3.4.1).

3.4.4 Standard Solvent-Based Wet Matrix Application

Standard solvent-based matrix can also be applied to sections that have been removed from –80 °C storage on dry ice then dried under nitrogen. For manual application of the matrix, typically a 10 mL solution of 10 mg/mL CHCA (50 % acetonitrile, 50 % water, with 0.1 % TFA) is homogeneously sprayed onto the tissues sections using a airbrush or thin layer chromatography sprayer (TLC). The matrix is applied in multiple single passes, spraying at a distance of approximately 30 cm from the sample. Each pass should last for less than a second and there should be sufficient drying time prior to next spray past. Drying can be aided by passing slide under a stream of nitrogen. The spray has to be such a fine mist that it appears to the eye that no wetting is occurring or droplets are forming on each individual pass (Fig. 3). Over 10 passes should occur before an observable matrix build up should be detected. The whole manual application process should take approximately 20–30 min. Automated matrix application systems are available and apply the matrix as a homogeneous coating or as discreet spots and should be applied as described by manufacturer's instructions. A combination of matrix applications, derivatization or processing might be required experimentally and can be performed on the same sample set (*see* **Note** 7).

3.5 Analysis

Following tissue processing (Sect. 3.4) samples can be stored under vacuum in a desiccator and wrapped in aluminum foil for approximately 72 h without undue deterioration of the sample. However, whenever possible samples should be analyzed immediately. The MALDI mass spectrometer used to perform the analysis will be determined by what is available locally. It should be noted that both the MALDI source and mass analyzer will affect a number of experimental factors; (1) the spatial resolution of data collected, (2) the sensitivity of detection of the target analytes, (3) the spectral resolution of the masses detected, (4) the speed at which the analysis is performed. Users will need to seek guidance on their specific mass spectrometer. However, there are a number of variables that are applicable to most systems when deciding on experiment parameters, now discussed briefly. Analysis can be either be a full spectrum analysis (Sect. 3.5.1) or selection of a target mass which is subsequently fragmented (MS/MS) (Sect. 3.5.2). Each mode has advantages; full spectrum analysis enables the simultaneous measurement of the target analyte and other endogenous or exogenous targets detectable during ionization. However, as other endogenous compounds may have the same mass as the target compound they can contribute to the abundance of the selected mass and produce a non-representative abundance or distribution image. To confirm the identification of a target mass requires the isolation and fragmentation of the compound and subsequent detection of the fragments of that molecule. This fragmentation MS/MS analysis confirms the identification and distribution of a target but does not allow simultaneous analyte detection. Quantification is ideally performed using MS/MS analysis to prevent any apparent abundance arising from background masses. It is ideal to perform both full spectrum MS analysis and fragmentation MS/MS analysis on adjacent brain sections to validate the data collected.

3.5.1 Standard Full Spectrum Analysis

MALDI analysis can be performed in positive or negative ionization mode, depending on the target analyte. For the exogenous analytes described here, positive ionization is typically performed resulting in detection of the $[M+H]^+$ ion. Positive ionization is also used for dopamine, detected as $[M]^+$. Optimization of mass spectrometer settings is best performed initially on manually spotted standards prepared under previously described methods for quantitation spots. Laser power, repetition rate, movement, mass analyzer specific trapping voltages or gases can then be “tuned” to optimum values. These optimized settings should then be applied to test tissue sections from drug treated brains. Once settings are optimized they should be fixed for the entire experiment, with any auto tune or laser power ramping options deactivated (*see* **Note 8**).

3.5.2 Fragmentation Analysis

As for standard full spectrum analysis, MS/MS target confirmation should have settings optimized by firstly using manually applied standards before moving onto tissue. This enables the appropriate fragmentation peaks to be identified resulting from the target analyte. For analytes described here, typically positive ionization is required and selection of the $[M+H]^+$ for fragmentation.

3.5.3 Data Processing

Due to fluctuation in individual spectra and in total ion count it is typical to perform some form of spectra baseline subtraction, peak picking and data normalization. This is instrument specific and beyond the scope of this chapter and researcher should refer to manufacture recommendations and seek expert user guidance.

3.6 Limitations

It is worth noting that there are limitation in the selectivity and sensitivity of the various ionization methods and mass analyzers employed for MSI analysis. The inability to detect a target endogenous or exogenous compound does not mean that a target is not present within the sample just that it was below the limit of detection of the analysis performed. Therefore, researchers should take particular care when interpreting data that indicate absence of a target. Weight can be added to MSI data where the absence of a target is recorded by use of additional bioanalytical assays, such as tissue homogenization and LC-MS quantitation or probe based assays. References provided throughout this chapter detail how researchers have validated the methods described, and the caveats associated with the data presented.

4 Conclusion

Using mass spectrometry imaging to perform label-free and multiplex analysis of compounds directly from brain tissue sections is a powerful new tool for neurobiological and pharmaceutical research. The protocols described here, by which reproducible and quantifiable data can be collected, are a starting point from which researchers can modify, refine and apply to a wide range of endogenous and exogenous targets. The technologies by which the analysis is performed will continue to evolve and improve but the requirement to collect, prepare and process the samples prior to analysis will continue to be a crucial component of any successful study.

5 Notes

1. Researchers may want to consider using heat-stabilization to deactivate endogenous proteases [47, 48]. This can aid proteomic and peptidomic analysis in tissue where enzymatic activity can cause large fluctuations in apparent analyte abundance. While no direct degradation of target analyte may occur when analyzing exogenous compounds, there can remain a risk of increased analyte suppression by such activity in non-stabilized tissues.
2. It is important that samples are frozen free floating or carefully supported, they should never be placed into tubes prior to freezing as the room temperature tissue will deform and take the shape of the container. This then limits the ability to cut anatomically determinable sections and can often prevent the

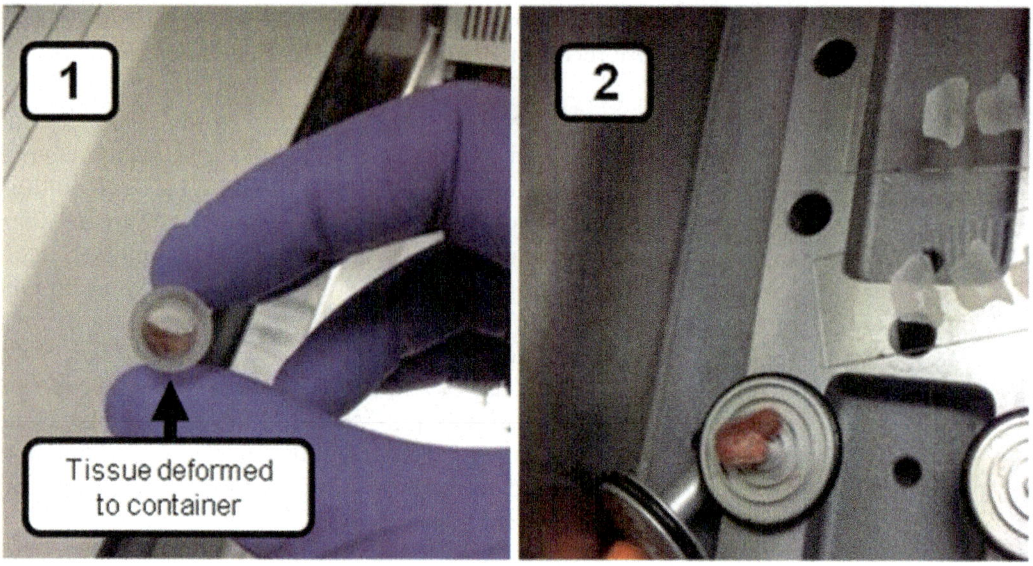

Fig. 4 Tissue deforming to shape of container. (1) Tissue was placed into container prior to snap-freezing and curved rather than retaining its original shape. (2) Highlighting difficulty in removing tissue from container and the deformation that causes poor quality tissue sections

sample being retrieved from the container without undue warming or even cutting (Fig. 4).

3. The process of formalin fixation and paraffin embedding (FFPE) is a method for traditional histology and is an effective way of preserving tissues. Unfortunately it is not compatible with standard MSI experiments as the formaldehyde causes disulfide cross-linking that prevents subsequent proteomic analysis. Small molecule analysis is prevented because the fixative solution can cause analyte delocalization. The process of paraffin embedding also has to be avoided as it contaminates the tissue section surface and suppresses ionization of target analytes. However, even with these caveats, researchers have developed methods that enable some analysis to be performed on tissues treated with FFPE to enable access to extensive archive biobanks. The harsh processing protocols act to enable antigen retrieval and subsequent MSI analysis [53].
4. A number embedding media are compatible with MSI if required. Gelatine, agarose and carboxymethyl cellulose are traditional media and have been successfully utilized, though more bespoke polymers have been developed specifically for MSI analysis [54].
5. It has been found that recently purchased CHCA matrix, when ground, does not adequately adhere to the brain tissue sections. CHCA matrix that has been stored at –20 °C for several

months does work. This is probably due to hydration of the matrix. This process can be speeded up for new CHCA matrix by removing the container lid and gently replacing it before storing in a −20 °C freezer. The matrix should then be taken out and have the lid removed to allow condensation to occur before returning to the freezer. Repeated daily for a week and combined with occasionally leaving the matrix at room temperature, this process has been found effective at aging the matrix sufficiently. The aged matrix should be darker mustard yellow prior to use.

6. While it is best to acquire optical images of samples prior to analysis and perform traditional histology and staining of adjacent tissue sections it is possible to remove the MALDI matrix following analysis and processing for subsequent image analysis such as hematoxylin and eosin (H and E) staining [36]. The slide with the sample should be rinsed by gently placing in a microscope slide washing rack or suitably large glass container containing 100 % ethanol for 3–5 s before removing and gently placing in a second bath. Excess ethanol should be gently shaken off and the slide dried under a stream of nitrogen. Care should be taken to not shake slide excessively while in ethanol bath as it risks removing the tissue sections from the slide surface.
7. Different processing can be performed to brain sections on the same slide by masking some tissue sections while processing the others (matrix coating or derivatization). Thaw-mounting the tissue sections onto slide a sufficient distance apart (a minimum of approximately 7.5 mm) allows 5 cm of Parafilm, (a plastic paraffin film) stretched and rolled to form a string to be wrapped around the slide between the tissue sections. A second piece of stretch film can then be placed the other side of the sections to be masked. A piece of non-stretched Parafilm can then be carefully used to mask samples, supported by the stretched film, without coming into contact with the tissues. Gentle pressing of the stretch and non-stretch film makes an effective temporary water and solvent barrier.
8. Some MALDI sources have fixed laser spot diameters but can generate higher spatial resolution images by setting the raster movement to a value lower than the laser spot diameter. Other MALDI sources have a variable laser spot diameter. The larger the laser spot diameter the greater the ionization area and usually the greater the sensitivity of analyte detection. Therefore, higher spatial resolution can result in decreased sensitivity of analyte detection if a smaller diameter laser spot is used or if oversampling is employed. An increase in spatial resolution from 100 μm to 50 μm will result in acquisition time increasing fourfold while the area ionized at each sampling positions

will be a reduced to a quarter. Therefore, the high spatial resolution imaging (~10 μm) should be focused to small regions of interest and not for multiple sections analysis.

9. An apparent loss of mass spectrometer sensitivity during an experimental is a common occurrence during MSI experiment. The abundance may appear sufficient at the point of pre-experimental optimization of mass spectrometer but be found to have gradually decreased during the run. This is typically due to the MALDI source of the mass spectrometer becoming heavily contaminated during previous experiments. The ionization processes causes the MALDI matrix on the tissue to coat the internal surfaces of the mass spectrometer. This is a gradual process that takes weeks to occur when operating the mass spectrometer in the standard spotted sample workflow. However, when operated in imaging mode the thicker matrix layer, higher laser energies and repetition rates means this source dirtying can occur within 24 h of acquisition.
10. If matrix appears to be either preferentially or insufficiently adhering to quantitation spots then it is an indication that the sample has not had an adequate amount of drying time. An alternative approach is to return sample to −80 °C freezer following quantitation spots added and leave for several hours. Once sample is subsequently processed as usual there should be even matrix adhesion.
11. If matrix crystals are too large and non-homogeneous it is an indication that the matrix solution has been applied too wet. Try increasing the drying time between matrix application passes or increase the gas flow.

References

1. Solon EG, Kraus L (2001) Quantitative whole-body autoradiography in the pharmaceutical industry: survey results on study design, methods, and regulatory compliance. J Pharmacol Toxicol Methods 46(2):73–81
2. Farde L, Hall H, Sedvall G (1986) Quantitative analysis of D2 dopamine receptor binding in the living human brain by PET. Science 17(231):258
3. Farde L et al (1987) PET analysis of human dopamine receptor subtypes using 11C-SCH 23390 and 11C-raclopride. Psychopharmacology (Berl) 92(3):6
4. Falck B et al (1962) Fluorescence of catechol amines and related compounds condensed with formaldehyde. J Histochem Cytochem 10(3):348
5. Jones BE, Beaudet A (1987) Distribution of acetylcholine and catecholamine neurons in the Cat brain-stem—a choline-acetyltransferase and tyrosine-hydroxylase immunohistochemical study. J Comp Neurol 261(1):15–32
6. de Jong LA et al (2005) Receptor-ligand binding assays: technologies and applications. J Chromatogr B Analyt Technol Biomed Life Sci 829(1–2):1–25
7. Keenan C, Hoopowitz H (1981) Limitations in identifying neurotransmitters within neurons by fluorescent histochemistry techniques. Science 4(214):1151
8. Chen W (2007) Clinical applications of PET in brain tumors. J Nucl Med 48(9):1468–1481
9. Niccolini F, Su P, Politis M (2014) Dopamine receptor mapping with PET imaging in Parkinson's disease. J Neurol 1–13
10. Talavage TM, Gonzalez-Castillo J, Scott SK (2014) Auditory neuroimaging with fMRI and PET. Hear Res 307:4–15

11. Pichler BJ et al (2010) PET/MRI: paving the Way for the next generation of clinical multimodality imaging applications. J Nucl Med 51(3):333–336
12. Hanrieder J et al (2013) Imaging mass spectrometry in neuroscience. ACS Chem Neurosci 4(5):666–679
13. Goodwin RJA, Pitt AR (2010) Mass spectrometry imaging of pharmacological compounds in tissue sections. Bioanalysis 2(2):279–293
14. Stoeckli M, Farmer T, Caprioli R (1999) Automated mass spectrometry imaging with a matrix-assisted laser desorption ionization time-of-flight instrument. J Am Soc Mass Spectrom 10(1):67–71
15. Hanrieder J et al (2013) Time-of-flight secondary Ion mass spectrometry based molecular histology of human spinal cord tissue and motor neurons. Anal Chem 85(18):8741–8748
16. Ye H et al (2013) Visualizing neurotransmitters and metabolites in the central nervous system by high resolution and high accuracy mass spectrometric imaging. ACS Chem Neurosci 4(7):1049–1056
17. Caprioli RM (2014) Imaging mass spectrometry: molecular microscopy for enabling a New Age of discovery. Proteomics 14(7–8):807–809
18. Amstalden van Hove ER, Smith DF, Heeren RM (2010) A concise review of mass spectrometry imaging. J Chromatogr A 1217(25): 3946–3954
19. Minerva L et al (2012) MALDI MS imaging as a tool for biomarker discovery: methodological challenges in a clinical setting. Proteomics 6(11–12):581–595
20. Bennet RV et al (2013) Imaging of biological tissues by desorption electrospray ionization mass spectrometry. J Vis Exp 77:e50575
21. Eberlin LS et al (2011) Desorption electrospray ionization mass spectrometry for lipid characterization and biological tissue imaging. Biochim Biophys Acta 1811(11):946–960
22. Vickerman JC (2011) Molecular imaging and depth profiling by mass spectrometry-SIMS, MALDI or DESI? Analyst 136(11):2199–2217
23. Weaver EM, Hummon AB (2013) Imaging mass spectrometry: from tissue sections to cell cultures. Adv Drug Deliv Rev 65(8):1039–1055
24. Bich C, Touboul D, Brunelle A (2014) Cluster TOF-SIMS imaging as a tool for micrometric histology of lipids in tissue. Mass Spectrom Rev. 33(6):442–451.
25. Ellis SR et al (2013) Surface analysis of lipids by mass spectrometry: more than just imaging. Prog Lipid Res 52(4):329–353
26. Blatherwick EQ et al (2011) Utility of spatially-resolved atmospheric pressure surface sampling and ionization techniques as alternatives to mass spectrometric imaging (MSI) in drug metabolism. Xenobiotica 41(8):720–734
27. Francese S et al (2013) Curcumin: a multipurpose matrix for MALDI mass spectrometry imaging applications. Anal Chem 85(10): 5240–5248
28. Jones EE et al (2014) MALDI imaging mass spectrometry profiling of proteins and lipids in clear cell renal cell carcinoma. Proteomics 14(7–8):924–935
29. Bradshaw R et al (2012) Separation of overlapping fingermarks by Matrix Assisted Laser Desorption Ionisation Mass Spectrometry Imaging. Forensic Sci Int 222(1–3):318–326
30. Ronci M et al (2013) MALDI MS imaging analysis of apolipoprotein E and lysyl oxidase-like 1 in human lens capsules affected by pseudoexfoliation syndrome. J Proteome 82:27–34
31. Reyzer ML et al (2003) Direct analysis of drug candidates in tissue by matrix-assisted laser desorption/ionization mass spectrometry. J Mass Spectrom 38(10):1081–1092
32. Shariatgorji M, Svenningsson P, Andren PE (2014) Mass spectrometry imaging, an emerging technology in neuropsychopharmacology. Neuropsychopharmacology 39(1):15
33. McDonnell L, Andrén PE, Corthals GL (2012) Preface. J Proteome 75(16):4881–4882
34. Prideaux B, Stoeckli M (2012) Mass spectrometry imaging for drug distribution studies. J Proteome 75(16):4999–5013
35. Goodwin RJA et al (2011) Qualitative and quantitative MALDI imaging of the positron emission tomography ligands raclopride (a D2 dopamine antagonist) and SCH 23390 (a D1 dopamine antagonist) in Rat brain tissue sections using a solvent-free Dry matrix application method. Anal Chem 83(24):9694–9701
36. Nilsson A et al (2010) Fine mapping the spatial distribution and concentration of unlabeled drugs within tissue micro-compartments using imaging mass spectrometry. PLoS ONE 5(7): e11411
37. Källback P et al (2012) Novel mass spectrometry imaging software assisting labeled normalization and quantitation of drugs and neuropeptides directly in tissue sections. J Proteome 75(16):4941–4951
38. Goodwin RJA (2012) Sample preparation for mass spectrometry imaging: small mistakes can lead to big consequences. J Proteome 75(16):4893–4911
39. Chacon A et al (2011) On-tissue chemical derivatization of 3-methoxysalicylamine for MALDI-imaging mass spectrometry. J Mass Spectrom 46(8):840–846
40. Cobice DF et al (2013) Mass spectrometry imaging for dissecting steroid intracrinology

within target tissues. Anal Chem 85(23): 11576–11584

41. Shariatgorji M, Nilsson A, Goodwin R, Zhang X, Schintu N, Svenningsson P, Andren PE (2013) MALDI-MS imaging and quantitation of primary amine neurotransmitters dopamine, GABA and glutamate directly in brain tissue sections. Proceedings for the 61st American Society for Mass Spectrometry Annual Conference. Minneapolis, MN, 9–13 June.
42. Khatib-Shahidi S et al (2006) Direct molecular analysis of whole-body animal tissue sections by imaging MALDI mass spectrometry. Anal Chem 78(18):6448–6456
43. Hamm G et al (2012) Quantitative mass spectrometry imaging of propranolol and olanzapine using tissue extinction calculation as normalization factor. J Proteome 75(16): 4952–4961
44. Spijker S (2011) Dissection of rodent brain regions. In: Li KW (ed) Neuroproteomics. Humana Press, Amsterdam, pp 13–26
45. Sturm RM et al (2012) Mass spectrometric evaluation of neuropeptidomic profiles upon heat stabilization treatment of neuroendocrine tissues in crustaceans. J Proteome Res 12(2): 743–752
46. Goodwin RJA et al (2008) Time-dependent evolution of tissue markers by MALDI-MS imaging. Proteomics 8(18):3801–3808
47. Goodwin RJA et al (2010) Stopping the clock on proteomic degradation by heat treatment at the point of tissue excision. Proteomics 10(9):1751–1761
48. Goodwin RJA, Iverson SL, Andren PE (2012) The significance of ambient-temperature on pharmaceutical and endogenous compound abundance and distribution in tissues sections when analyzed by matrix-assisted laser desorption/ionization mass spectrometry imaging. Rapid Commun Mass Spectrom 26(5):494–498
49. Gregson C (2009) Optimization of MALDI tissue imaging and correlation with immunohistochemistry in rat kidney sections. Biosci Horizons 2(2):134–146
50. Seeley EH et al (2008) Enhancement of protein sensitivity for MALDI imaging mass spectrometry after chemical treatment of tissue sections. J Am Soc Mass Spectrom 19(8):1069–1077
51. Goodwin RJA et al (2010) Use of a solvent-free Dry matrix coating for quantitative matrix-assisted laser desorption ionization imaging of 4-bromophenyl-1,4-diazabicyclo(3.2.2)nonane-4-carboxylate in Rat brain and quantitative analysis of the drug from laser microdissected tissue regions. Anal Chem 82(9):3868–3873
52. Puolitaival S et al (2008) Solvent-free matrix dry-coating for MALDI imaging of phospholipids. J Am Soc Mass Spectrom 19(6): 882–886
53. Casadonte R, Caprioli RM (2011) Proteomic analysis of formalin-fixed paraffin-embedded tissue by MALDI imaging mass spectrometry. Nat Protoc 6(11):14
54. Strohalm M et al (2011) Poly[N-(2-hydroxypropyl)methacrylamide]-based tissue-embedding medium compatible with MALDI mass spectrometry imaging experiments. Anal Chem 83(13):5458–5462

Chapter 11

Positron Emission Tomography Imaging of Dopaminergic Receptors in Rats

Boguslaw Szczupak and Abraham Martín

Abstract

Positron emission tomography (PET) is an imaging technique able to provide detailed spatial-temporal functional data of the cerebral neurotransmission system. The dopaminergic pathway has been largely characterized with PET imaging due to the existence of large number of radiotracers available to bind to the different targets of this system. Thus, the selection of the radiotracer depends on the particular aspect of the dopaminergic system to be studied. In this chapter, a methodological PET study of the D_2/D_3 receptor bioavailability with [^{11}C]raclopride is described. Likewise, the preparation steps for the synthesis of [^{11}C]raclopride as well as the protocols of PET/CT acquisition, reconstruction, and quantification of the data are presented.

Key words Positron emission tomography (PET), Dopaminergic neurotransmission, [^{11}C]Raclopride, Computed tomography (CT)

1 Introduction

Positron emission tomography (PET) has played an important role in the in vivo characterization of the dopaminergic neurotransmission [1–5]. PET provides high sensitivity achieved by the administration of trace doses of selective and high-affinity radioligands to the evaluation of the dopaminergic system. A large number of PET radiopharmaceuticals have been developed to bind to the different targets of the dopaminergic neurotransmission: (1) L-DOPA decarboxylase (the enzyme that produces dopamine from L-DOPA), (2) storage vesicles (vesicular monoamine transporter), (3) presynaptic transporters, and (4) postsynaptic receptors [6] (Fig. 1). The main dopaminergic pathway is the nigrostriatal that originates in the substantia nigra pars compacta and projects to the striatum. Dopaminergic neurons are also found in groups of cells in the ventral tegmentum of the midbrain and dorsal hypothalamus projecting to the cerebral cortex, limbic areas, and spinal cord [7]. PET radiotracers that measure dopamine synthesis and transport

Mario Tiberi (ed.), *Dopamine Receptor Technologies*, Neuromethods, vol. 96,
DOI 10.1007/978-1-4939-2196-6_11, © Springer Science+Business Media New York 2015

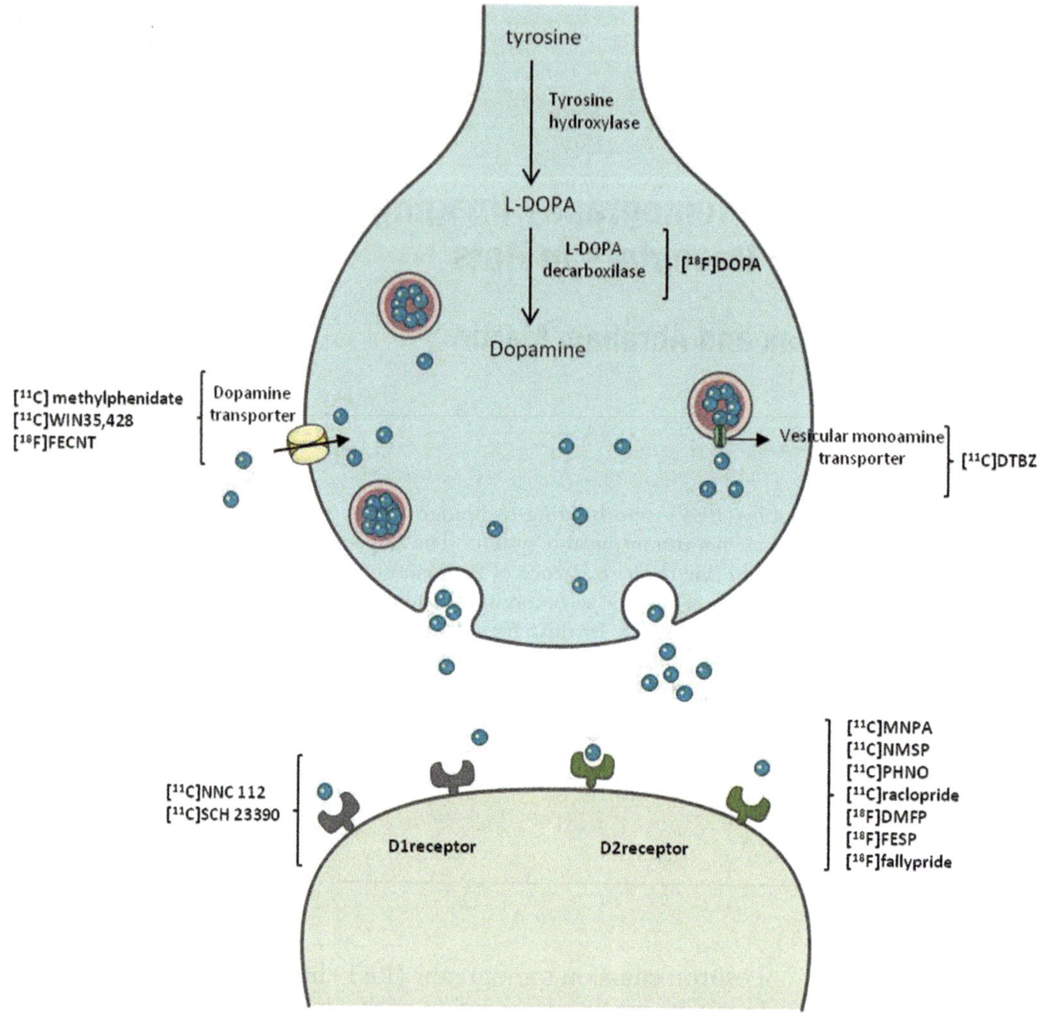

Fig. 1 The diagram shows the dopaminergic synapse. The tyrosine is transformed to L-dihydroxyphenylalanine (L-DOPA) and then to free dopamine. The vesicular monoamine transporter introduces the dopamine into vesicles. The dopamine remains stored until a nerve impulse promotes the release of vesicular dopamine into the synaptic cleft. Dopamine binds to the dopamine receptors (D_1R or D_2R), and the excess is transported back to the presynaptic neuron through dopamine transporters. WIN35,428 (2β-carbomethoxy-3β-(4-fluorophenyl)-tropane); FECNT (2β-carbomethoxy-3β-(4-chlorophenyl)-8-(2-fluoroethyl)nortropane); DTBZ (dihydrotetrabenazine); NNC112 ((+)-5-(7-benzofuranyl)-8-chloro-7-hydroxy-3-methyl-2,3,4,5-tetrahydro-1H-3 benzazepine); SCH 23390 ((R)-(+)-8-chloro-2,3,4,5-tetrahydro-3-methyl-5-phenyl-1H-3-benzazepin-7-ol); MNPA ((R)-2-CH_3O-*N*-n-propylnorapomorphine); NMSP (*N*-methylspiperone); PHNO (4-propyl-9-hydroxynaphthoxazine); DMFP (dimethoxyfallypride); FESP (fluoroethylspiperone). The diagram was produced using Servier Medical Art (www.servier.com)

have been developed to evaluate the presynaptic function of the dopaminergic system. The most commonly used PET radiotracers are [^{18}F]DOPA [8] to measure presynaptic aromatic amino acid decarboxylase, [^{11}C]DTBZ [9] to target the type 2 vesicular monoamine transporter (VMAT-2), and [^{11}C]methylphenidate [10],

[^{11}C]WIN35,428 [11], and [^{18}F]FECNT [12] to assess expression and activity of dopamine transporter. Postsynaptic dopaminergic radioligands can target either the D_1-like (D_1R, D_5R) or D_2-like (D_2R, D_3R, D_4R) receptors. The radioligands that evaluate the D_1 receptor availability are [^{11}C]NNC 112 [13] and [^{11}C]SCH 23390 [14], and those for the D_2/D_3 receptors are [^{11}C]MNPA [15], [^{11}C]NMSP [16], [^{11}C]raclopride [17], [^{18}F]DMFP [18], [^{18}F] FESP [19], [^{18}F]fallypride [20], and [^{11}C]PHNO [21], among others (Fig. 1). Therefore, the choice of the radioligand with which to assess a particular target of the neurotransmitter system depends on the aspects to be studied. In the present chapter, a methodological PET study of the dopaminergic receptors in rats has been performed using [^{11}C]raclopride because it is a widely used radiotracer to study changes in D_2/D_3 receptors [22]. The essential preparation steps of [^{11}C]raclopride and corresponding PET imaging acquisition protocols are presented together with their resulting PET/CT fusion rat brain images and quantification.

2 Materials

2.1 Radiochemistry

2.1.1 IBA Cyclone 189 Cyclotron (IBA Molecular, Belgium)

For the production of [^{11}C] raclopride, [^{11}c]CH_4 was directly generated in an IBA Cyclone 189 Cyclotron.

2.1.2 Lead-Shielded Automated Synthesis Module

Once the activity is generated in the cyclotron, it is transferred to the automated synthesis module, TRACERlab FXCPro (GE Healthcare, WI, USA). The module is placed inside a lead-shielded hot cell (Commercer, Italy).

2.1.3 Quality Control Equipment

Radio-HPLC (Agilent Technologies, Spain).

2.1.4 Shielded Syringes and Activimeter (Commercer, Italy)

2.2 Animal Preparation and Monitoring

2.2.1 Anesthesia

1. Isoflurane and oxygen.
2. Isoflurane-/oxygen-based anesthesia system fitted with an induction chamber and inhalation masks for rats (La Bouvet, Spain).

2.2.2 Preparation for Radiotracer Administration

1. Infrared heating lamp.
2. 24-G catheter, 1 ml syringes (BD insyte, Spain).
3. Physiologic saline: 0.9 % NaCl.

2.2.3 Preparation for Animal Monitoring

1. Respiration monitoring system: small animal monitoring and gating system (Mo10255; SA Instruments, Inc, USA).
2. Temperature control and monitoring:
   (a) Water-based heating pad and heating pump to maintain temperature (Huber UK, UK).
   (b) Rectal probe (SA Instruments, Inc, USA).
   (c) Monitoring software (SA Instruments, Inc, USA).

2.3 PET/CT Imaging Instrumentation, Acquisition, Reconstruction, and Analysis Software

2.3.1 PET/CT Imaging Instrumentation

PET/CT images are acquired using the eXplore Vista-CT dedicated preclinical imaging system (GE Healthcare, Waukesha, USA). The system provides functional imaging (PET) along with anatomical images (CT) within a single instrument (Fig. 2). The 7 cm bore is suitable for mice and rats up to 400 g. Both static and dynamic studies are possible with time-uptake analysis.

2.3.2 PET/CT Image Reconstruction Software

PET and CT images are acquired and reconstructed by the MMWKS Vista-CT software provided by the scanner manufacturer. The software's features include FBP and 2DOSEM reconstruction, random, scatter, and CT-based attenuation correction. The software was installed on a Dell Precision 690 workstation with 1.6 GHz CPU and 8 GB RAM.

2.3.3 PET/CT Image Analysis Software

PMOD software version 3.4 (PMOD Technologies, Zurich, Switzerland) with image fusion and general kinetic modeling tools is used for image quantification. The software was installed on Hewlett-Packard ZR 420 workstation equipped with 3.6 GHz CPU and 16 GB RAM memory.

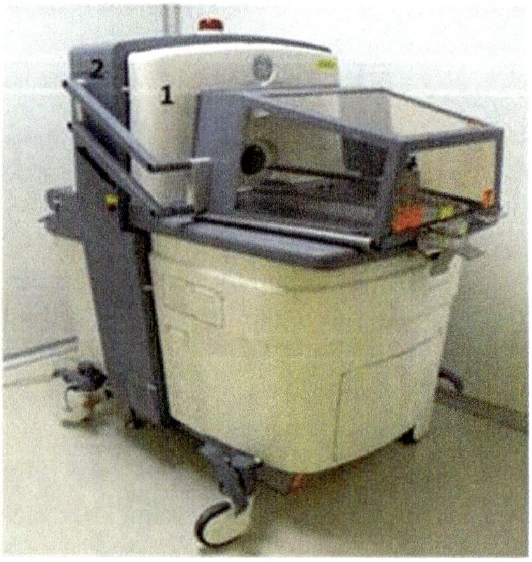

Fig. 2 eXplore Vista-CT scanner with PET compartment (*1*) in the front and CT (*2*) in the back

3 Methods

3.1 Preparation of [^{11}C]Raclopride

1. Generate [^{11}C]CH_4 using the nuclear reaction ^{14}N(p,He)^{11}C in 5 % H2 in an IBA Cyclone 18/9 cyclotron and transfer to a TRACERlab FXC Pro synthesis module (GE Healthcare, Waukesha, WI, USA) to generate [^{11}C]methyl iodide.
2. Distill [^{11}C]CH3I under continuous helium flow (20 ml/min) and introduce in a 2 ml stainless steel reaction loop, pre-charged with a solution of O-desmethyl-raclopride (free base, 1 mg, ABX) in dimethylsulfoxide (80 ml) and aqueous 5 M sodium hydroxide solution (3 ml).
3. Purify the reaction mixture by means of high-performance liquid chromatography.
4. Formulate the collected fraction by retention on a C-18 cartridge (Sep-Pak Light, Waters, Milford, MA, USA).
5. Elute with ethanol (1 ml) and saline (9 ml).
6. Filtrate twice with 0.22 mm sterile filters to obtain the final [^{11}C]raclopride solution.
7. In our study [17], typical radiochemical yields and specific activities were 51.3 ± 11.2 % (end of bombardment) and 109 ± 20 GBq/μmol (end of synthesis), respectively.

3.2 PET/CT Scanner Preparation

1. Set up the scanner for the acquisition according to the user's manual.
2. Perform a blank scan (Note 1) to check the sensitivity and correct functionality of the scanner. Compare the results with the specification and previous readings.
3. Attach the rat gas anesthesia mask with associated tubing, heating pad, and animal monitoring system to the scanner bed.
4. Cover the heating pad with the blotter (Fig. 3) and set the temperature of the heating pad to 37–39 °C (Note 2).
5. Set up the PET protocol for dynamic acquisition by defining the number of frames and their durations (Note 3). In our study [17], 18 frames (4×5 s, 2×30 s, 2×60 s, 2×150 s, 2×250 s, 2×450 s, and 2×600 s) were acquired, with a total acquisition time of 63 min.

3.3 Animal Preparation and Transfer to the PET/CT Scanner

1. Obtain the Institution's animal care and ethic committee's approval (prior to performing studies in animals).
2. Anesthetize the rat with 2–2.5 % isoflurane in 100 % O_2 in an anesthesia induction chamber and maintain by 1–2 % of isoflurane with anesthetic mask.
3. Heat the tail with an infrared lamp during 3–4 min to dilate the tail veins.

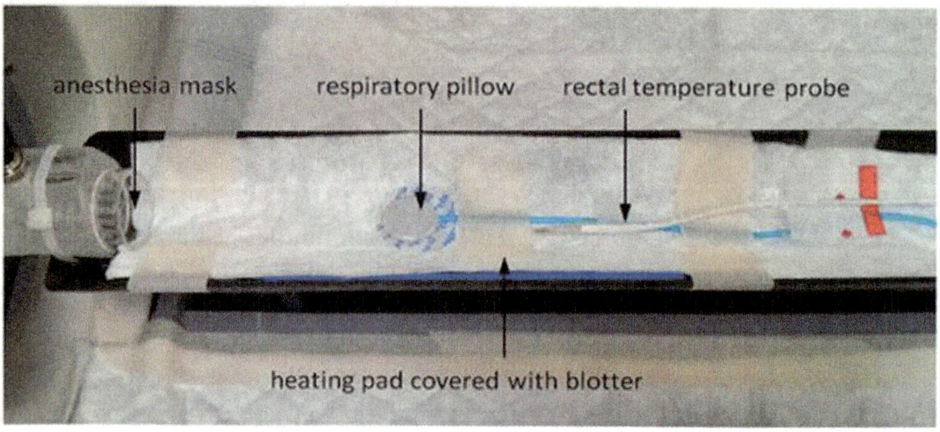

Fig. 3 Setup of the scanner bed for the PET/CT experiment

4. Place a 24-G catheter in the tail vein for intravenous administration. This is the most common way for radiotracer administration in rats.
5. Fix the catheter to the tail with surgical tape.

3.4 PET/CT Imaging

1. Transfer the rat to the scanner bed.
2. Fix the animal in prone position using surgical tape and attach vital sign monitoring probes (Note 4). Position the respiratory pillow sensor near the abdomen of the rat and secure it with surgical tape (Fig. 3). Apply a small amount of lubricant over the temperature probe before insertion into the rectum.
3. In order to adjust precisely the scanning area, perform a scout acquisition. Place the field of view (FOV) of the scanner over the entire brain area (Note 5; Fig. 4).
4. Start the PET acquisition simultaneously with the radiotracer injection (Note 6; Fig. 5). Inject around 20 MBq of [^{11}C] raclopride in a volume of 0.2–0.3 ml as a bolus and flush it with 0.1 ml of saline (Note 7). Note the time of injection (acquisition start).
5. In order to exactly calculate the injected dose, measure syringe dosages before, after, and at the time of the injection in the dose calibrator together with the times of these measures (Note 8).
6. After completion of the PET acquisition, set up the parameters for the CT (Note 9) and perform the scan. Use the previously obtained scout to determine the area of interest.
7. At the end of experiment, turn off the anesthesia gas, remove the animal from the scanner, and clean the scanner bed and animal monitoring probes with appropriate cleaning products.

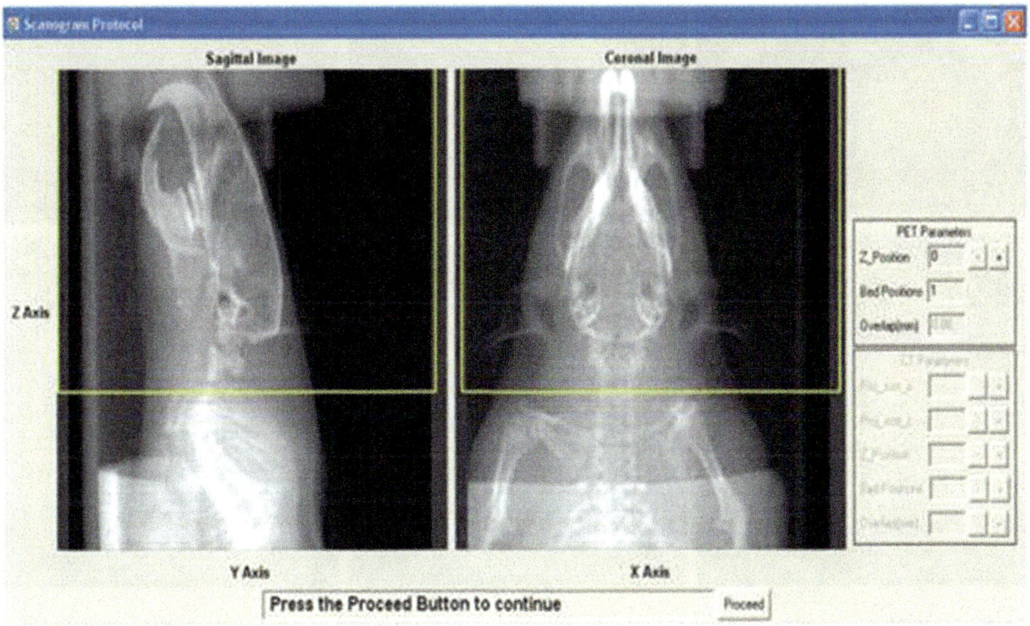

Fig. 4 Scout scan acquired prior to the PET and CT acquisitions. The field of view (FOV) of the PET/CT is represented by the *yellow rectangle*

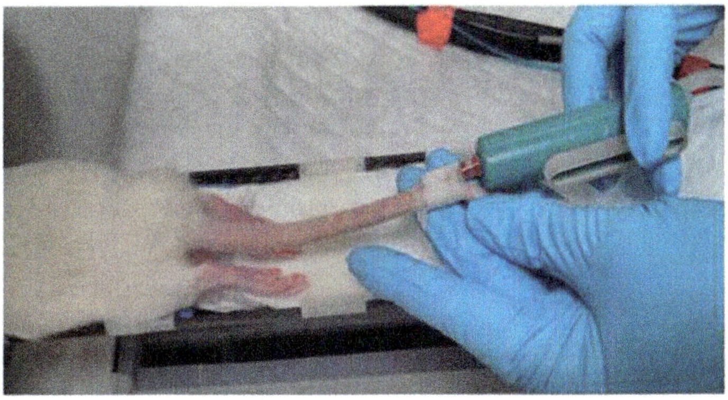

Fig. 5 For the dynamic study, the PET acquisition should start simultaneously with the tracer injection

3.5 PET/CT Image Reconstruction

1. Reconstruct CT data using cone-beam filtered back-projection Feldkamp-David-Kress (FDK) algorithm [23] (Note 10).
2. Reconstruct PET data using an appropriate reconstruction algorithm with corrections for random and scatter coincidences and photon attenuation (Note 10).

3.6 Positron Emission Tomography Image Analysis

1. In our study [17], the images were visualized and analyzed using PMOD (Note 11).
2. Fuse the frames of the reconstructed PET data to better interpret the radiotracer uptake in the brain.

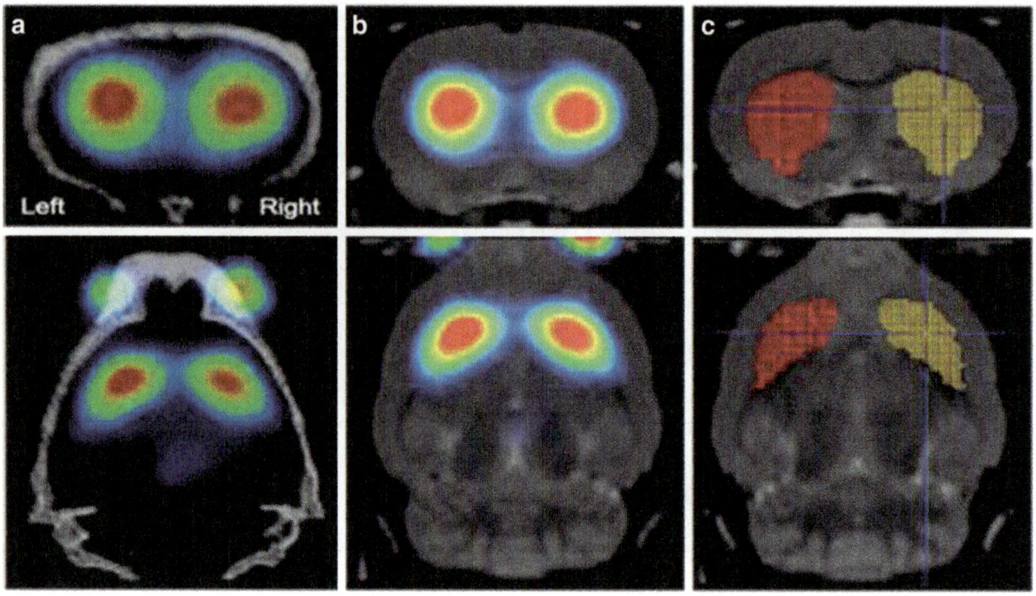

Fig. 6 Co-registered positron emission tomography axial (*upper row*) and coronal (*lower row*) images of [^{11}C] raclopride with the computed tomography image of the same rat (**a**) and the magnetic resonance imaging (T2 weighting MRI) rat brain template (**b**) at the level of the right and left striata. (**c**) Example of the ROI of cerebral striata proposed by PMOD and generated over the MRI-T2W rat brain template

3. Co-register the PET images either with the anatomical data provided by the CT at the time of the PET scan or by an MRI brain template provided by the software of analysis.
4. Define the volumes of interest (VOIs) either manually or with a brain template of VOIs generated over an MRI brain template (PMOD) (Note12; Fig. 6).
5. Calculate the radiotracer uptake as (1) the average of the percentage of the injected dose per cubic centimeter or gram of tissue (%ID/cc-g) during the last 10–15 min of the PET acquisition or (2) the binding potential non-displaceable (BP_{ND}) of [^{11}C]raclopride in the brain (Note 13; Fig. 7).

4 Notes

1. In general, a few-minute-long static emission scan without any radioactivity is good enough to identify the problem.
2. Animal handling has a dramatic effect on the biodistribution of the radiotracers and might significantly influence the results of the study [24]. To maintain normothermic conditions, the temperature of an anesthetized rat has to be controlled during the experiment. This can be done by using a water-based pad with temperature control provided by a rectal temperature probe.

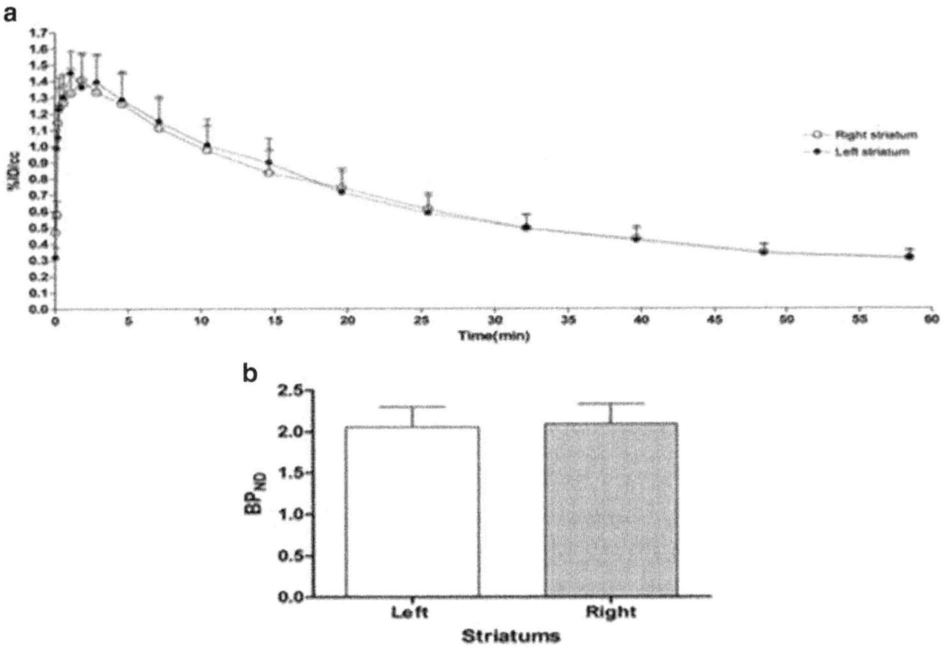

Fig. 7 Time-activity curves (**a**) and BP_{ND} (**b**) for [^{11}C]raclopride in the right and left striata ($N = 4$)

3. Number of frames and their durations should be set up to better reflect the radiotracer dynamics and the biological process to be studied. Short time frames should be acquired at the beginning of the experiment to follow fast uptake dynamics. Longer time frames are used at the end of the study when the steady state of the radiotracer binding is expected. Data can be collected either in given time frames or list mode where the framing process takes place after the completion of the study. Data collected in given time frames cannot be reframed. For the acquisition in list mode, the framing process takes place after the completion of the study; thus, it can be repeated with different settings if needed. The list mode acquisition should always be used when studying a new radiotracer or no previous information on its dynamics is available.
4. Some manufacturers offer dedicated scanner beds with an adjustable incisor bar nose cone system and stereotaxic ear bars for quick and precision animal loading. Those beds usually provide also connections for physiological monitoring probes and anesthesia gas management.
5. It is advisable to place the FOV only in the brain area and to avoid the pulmonary area that exhibits a high radiotracer uptake in lung.
6. For those radiotracers that need cellular uptake to be visualized, the administration is performed for a total of 30–60 min

before the PET static acquisition. In this case, a single frame of variable duration will be taken to provide a static image of the radiotracer distribution in the animal body.

7. To ensure that the whole radiotracer has been administrated is advisable to flush the catheter with saline immediately after the radiotracer administration.

8. To ensure that the data are reproducible and to decrease variability, it is advisable to use the same (or similar) dosage and volume in all PET acquisitions.

9. The CT scan not only provides detailed information about the animal anatomy, but also it is use to generate the attenuation correction map for the PET images [25]. The choice of acquisition parameters depends largely on the type and model of the scanner used. For example, in the case of our scanner, the parameters for standard CT acquisition are as follows: tube current (the number of electrons accelerated across an X-ray tube per unit time, expressed in units of microampere), 140 μA; tube potential (the electric potential applied across an X-ray tube to accelerate electrons toward a target material, expressed in units of kilovolts), 40 kV; number of shoots, 4; resolution, standard (200 μm), and number of projections (the number of rotations of the X-ray beam around the animal, expressed in sexagesimal degrees), 360°.

10. In general, two major categories of reconstruction methods exist, analytical and iterative. Analytical reconstruction utilizes the mathematics of computed tomography that relates line integral measurements to the activity distribution in the object [26]. In other words, it tries to formulate the exact solution in a closed-form equation. Most of commonly used analytical algorithms are based on filtered back projection (FBP) and Fourier reconstruction. In comparison, iterative methods model the data collection process and attempt, in a series of successive iterations, to find the solution that is most consistent with the measured data [27]. The reconstruction algorithms differ in the compromise between noise handling, image resolution, and computational expense.

    In our study [17] CT data were reconstructed using a modified version of Feldkamp-David-Kress (FDK) algorithm for cone-beam reconstruction [27] with binning set to 2. PET data were reconstructed with filtered back projection (FBP) using a Ramp filter with a cutoff frequency of 1 Hz. Recently, the FBP has been recommended as a reconstruction method for studies involving an evaluation of binding potential in rats with [^{11}C]raclopride [28]. PET images were reconstructed to a nominal voxel size of $0.3875 \times 0.3875 \times 0.775$ mm and were corrected for radioactive decay, detector dead time, random and scatter coincidences, and photon attenuation (CT-based correction). Both reconstruction algorithms are included in MMWKS Vista-CT software provided by the scanner manufacturer.

11. In the present study PMOD software was used to visualize and quantify the PET data. However, several other commercial and open source packages for image quantification and processing, such as BrainVISA/Anatomist (http://brainvisa.info/), VINCI (http://www.nf.mpg.de/vinci3/doc/vinci-about.html), or AMIDE (http://amide.sourceforge.net/), are available.
12. It is advisable to use large VOIs to reduce both the partial volume and spillover effects.
13. In our study [17], the binding potential of [^{11}C]raclopride uptake was evaluated with the simplified reference tissue model (SRTM) [29] that relies on a two-tissue reversible compartment for a target region (here the ipsilateral and contralateral striata) and a single compartment for a reference region without target to the radioligand to be studied (here cerebellum). The binding potential values calculated by the SRTM are considered as the binding potential non-displaceable (BPND), which is the BP of the tracer to the tissue and refers to the ratio of specifically bound radioligand to that of non-displaceable radioligand in tissue at equilibrium. The SRTM also provides the R_1 (tracer delivery from vascular compartment to the target tissue) and K_2 (clearance from the target tissue back to the vascular compartment).

References

1. Farde L, Hall H, Ehrin E, Sedvall G (1986) Quantitative analysis of D2 dopamine receptor binding in the living human brain by PET. Science 231:258–261
2. Pavese N, Andrews TC, Brooks DJ, Ho AK, Rosser AE, Barker RA, Robbins TW, Sahakian BJ, Dunnett SB, Piccini P (2003) Progressive striatal and cortical dopamine receptor dysfunction in Huntington's disease: a PET study. Brain 126:1127–1135
3. Rinne JO, Laihinen A, Ruottinen H, Ruotsalainen U, Nagren K, Lehikoinen P, Oikonen V, Rinne UK (1995) Increased density of dopamine D2 receptors in the putamen, but not in the caudate nucleus in early Parkinson's disease: a PET study with [11C] raclopride. J Neurol Sci 132:156–161
4. Farde L, Wiesel FA, Stone-Elander S, Halldin C, Nordstrom AL, Hall H, Sedvall G (1990) D2 dopamine receptors in neuroleptic-naive schizophrenic patients. A positron emission tomography study with [11C]raclopride. Arch Gen Psychiatry 47:213–219
5. Perani D, Garibotto V, Gorini A, Moresco RM, Henin M, Panzacchi A, Matarrese M, Carpielli A, Bellodi L, Fazio F (2008) In vivo PET study of 5HT(2A) serotonin and D(2) dopamine dysfunction in drug-naive obsessive-compulsive disorder. Neuroimage 42:306–314
6. Hammoud DA, Hoffman JM, Pomper MG (2007) Molecular neuroimaging: from conventional to emerging techniques. Radiology 245:21–42
7. Kandel E, Schwartz J, Jessell T, Siegelbaum SA, Hudspeth AJ (2013) Principles of neural science, 5th edn. McGraw-Hill, New York
8. Nagasawa H, Tanji H, Itoyama Y, Saito H, Kimura I, Fujiwara T, Iwata R, Itoh M, Ido T (1996) Brain 6-[18F]fluorodopa metabolism in early and late onset of Parkinson's disease studied by positron emission tomography. J Neurol Sci 144:70–76

9. Bohnen NI, Albin RL, Frey KA, Fink JK (1999) (+)-alpha-[11C]Dihydrotetrabenazine PET imaging in familial paroxysmal dystonic choreoathetosis. Neurology 52:1067–1069
10. Fischer K, Sossi V, von Ameln-Mayerhofer A, Reischl G, Pichler BJ (2012) In vivo quantification of dopamine transporters in mice with unilateral 6-OHDA lesions using [11C] methylphenidate and PET. Neuroimage 59: 2413–2422
11. Villemagne VL, Rothman RB, Yokoi F, Rice KC, Matecka D, Dannals RF, Wong DF (1999) Doses of GBR12909 that suppress cocaine self-administration in non-human primates substantially occupy dopamine transporters as measured by [11C] WIN35, 428 PET scans. Synapse 32:44–50
12. Goodman MM, Kilts CD, Keil R, Shi B, Martarello L, Xing D, Votaw J, Ely TD, Lambert P, Owens MJ, Camp VM, Malveaux E, Hoffman JM (2000) 18F-labeled FECNT: a selective radioligand for PET imaging of brain dopamine transporters. Nucl Med Biol 27:1–12
13. Halldin C, Foged C, Chou YH, Karlsson P, Swahn CG, Sandell J, Sedvall G, Farde L (1998) Carbon-11-NNC 112: a radioligand for PET examination of striatal and neocortical D1-dopamine receptors. J Nucl Med 39: 2061–2068
14. Halldin C, Stone-Elander S, Farde L, Ehrin E, Fasth KJ, Långström B, Sedvall G (1986) Preparation of 11C-labelled SCH 23390 for the in vivo study of dopamine D-1 receptors using positron emission tomography. Int J Rad Appl Instrum A 37:1039–1043
15. Seneca N, Zoghbi SS, Skinbjerg M, Liow JS, Hong J, Sibley DR, Pike VW, Halldin C, Innis RB (2008) Occupancy of dopamine D2/3 receptors in rat brain by endogenous dopamine measured with the agonist positron emission tomography radioligand [11C]MNPA. Synapse 62:756–763
16. Maggos CE, Tsukada H, Kakiuchi T, Nishiyama S, Myers JE, Kreuter J, Schlussman SD, Unterwald EM, Ho A, Kreek MJ (1998) Sustained withdrawal allows normalization of in vivo [11C]N-methylspiperone dopamine D2 receptor binding after chronic binge cocaine: a positron emission tomography study in rats. Neuropsychopharmacology 19: 146–153
17. Martin A, Gomez-Vallejo V, San SE, Pedro D, Markuerkiaga I, Llarena I, Llop J (2013) In vivo imaging of dopaminergic neurotransmission after transient focal ischemia in rats. J Cereb Blood Flow Metab 33:244–252
18. Döbrössy MD, Braun F, Klein S, Garcia J, Langen KJ, Weber WA, Nikkhah G, Meyer PT (2012) [18F]desmethoxyfallypride as a novel PET radiotracer for quantitative in vivo dopamine D2/D3 receptor imaging in rat models of neurodegenerative diseases. Nucl Med Biol 39:1077–1080
19. Jovkar S, Wienhard K, Pawlik G, Coenen HH (1990) The quantitative analysis of D2-dopamine receptors in baboon striatum in vivo with 3-N-[2′-18F]fluoroethylspiperone using positron emission tomography. J Cereb Blood Flow Metab 10:720–726
20. Cropley VL, Innis RB, Nathan PJ, Brown AK, Sangare JL, Lerner A, Ryu YH, Sprague KE, Pike VW, Fujita M (2008) Small effect of dopamine release and no effect of dopamine depletion on [18F]fallypride binding in healthy humans. Synapse 62:399–408
21. Graff-Guerrero A, Willeit M, Ginovart N, Mamo D, Mizrahi R, Rusjan P, Vitcu I, Seeman P, Wilson AA, Kapur S (2008) Brain region binding of the D2/3 agonist [11C]-(+)-PHNO and the D2/3 antagonist [11C]raclopride in healthy humans. Hum Brain Mapp 29: 400–410
22. Ikoma Y, Watabe H, Hayashi T, Miyake Y, Teramoto N, Minato K, Iida H (2009) Quantitative evaluation of changes in binding potential with a simplified reference tissue model and multiple injections of [11C]raclopride. Neuroimage 47:1639–1648
23. Feldkamp LA, Davis LC, Kress JW (1984) Practical cone-beam algorithm. J Opt Soc Am 1:612–619
24. Fueger BJ, Czernin J, Hildebrandt I, Tran C, Halpern BS, Stout D, Phelps ME, Weber WA (2006) Impact of animal handling on the results of ^{18}F-FDG PET studies in mice. J Nucl Med 47:999–1006
25. Christian PE, Waterstram-Rich KM (2013) Nuclear medicine and PET/CT: technology and techniques, 7th edn. Elsevier Health Sciences, Amsterdam
26. Phelps ME (2006) Positron-emission tomography: physics, instrumentation, and scanners. Springer Science, New York
27. Vaquero JJ, Redondo S, Lage E, Abella M, Sisniega A, Tapias G, Desco M (2008) Assessment of a new high-performance small-animal X-ray tomography. IEEE Trans Nucl Sci 55:898–905
28. Torrent E, Farre M, Abasolo I, Millan O, Llop J, Gispert D, Ruiz A, Pareto D (2013) Optimization of [11C]raclopride positron emission tomographic rat studies: comparison of methods for image quantification. Mol Imaging 4:257–262
29. Lammertsma AA, Hume SP (1996) Simplified reference tissue model for PET receptor studies. Neuroimage 4:153–158

Part IV

Molecular and Cell Biological Approaches in the Study of Dopamine Receptor Function

Chapter 12

Transactivation of Receptor Tyrosine Kinases by Dopamine Receptors

Jeff S. Kruk, Azita Kouchmeshky, Nicholas Grimberg, Marina Rezkella, and Michael A. Beazely

Abstract

As our understanding of G protein-coupled receptor (GPCR) signaling grows, it is clear that the arsenal of GPCR effectors is far greater than their classical second messenger signaling pathways. The *transactivation*, or GPCR-induced activation of receptor tyrosine kinases (RTKs), presents an avenue for GPCRs to affect signaling pathways that have previously been attributed to growth factors and opens the door for modulation of RTK activity with small-molecule GPCR ligands. Several RTK transactivation pathways initiated by dopamine receptors have been described. One of the best characterized is the D2-class dopamine receptor transactivation of the platelet-derived growth factor (PDGF) receptor: a ligand-independent, intracellular signaling pathway. Dopamine receptors can also transactivate epidermal growth factor (EGF) receptors, often via the metalloproteinase-dependent cleavage of EGF itself from the cell surface. Although the discovery of RTK transactivation is relatively recent, a growing body of research has identified these pathways in several cell line, primary cell, and in vivo systems. These studies have characterized the time course and magnitude of RTK transactivation and have identified several common effectors involved. The choice of the primary readout in transactivation studies, i.e., RTK activation/phosphorylation or a downstream RTK effector such as extracellular signal-regulated kinases (ERK) or Akt, is an important consideration for transactivation studies. Equally important is identifying whether the transactivation is ligand (growth factor) dependent or independent. These and other considerations are described, not only with a focus on dopamine receptor-initiated transactivation pathways but also with a discussion of general considerations for the study of GPCR–RTK transactivation.

Key words Dopamine, GPCR, RTK, PDGF, EGF, Transactivation, ERK, Akt, Phosphorylation

1 Introduction

The binding of agonists to the extracellular portion of dopaminergic G protein-coupled receptors (GPCRs) results in the activation of heterotrimeric GTP-binding proteins (G proteins) on the intracellular side [1]. The α-subunit exchanges GDP for GTP and dissociates from the receptor as well as the βγ-subunits. Both the α- and βγ-subunits are able to activate effector enzymes, resulting in changes in intracellular second messengers such as

Mario Tiberi (ed.), *Dopamine Receptor Technologies*, Neuromethods, vol. 96,
DOI 10.1007/978-1-4939-2196-6_12, © Springer Science+Business Media New York 2015

cyclic AMP and calcium [1]. Receptor tyrosine kinases (RTKs) are composed of a single transmembrane domain, an extracellular ligand-binding domain, and an intracellular tyrosine kinase domain [2]. Many RTKs bind growth factors that are small proteins/ large polypeptides and promote growth and development [2]. Growth factor binding brings two RTKs into close proximity and allows for a transautophosphorylation of intracellular tyrosines that serve as docking points for several adaptor proteins and effector enzymes [2]. As such, this state of increased phosphorylation indicates an activated RTK.

As GPCR and RTK signaling and function were elucidated, they were considered to be relatively siloed, independent signaling pathways. However, RTKs can be activated through intracellular signaling cascades via cross-talk with other receptors such as GPCRs, a process called transactivation first described by Daub [3]. Within the last 25 years, several examples of GPCR–RTK transactivation have been described [4, 5]. This method of RTK activation is termed "transactivation," but within that term, there are at least two very distinct and well-characterized signaling mechanisms for GPCR activation of RTKs: ligand-dependent and ligand-independent RTK transactivation.

The best-characterized ligand-dependent transactivation pathway involves GPCR-induced activation of epidermal growth factor (EGF) receptors. Briefly, the activation of a GPCR initiates a series of intracellular steps that result in the activation of matrix metalloproteinases (MMPs) (members of the A Disintegrin and Metalloproteinase (ADAM) family of peptidase proteins) that cleave membrane-tethered growth factor [6]. The released growth factor is then able to bind and activate its receptors on the same or on adjacent cells [6]. This type of transactivation could be described as "triple-membrane-passing" signaling as it is initiated with an extracellular stimulus (GPCR ligand), resulting in an intracellular signaling pathway that releases another extracellular stimulus (the growth factor), which then binds and activates an RTK and its multiple intracellular pathways [6–8].

In contrast to triple-membrane-passing transactivation, ligand-independent transactivation is entirely intracellular. One of the first and best-characterized ligand-independent pathways was elucidated in the laboratory of the late Hubert van Tol and involves D2-class dopamine receptor transactivation of the platelet-derived growth factor (PDGF)β receptor [9]. As its name suggests, this and similar pathways involve a series of intracellular signaling steps but at no time is growth factor released extracellularly or involved with RTK activation. Dopamine receptors are grouped into the D2 family (including D2S (short isoform), D2L (long isoform), as well as D3 and D4 receptors and isoforms) and the D1 family (D1 and D5). The activation of either D2L or D4 dopamine receptors in Chinese hamster ovary (CHO) cells results in the phosphorylation of extracellular signal-regulated kinases (ERK1/2) downstream of PDGFβ receptor transactivated in a $G\alpha_{i/o}$- and Src-dependent manner (note that here and in many other studies, the Src-family kinase inhibitor, PP2, is used to implicate Src, but the specific Src-family kinase involved is not identified) [9]. This transactivation is ligand independent and does not even require PDGFβ receptor dimerization [10], a step that is required for growth factor-induced PDGFβ receptor activation [11]. Cell surface-expressed PDGFβ receptors possess *N*-linked and *O*-linked glycosyl chains that raise its molecular weight by about 40 kDa [12, 13]. Interestingly, prevention of PDGFβ receptor glycosylation did not prevent GPCR-induced transactivation of intracellular PDGFβ receptors, suggesting that intracellular RTKs can be transactivated [13], likely because their C-terminal tails containing the phosphorylation sites are still exposed to the cytoplasm. D2 dopamine receptors also transactivate the PDGF receptor in human embryonic kidney (HEK) 293 cells [14]. In electrophysiology studies, activation of the PDGFβ receptor inhibits *N*-methyl-D-aspartate (NMDA) receptor currents [15–17]. Both D4 and D2/3 dopamine receptors similarly inhibit NMDA-evoked currents in a PDGFβ receptor-dependent manner in hippocampal [18] and cortical [19] neurons, respectively, suggesting that ligand-activated and transactivated PDGFβ receptors have similar downstream signaling consequences.

D2-class dopamine receptors are also able to transactivate the EGF receptor. In PC12 cells stably transfected with the D2 receptor, the D2 receptor agonist bromocriptine protected neurons by increasing Akt activation in an EGF receptor-dependent manner [20]. D2 dopamine receptor activation also results in an increase in EGF receptor phosphorylation in NS20Y neuroblastoma cells [14]. In the latter model, the inhibition of metalloproteinases attenuated the transactivation of the EGF receptor, suggesting that this pathway is of the triple-membrane-passing variety [14]. Dopamine promotes the proliferation of cells in the subventricular zone in the CNS via an increase in EGF release [21] and this effect was also metalloproteinase dependent [22]. See Ohtsu et al. for review including a summary of the role of metalloproteinases in EGF receptor transactivation and a discussion of the most common ADAM subtypes involved: 17 is the most often identified, but types 10, 12, and 15 are also involved [23]. A similar D2 dopamine–EGF receptor transactivation pathway is observed in primary mesencephalic cultures [24]. Intriguingly, differences have been observed within the D2-class receptor family. For example, while both D2 and D3 receptor activation increase ERK1/2 phosphorylation in HEK293 cells, only D2 receptor-induced phosphorylation of ERK1/2 was EGF receptor dependent [25].

Differences between D2 and D3 receptor-mediated pathways were also observed in the transactivation of the insulin-like growth factor 1 (IGF-1) receptor and subsequent increased phosphorylation of Akt (threonine 308 and serine 473) and glycogen synthase kinase-3β (GSK-3β, serine 9) [26]. Both D2L and D3 receptor-induced Akt/GSK-3β phosphorylation was blocked by pertussis toxin (PTX) and the Src tyrosine kinase inhibitor PP2 (*see* Table 1), but whereas D2L receptor-induced

GSK-3β phosphorylation required calmodulin, Akt, and ERK1/2, D3 receptor-induced phosphorylation required phospholipase C (PLC)γ activity but not ERK1/2 activity [26]. Many of the steps in the D3 receptor to Akt pathway were recently reproduced in vivo by injecting the D2/3 receptor agonist, quinelorane, and measuring changes in phosphorylation of proteins in the

Table 1
Effectors identified in GPCR–RTK transactivation pathways

GPCR	RTK	Transactivation effectors	Reference
D4.4	PDGFβ	Gαi/o (200 ng/mL pertussis toxin (PTX) overnight)	[9][a]
		βARKct (Lipofectamine transfection, 24 h)	
		Src (10–100 μM PP2 for 1 h)	
D4	PDGFβ	Gαi/o (5 μg/mL PTX in patch electrode[a])	[18]
		βARKct (5 μM in patch electrode)	
		GDPβS (in patch electrode)	
D2S	PDGFβ	β-Arrestins 1 and 2 promoted D2S receptor-induced ERK1/2 phosphorylation	[28]
		Dominant-negative β-arrestin inhibited D2S receptor to ERK signaling (note only D2L receptor-induced ERK1/2 phosphorylation was RTK dependent)	
D2L	EGF	Src (1 μM PP2 for 30 min)	[20][b]
D2	EGFR	Gαi only (100 ng/mL pertussis toxin (PTX) overnight)	[25]
D1	EGF	PKC (1 μM calphostin C), PMA (100 ng/mL for 15 min increased EGF shedding)	[30]
		Intracellular calcium (10 μM BAPTA-AM for 30 min)	
D2/D3	IGF-1	Gαi/o (100 ng/mL PTX overnight)	[26]
		Src (10 μM PP2 for 30 min blocked DA-induced Akt and GSK-3β phosphorylation, not clear if Src is up- or downstream of the IGF-1 receptor)	
D1	TrkB	Intracellular calcium (10 μM BAPTA and BAPTA-AM for 30 min)	[31]

Effectors/inhibitors listed are involved in GPCR-induced transactivation of RTKs as measured by RTK phosphorylation or other readout measures. Note that effectors downstream of the RTK (after transactivation) are not listed in this table

[a]ERK phosphorylation was the readout for most experiments

[b]In this manuscript, the readout was a PDGFβ receptor-dependent D4 dopamine receptor inhibition of NMDA receptor currents

nucleus accumbens and dorsal striatum; however, it remains unknown whether RTK transactivation is involved [27]. Even within isoforms of the same D2 receptor subtype, differences in ERK1/2 phosphorylation have been observed [28], *see* Sect. 2.5). See Beaulieu et al. for a recent review of dopamine receptor to Akt signaling [29].

Although the transactivation literature primarily involves reports on D2-class dopamine receptors, treatment of neuron-enriched striatal cultures with the D1 receptor agonist, SKF38393, promoted EGF shedding, whereas the D2 agonist, quinpirole, promoted shedding from non-neuronal cells [30]. D1 receptor activation also promotes the phosphorylation of tropomyosin-related kinase B (TrkB) receptor in striatal neurons via promoting an increase in TrkB receptor membrane localization [31].

The techniques for studying transactivation are for the most part standard biochemical techniques such as Western blotting for RTK tyrosine phosphorylation, immunoprecipitation, and immunofluorescence. In the next section, we will outline considerations for experimental setup, choose a readout to measure, determine whether the transactivation pathway depends on extracellular growth factor release, and highlight common enzymes and proteins involved in transactivation pathways. In the Sect. 3, we will discuss some unique aspects of dopamine receptor pathways and unanswered questions and briefly highlight novel approaches to GPCR–RTK signaling.

2 Materials and Methods

2.1 Model Systems

Initial considerations when designing a transactivation experiment include the model system and cell culturing protocols, the choice of ligand, the concentration, and the duration of treatment. Dopamine receptor-initiated transactivation studies have been performed by stable transfection of D2L or D4 receptors in CHO cells [9, 10, 13], PC12 and SN4741 cells [20], and HEK293 and NS20Y cells [14] as well as D1 transfection in HEK293 cells [31]; however, dopamine is able to transactivate the PDGFβ receptor in non-transfected SH-SY5Y cells [32]. Dopamine receptor-induced RTK transactivation has also been observed in primary striatal [14, 30, 31], mesencephalic [24], and subventricular zone (SVZ) precursor cell cultures [21] as well as acutely dissected CA1 hippocampal [18, 19] and cortical slices [19]. Typically, transactivation studies involve Western blotting and 6-well culture plates are employed. However, for studies involving readouts such as cell survival, 48- or 96-well plates are used. To avoid the presence of growth factors interfering with the transactivation experiment, cells are often incubated in serum-free media for 24 h prior to treatment. However, if serum starvation adversely affects the cells to a high degree, a shorter serum-free incubation time may be sufficient. In certain combinations of cell lines/serum and medium, the contribution of serum growth factor to basal RTK activation levels may be so minimal that serum starvation is not required.

2.2 Timing and Choice of Ligand

One of the first experiments that should be employed in a novel transactivation system is a time-course and/or dose–response investigation determining a phosphorylation profile of the RTK by the GPCR ligand of choice. GPCR transactivation is generally an "acute" phenomenon, with a peak increase in RTK phosphorylation usually observed between 2 and 15 min and a return to baseline levels by 30 min. This can then be compared to analogous experiments using growth factors in place of GPCR ligands, where the growth factor-induced phosphorylation of the RTK is typically much more intense and longer lasting than that seen with the GPCR agonist [32]. Most studies employ either dopamine itself (10–100 μM) or quinpirole (1–10 μM) to activate D2-class dopamine receptors as well as 10 μM PD168077 to activate D4 dopamine receptors (purchased from Sigma and/or Tocris). Oak et al. performed full time-course and dose–response experiments, although their readout was ERK1/2 phosphorylation rather than RTK phosphorylation (*see* Sect. 2.4, [9]).

On the other hand, there are several exceptions to the "acute RTK transactivation rule" that should be considered, as the time at which peak phosphorylation is seen can be variable between the systems and receptors being studied. For example, activation of adenosine receptors or D1 dopamine receptors (using 1 μM SKF38393) increases TrkB receptor phosphorylation, but the phosphorylation of TrkB receptor peaks at 3 h and stays elevated up to 6 h [31]. See Lee et al. for a discussion on some of the unique aspects of Trk family RTK transactivation [33]. Similarly, when Yoon and Baik applied quinpirole to primary mesencephalic cultures, they measured a sustained release of EGF that peaked at 2 h but remained significantly elevated above baseline for up to 12 h [24]. D2L and D3 receptor-induced Akt/GSK-3β phosphorylation (IGF-1 receptor dependent) remained elevated for more than 2 h post-treatment [26]. Thus, although RTK transactivation is often a very short-lived event, there are several exceptions (*see* **Note 1**).

2.3 Common Proteins Identified in Transactivation Pathways

Although technically GPCR–RTK transactivation describes an activation of the RTK, usually using tyrosine phosphorylation of the RTK as the readout, a significant amount of data with respect to dopamine receptor–RTK pathways uses other readouts (*see* Sect. 2.4, **Note 2**). After determining a concentration and duration of GPCR agonist application, the suspected effector proteins involved can be knocked out, knocked down, or pharmacologically inhibited to determine whether they play a role in a transactivation pathway.

To identify effector proteins involved in transactivation pathways in cases where the primary readout is downstream of the RTK rather than the RTK itself, researchers must be careful to place the effector either in between the GPCR and RTK or downstream of the RTK. Table 1 summarizes data collected from experiments to identify effector proteins between the GPCR and RTK (but not downstream of the RTK).

2.4 Choice of Readout

As mentioned above, particularly with many of the early transactivation studies, the primary readout was to measure ERK1/2 phosphorylation rather RTK phosphorylation (*see* **Note 2**, [9]). In addition to ERK1/2, other readouts include Akt phosphorylation [20] and a PDGFβ receptor-dependent inhibition of NMDA-evoked currents [18, 19]. When RTK phosphorylation was measured, protocols often involved immunoprecipitating the RTK and measuring total tyrosine phosphorylation [9]. However, as specific anti-phosphotyrosine antibodies become available, researches will be able to (1) identify which specific RTK residues are phosphorylated after a transactivating stimulus and (2) compare and contrast the phosphorylation profile of transactivated vs. ligand-activated RTKs (*see* **Note 3**, [10, 32]). With a focus on ERK1/2 phosphorylation as the primary readout, effectors identified were frequently downstream of the transactivated RTK and are similar to those effectors downstream of ligand-activated RTKs: PLCγ, phosphoinositide 3-kinase (PI3-kinase), calcium, Src, Ras, mitogen-activated protein kinase kinase (MEK), and others. In the case of intracellular transactivation pathways, while it is clear that effectors downstream of a transactivated RTK are similar to those observed when an RTK is directly activated with ligand, it remains to be seen whether there are small but significant differences between direct ligand and transactivated RTK downstream signaling (*see* **Note 3**).

2.5 Role of β-Arrestins

β-Arrestin overexpression promoted D2S receptor-induced ERK1/2 phosphorylation in CHO cells, whereas D2L-induced phosphorylation was not affected [28]. Conversely, dominant-negative β-arrestins (β-arrestin 2 (319–418) as well as a dominant-negative dynamin construct, dynamin I (K44A)) inhibited D2S, but not D2L receptor-induced ERK1/2 phosphorylation [28], despite both β-arrestin and dominant-negative constructs affecting D2S and D2L receptor trafficking in a similar manner [28]. Interestingly, the (relatively nonselective) PDGFβ receptor kinase inhibitor, tyrphostin A9, blocked D2L, but not D2S receptor-induced ERK1/2 phosphorylation, suggesting that only the D2L receptor required RTK transactivation to signal to ERK [28]. In HEK293 cells, siRNA against $G\alpha_{i2}$, but not β-arrestin 1 or 2, decreased D2S receptor-induced ERK1/2 phosphorylation [34]. This is one of the few studies to go beyond using PTX to broadly inactivate $G\alpha_{i/o}$ proteins (Table 1) and to identify the specific Gα subunit involved.

2.6 Involvement of Extracellular Growth Factor Release

An important question to address in transactivation studies is whether the pathway is completely intracellular or of the triple-membrane-passing type. As noted in the Introduction, the best-characterized transactivation pathways involve D2-class dopamine receptor transactivation of PDGF receptors via intracellular pathways and either D1- or D2-class transactivation of EGF receptors via triple-membrane-passing signaling, although D2 receptor transactivation of EGF receptors is not necessarily extracellular in all cases [20]. To determine if the transactivation pathway under investigation involves shedding/release of growth factor, researchers first measure and/or block any released growth factor. Preventing growth factor release can be accomplished by first applying anti-growth factor antibodies (which must bind the native protein conformation) to the culture media just prior to the addition of the GPCR agonist. The concentration of antibody to add should be at least tenfold higher than an equivalent concentration of growth factor that elicits a level of phosphorylation similar to that observed with GPCR application. These neutralizing antibodies should successfully inhibit binding of released growth factor to its receptor, thus preventing transactivation. Growth factor release can also be detected by collecting conditioned cell culture media and performing an ELISA or Western blot [22, 30]. Concentrating the cell culture media using centrifugation-based filters may be necessary if released growth factor quantities are judged to be too dilute. If growth factor release is detected and suspected to be metalloproteinase dependent, the next step is to use metalloproteinase inhibitors or to knock down metalloproteinase expression.

In their D4-CHO cell line, Chi et al. demonstrated that they could not even detect mRNA for the PDGF-B chain (PDGF-BB dimers are the primary ligand for PDGFβ receptor) as evidence for a ligand-independent PDGFβ receptor-mediated ERK1/2 phosphorylation [10]. To rule out an EGF receptor transactivation that involved extracellular release of EGF, they preincubated their cultures for 30 min with the broad-spectrum metalloproteinase inhibitor, GM6001 (5 μM), as well as 10 μg/mL diphtheria toxin CRM197 to bind membrane-bound EGF and prevent shedding [10, 35]. Using GM6001 (purchased from Calbiochem) is the most common molecular approach in determining whether the transactivation pathway is metalloproteinase dependent – 200 nM to 10 μM, 15 min pretreatment [14]; 100 nM, 1 h pretreatment [30]; and 10 μM, 1 h pretreatment [24] – although BB-94 (British Biotech Products, UK [22]) and botulinum neurotoxin type A (BoNT/A, 100 nM, 8 h) [30] have also been employed.

A novel approach by Pierce et al. was to co-culture α2A receptor-expressing cells with cells that do not express the α2A receptor [36]. Each cell type expressed a different epitope-tagged ERK2 that allowed for a distinction between ERK activation. The authors demonstrated that in their co-culture, α2A receptor activation was able to increase ERK phosphorylation in cells lacking the α2A receptor, via EGF release and EGF receptor activation on adjacent, non-α2A receptor-expressing cells [36]. A related approach involves treating the cells with agonist to promote growth factor shedding, collecting the conditioned media, and using it to treat separate cultures to induce growth factor receptor activation. Iwakura et al. used this approach to show that D1–TrkB receptor transactivation was not the result of glial cell release of BDNF [31]; nevertheless, it is important to take into account that the GPCR agonist may remain in the conditioned media when selecting the second culture. Recently, investigators have begun to directly measure metalloproteinase enzymatic activity (using, e.g., the SensoLyte 520 MMP Substrate Sampler Kit; AnaSpec [30]) and have even begun to identify the specific subtypes involved (e.g., ADAM 10, 17) using siRNA knockdown [24].

2.7 RTK Dimerization, Localization, and Complex Formation

In the case of triple-membrane-passing signaling to RTKs, it is the endogenous ligand itself that, after being released downstream of a GPCR, directly activates the RTK [14]. Similar to exogenous ligand-mediated activation, EGF receptor transactivation-mediated ligand release results in receptor dimerization and transautophosphorylation at several tyrosine residues [37]. For ligand-independent transactivation, Chi et al. investigated whether RTK dimerization was required for transactivation of the RTK and subsequent downstream ERK1/2 phosphorylation using a C-terminal deletion mutant of the PDGFβ receptor by truncating the human receptor at residue 615 [10, 38, 39]. After transfection of the truncated RTK, D4 receptor activation still resulted in transactivation and activation of ERK1/2, suggesting that at least for the PDGFβ receptor, dimerization was not required for transactivation [10]. If demonstrated to be true for all cases of ligand-independent transactivation, this would represent a significant deviation from classical ligand-induced RTK activation.

For triple-membrane-passing signaling, the released ligand interacts with cell surface-localized receptors. For intracellular transactivation pathways, this is not necessarily the case. Using tunicamycin (2 μg/mL, overnight) to inhibit the glycosylation of the PDGFβ receptor (a requirement for its cell surface expression), Gill et al. demonstrated that intracellular, immature receptor could be transactivated by D4 dopamine receptors and that these transactivated receptors retained their ability to signal to ERK1/2 [13]. Similarly, adenosine and pituitary adenylate cyclase-activating peptide are each able to increase the phosphorylation of intracellular TrkA receptors [40].

Several GPCR–RTK transactivation systems involve physical interactions between the GPCR and the RTK: e.g., sphingosine-1-phosphate and lysophosphatidic acid both physically interact with the PDGFβ receptor [41]. For dopamine receptors, D2 receptors expressed in PC12 cells form a complex with Src and the EGF receptor, and this complex is required for dopamine-induced Akt phosphorylation [20], one of the few examples of an exclusively intracellular EGF transactivation system. Determining these protein–protein interactions typically involves common molecular techniques such as fluorescent labeling, BRET/FRET, or co-immunoprecipitation.

3 Notes

1. Defining transactivation.
   The term transactivation is used to describe GPCR-induced activation of RTKs. Within that term, there are two major common types of transactivation pathways, one intracellular and one requiring extracellular growth factor release. The term triple-membrane-passing is useful to distinguish the transactivation pathways that require extracellular ligand shedding from intracellular pathways. As discussed in Sect. 2.2, many RTK transactivation events are short lived, but there are several exceptions to this rule: the most well-described exception being the transactivation of Trk receptors that is observed over hours, rather than minutes [42]. We have recently described a signaling pathway in which very long-term (24 h) activation of the serotonin (5-HT) type 7 receptor increases the expression and phosphorylation of the PDGFβ receptor and promotes its neuroprotective effects against NMDA-induced neurotoxicity [43, 44]. Although there is a change in expression, the "basal" i.e., ligand-independent, activity of the PDGFβ receptor is increased. Given the time frame, the fact that RTK protein levels are changing in addition to the ligand-independent increase in activity, would it be appropriate to describe this pathway as "transactivation"? One other important note is that the term transactivation is actually more frequently used to describe transcriptional activation/regulation, and this can be problematic when searching for GPCR–RTK transactivation studies in the literature.
2. Using ERK1/2 to measure GPCR–RTK transactivation.
   Several transactivation studies, including many of those described above, measured ERK1/2 either primarily or exclusively as the primary readout of dopamine receptor transactivation of RTKs. In some systems, such as the well-characterized D4–PDGFβ receptor transactivation pathway, ERK1/2 phosphorylation is consistently dependent upon PDGFβ receptor activity [9, 10, 13]. In other systems, such as the co-culture system used by Prenzel et al., some ERK1/2 phosphorylation in non-α2A receptor-expressing cells was

dependent upon EGF shedding and EGF receptor activation; however, an intracellular α2A receptor-induced ERK1/2 was also detected [35]. Recently, we reported a 5-HT–PDGFβ receptor transactivation pathway where the pathways downstream of 5-HT receptor activation for PDGFβ receptor and ERK1/2 were parallel but diverged at the level of NADPH oxidase (*see* Fig. 1). In other words, 5-HT treatment resulted in ERK1/2 phosphorylation but it was PDGFβ receptor independent [45]. Transactivation studies should take into consideration these examples when measuring ERK1/2 phosphorylation after GPCR activation. When using inhibitors of effectors thought to be involved in transactivation pathways, measuring both RTK and downstream effector phosphorylation/activity allows the researcher to place the effector either upstream or downstream of the RTK.

Note that for dopamine receptors, descriptions of dopamine receptor–ERK pathways outnumber dopamine receptor–RTK–ERK pathways by a ratio of ~ 3:1 in the literature. In many of those publications, RTK transactivation or a dependence on RTK activity for ERK1/2 phosphorylation was not explored; however, PTX [46–48], β-adrenergic receptor kinase 1, C-terminus (βARK-ct) [47, 48], and Src sensitivity [46] are often reported, and the time courses for ERK1/2 phosphorylation are very similar to what is observed for RTK transactivation [49, 50]. See Lutrell or Rozengurt for reviews of all the potential signaling pathways from GPCRs to ERK [51, 52].

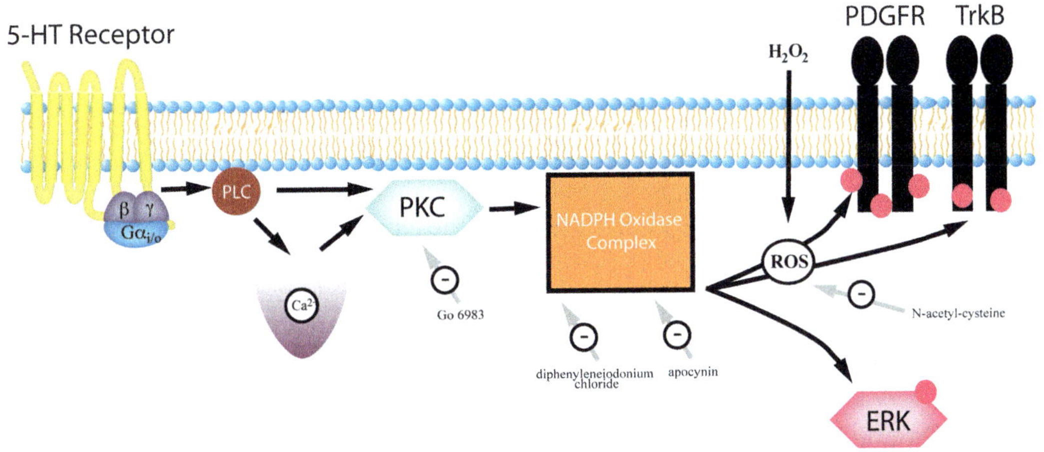

Fig. 1 An example of an intracellular transactivation pathway: 5-HT-induced ERK1/2 phosphorylation and PDGFβ receptor transactivation in SH-SY5Y cells. Application of 5-HT to SH-SY5Y cells results in a transient increase in ERK1/2 phosphorylation as well as PDGFβ and TrkB receptor phosphorylation. The intracellular signaling cascade for all three phosphorylation events runs parallel until NADPH oxidase and then diverges. RTK phosphorylation is ROS dependent, but ERK1/2 phosphorylation is ROS independent. Many of the components identified above are commonly reported throughout the dopamine-induced intracellular transactivation pathways (Figure reproduced from [45])

3. Comparing transactivated vs. ligand-activated RTK phosphorylation and signaling.
   Many transactivation studies use either total tyrosine phosphorylation or one phospho-specific antibody when measuring GPCR-induced RTK phosphorylation. For D2 receptor–EGF receptor transactivation in PC12 cells, the D2 agonist bromocriptine increased the tyrosine phosphorylation of tyrosines 845, 992, and 1068 (as well as Akt), in all cases to a lower extent than EGF itself (100 ng/mL) [20]. For the PDGFβ receptor, dopamine (1 μM) increased the phosphorylation of tyrosines 716, 740/751, and 1021 but not 857, whereas PDGF-BB (10 ng/mL) increased the levels of all phosphotyrosine sites and to a greater extent [10]. Although the phosphorylation levels of transactivated RTKs are generally much lower than ligand-stimulated RTKs, to date, a systematic examination of the differences in the dose–response, time-course, and downstream consequences of possible differential phosphorylation profiles of RTKs after GPCR activation has not been completed.
4. $G\alpha_{i/o}$ and GPCR kinases.
   As noted above and in Table 1, a common effector, not surprisingly, downstream of D2-class dopamine receptors is the $G\alpha_{i/o}$ proteins. Interestingly, $G\alpha_{i/o}$ proteins are involved not only in the transactivation of RTKs but also in the ligand activation of RTKs. For example, PDGF receptor-induced ERK1/2 activation in cultured airway smooth muscle cells is partially blocked by PTX [53]. Furthermore, PDGFβ receptors can physically interact with GPCRs and phosphorylate Gα subunits [54]. See Pyne and Pyne [55] and Pyne et al. [56] for reviews (and methods) to study the involvement of G proteins in RTK signaling.

   There is also a growing body of research into the regulation of RTKs by G protein-coupled receptor kinases (GRKs). For example, GRK2 phosphorylates the PDGFβ receptor on serine residues after activation of the receptor, and this promotes desensitization of the RTK [57, 58]. GRK5 is also able to serine phosphorylate the PDGFβ receptor in vascular smooth muscle cells, and this regulatory pathway may be involved in the pathogenesis of atherosclerosis [59, 60]. See Hupfeld and Olefsky [61] for a review on the role of GRKs and β-arrestins in RTK signaling.
5. Reactive oxygen species.
   We recently reported that the reactive oxygen species (ROS) scavenger *N*-acetylcysteine (1 mM, 45 min pretreatment) inhibited 5-HT-induced transactivation of the PDGFβ receptor [45]. One potential mechanism for ROS promotion of RTK phosphorylation is the oxidation of cysteine residues on

tyrosine phosphatase enzymes. This is the proposed mechanism of endothelin 1 receptor-induced EGF receptor transactivation: a transient oxidation of SHP-2 cysteine residues to temporarily inactivate the phosphatase results in a relative net increase in tyrosine phosphorylation of the RTK [62]. One potential source for ROS is the NADPH complex. Inhibitors of NADPH activity, diphenyleneiodonium chloride (DPI) and apocynin, abolished 5-HT-induced PDGFβ receptor transactivation and ERK1/2 phosphorylation in SH-SY5Y cells [45]. Note that while 0.1 μM H_2O_2 promoted PDGFβ receptor phosphorylation in SH-SY5Y cells and primary cortical mouse neurons, no change in ERK1/2 phosphorylation was observed [45]. This is contrast to previous reports that used much higher (0.1–2 mM H_2O_2) to promote ERK1/2 phosphorylation [63–65].

6. Agonist differences in transactivation systems.
   As noted in Sect. 2.2, most transactivation studies involving dopamine receptors employ either dopamine or a single D2- or D1-class agonist to initiate the transactivation pathway. To date, it remains unknown whether screening libraries of dopamine receptor agonists would identify ligands that preferentially initiate transactivation pathways vs. classical second messenger signaling. Interestingly, evidence that this might be the case comes from one of the first D2–EGF receptor transactivation reports. In PC12-D2 cells, treatment with bromocriptine, but not pramipexole, promoted the formation of the D2–Src–EGF receptor complex, transactivation of the EGF receptor, activation of Akt, and pro-survival effects on the cells [20].

7. Role of membrane microdomains.
   Cholesterol-rich membrane microdomains such as lipid rafts may be important for the transactivation of RTKs [66, 67]. The use of β-cyclodextrin or filipin to disrupt cholesterol greatly reduces sphingosine-1-phosphate (S1P)-induced, but not PDGF-BB-induced, PDGFβ receptor activation [66]. This is remarkable since PDGF receptors as well as $S1P_1$ receptors are localized in caveolae [11, 66, 68], a type of lipid raft. This suggests that caveolae may be an environment required for S1P and PDGF receptors to be in close proximity for signaling to occur or for signaling complexes to form, and indeed they have been shown to be physically associated [14, 69]. Stimulation with PDGF ligand results in PDGF receptor internalization [11], and $S1P_1$ receptors complexed with PDGF receptors are also co-internalized into the same vesicle in a clathrin-dependent manner [70]. There is increasing evidence that certain GPCR signaling complexes are compartmentalized in lipid rafts [71], which may be necessary for

signal transduction. Caveolae have also been associated with 5-HT receptors, dopamine receptors, angiotensin II receptors, and endothelin receptors [67, 72–78].

8. Novel approaches.
   Transactivation studies have primarily employed protein-based techniques such as Western blotting, immunoprecipitation, etc. Recently, the Beaulieu group has employed spatial intensity distribution analysis [79] to quantify both ligand-induced activation and transactivation of RTKs. This technique has demonstrated that RTK dimerization (for the EGF receptor and TrkB receptor) is similar for both ligand-induced activation and transactivation (including by dopamine receptors) of these RTKs, although differences were noted with respect to the level of internalization [37]. George et al. have recently proposed a functional genomics approach to achieve a more complete understanding of the components of RTK transactivation [80]. Both approaches provide the much needed opportunity to move transactivation studies beyond one GPCR–one RTK approach and to fully appreciate the full scope of RTK transactivation.

References

1. Beaulieu JM, Gainetdinov RR (2011) The physiology, signaling, and pharmacology of dopamine receptors. Pharmacol Rev 63: 182–217
2. Hubbard SR (1999) Structural analysis of receptor tyrosine kinases. Prog Biophys Mol Biol 71:343–358
3. Daub H, Weiss FU, Wallasch C, Ullrich A (1996) Role of transactivation of the EGF receptor in signalling by G-protein-coupled receptors. Nature 379:557–560
4. Almendro V, Garcia-Recio S, Gascon P (2010) Tyrosine kinase receptor transactivation associated to G protein-coupled receptors. Curr Drug Targets 11:1169–1180
5. Delcourt N, Bockaert J, Marin P (2007) GPCR-jacking: from a new route in RTK signalling to a new concept in GPCR activation. Trends Pharmacol Sci 28:602–607
6. Wetzker R, Bohmer FD (2003) Transactivation joins multiple tracks to the ERK/MAPK cascade. Nat Rev Mol Cell Biol 4:651–657
7. Maretzky T, Evers A, Zhou W et al (2011) Migration of growth factor-stimulated epithelial and endothelial cells depends on EGFR transactivation by ADAM17. Nat Commun 2:229
8. Werry TD, Sexton PM, Christopoulos A (2005) "Ins and outs" of seven-transmembrane receptor signalling to ERK. Trends Endocrinol Metab 16:26–33
9. Oak JN, Lavine N, Van Tol HH (2001) Dopamine D(4) and D(2 L) Receptor Stimulation of the Mitogen-Activated Protein Kinase Pathway Is Dependent on trans-Activation of the Platelet-Derived Growth Factor Receptor. Mol Pharmacol 60:92–103
10. Chi SS, Vetiska SM, Gill RS, Hsiung MS, Liu F, Van Tol HH (2010) Transactivation of PDGFRbeta by dopamine D4 receptor does not require PDGFRbeta dimerization. Mol Brain 3:22
11. Heldin CH, Westermark B (1999) Mechanism of action and in vivo role of platelet-derived growth factor. Physiol Rev 79:1283–1316
12. Daniel TO, Milfay DF, Escobedo J, Williams LT (1987) Biosynthetic and glycosylation studies of cell surface platelet-derived growth factor receptors. J Biol Chem 262:9778–9784
13. Gill RS, Hsiung MS, Sum CS, Lavine N, Clark SD, Van Tol HH (2010) The dopamine D4 receptor activates intracellular platelet-derived growth factor receptor beta to stimulate ERK1/2. Cell Signal 22:285–290

14. Wang C, Buck DC, Yang R, Macey TA, Neve KA (2005) Dopamine D2 receptor stimulation of mitogen-activated protein kinases mediated by cell type-dependent transactivation of receptor tyrosine kinases. J Neurochem 93:899–909
15. Lei S, Lu WY, Xiong ZG, Orser BA, Valenzuela CF, MacDonald JF (1999) Platelet-derived growth factor receptor-induced feed-forward inhibition of excitatory transmission between hippocampal pyramidal neurons. J Biol Chem 274:30617–30623
16. Valenzuela CF, Xiong Z, MacDonald JF et al (1996) Platelet-derived growth factor induces a long-term inhibition of N-methyl-D-aspartate receptor function. J Biol Chem 271: 16151–16159
17. Beazely MA, Lim A, Li H et al (2009) Platelet-derived growth factor selectively inhibits NR2B-containing N-methyl-D-aspartate receptors in CA1 hippocampal neurons. J Biol Chem 284: 8054–8063
18. Kotecha SA, Oak JN, Jackson MF et al (2002) A D2 class dopamine receptor transactivates a receptor tyrosine kinase to inhibit NMDA receptor transmission. Neuron 35:1111–1122
19. Beazely MA, Tong A, Wei WL, Van Tol H, Sidhu B, MacDonald JF (2006) D2-class dopamine receptor inhibition of NMDA currents in prefrontal cortical neurons is platelet-derived growth factor receptor-dependent. J Neurochem 98:1657–1663
20. Nair VD, Sealfon SC (2003) Agonist-specific transactivation of phosphoinositide 3-kinase signaling pathway mediated by the dopamine D2 receptor. J Biol Chem 278:47053–47061
21. O'Keeffe GC, Tyers P, Aarsland D, Dalley JW, Barker RA, Caldwell MA (2009) Dopamine-induced proliferation of adult neural precursor cells in the mammalian subventricular zone is mediated through EGF. Proc Natl Acad Sci U S A 106:8754–8759
22. O'Keeffe GC, Barker RA (2011) Dopamine stimulates epidermal growth factor release from adult neural precursor cells derived from the subventricular zone by a disintegrin and metalloprotease. Neuroreport 22:956–958
23. Ohtsu H, Dempsey PJ, Eguchi S (2006) ADAMs as mediators of EGF receptor transactivation by G protein-coupled receptors. Am J Physiol Cell Physiol 291:C1–C10
24. Yoon S, Baik JH (2013) Dopamine D2 receptor-mediated epidermal growth factor receptor transactivation through a disintegrin and metalloprotease regulates dopaminergic neuron development via extracellular signal-related kinase activation. J Biol Chem 288: 28435–28446
25. Beom S, Cheong D, Torres G, Caron MG, Kim KM (2004) Comparative studies of molecular mechanisms of dopamine D2 and D3 receptors for the activation of extracellular signal-regulated kinase. J Biol Chem 279: 28304–28314
26. Mannoury la Cour C, Salles MJ, Pasteau V, Millan MJ (2011) Signaling pathways leading to phosphorylation of Akt and GSK-3beta by activation of cloned human and rat cerebral D(2)and D(3) receptors. Mol Pharmacol 79:91–105
27. Salles MJ, Herve D, Rivet JM et al (2013) Transient and rapid activation of Akt/GSK-3beta and mTORC1 signaling by D3 dopamine receptor stimulation in dorsal striatum and nucleus accumbens. J Neurochem 125: 532–544
28. Kim SJ, Kim MY, Lee EJ, Ahn YS, Baik JH (2004) Distinct regulation of internalization and mitogen-activated protein kinase activation by two isoforms of the dopamine D2 receptor. Mol Endocrinol 18:640–652
29. Beaulieu JM, Del'guidice T, Sotnikova TD, Lemasson M, Gainetdinov RR (2011) Beyond cAMP: The Regulation of Akt and GSK3 by Dopamine Receptors. Front Mol Neurosci 4:38
30. Iwakura Y, Wang R, Abe Y et al (2011) Dopamine-dependent ectodomain shedding and release of epidermal growth factor in developing striatum: target-derived neurotrophic signaling (Part 2). J Neurochem 118:57–68
31. Iwakura Y, Nawa H, Sora I, Chao MV (2008) Dopamine D1 receptor-induced signaling through TrkB receptors in striatal neurons. J Biol Chem 283:15799–15806
32. Kruk JS, Vasefi MS, Liu H, Heikkila JJ, Beazely MA (2013) 5-HT(1A) receptors transactivate the platelet-derived growth factor receptor type beta in neuronal cells. Cell Signal 25:133–143
33. Lee FS, Rajagopal R, Chao MV (2002) Distinctive features of Trk neurotrophin receptor transactivation by G protein-coupled receptors. Cytokine Growth Factor Rev 13:11–17
34. Quan W, Kim JH, Albert PR, Choi H, Kim KM (2008) Roles of G protein and beta-arrestin in dopamine D2 receptor-mediated ERK activation. Biochem Biophys Res Commun 377:705–709
35. Prenzel N, Zwick E, Daub H et al (1999) EGF receptor transactivation by G-protein-coupled receptors requires metalloproteinase cleavage of proHB-EGF. Nature 402:884–888
36. Pierce KL, Tohgo A, Ahn S, Field ME, Luttrell LM, Lefkowitz RJ (2001) Epidermal growth factor (EGF) receptor-dependent ERK activation by G protein-coupled receptors: a co-culture system for identifying intermediates upstream

and downstream of heparin-binding EGF shedding. J Biol Chem 276:23155–23160

37. Swift JL, Godin AG, Dore K et al (2011) Quantification of receptor tyrosine kinase transactivation through direct dimerization and surface density measurements in single cells. Proc Natl Acad Sci U S A 108:7016–7021
38. Ueno H, Colbert H, Escobedo JA, Williams LT (1991) Inhibition of PDGF beta receptor signal transduction by coexpression of a truncated receptor. Science 252:844–848
39. Gronwald RG, Grant FJ, Haldeman BA et al (1988) Cloning and expression of a cDNA coding for the human platelet-derived growth factor receptor: evidence for more than one receptor class. Proc Natl Acad Sci U S A 85:3435–3439
40. Rajagopal R, Chen ZY, Lee FS, Chao MV (2004) Transactivation of Trk neurotrophin receptors by G-protein-coupled receptor ligands occurs on intracellular membranes. J Neurosci 24:6650–6658
41. Pyne NJ, Waters C, Moughal NA, Sambi BS, Pyne S (2003) Receptor tyrosine kinase-GPCR signal complexes. Biochem Soc Trans 31:1220–1225
42. Lee FS, Chao MV (2001) Activation of Trk neurotrophin receptors in the absence of neurotrophins. Proc Natl Acad Sci U S A 98: 3555–3560
43. Vasefi MS, Kruk JS, Liu H, Heikkila JJ, Beazely MA (2012) Activation of 5-HT7 receptors increases neuronal platelet-derived growth factor beta receptor expression. Neurosci Lett 511:65–69
44. Vasefi MS, Kruk JS, Heikkila JJ, Beazely MA (2013) 5-Hydroxytryptamine type 7 receptor neuroprotection against NMDA-induced excitotoxicity is PDGFbeta receptor dependent. J Neurochem 125:26–36
45. Kruk JS, Vasefi MS, Heikkila JJ, Beazely MA (2013) Reactive oxygen species are required for 5-HT-induced transactivation of neuronal platelet-derived growth factor and TrkB receptors, but not for ERK1/2 activation. PLoS One 8:e77027
46. Zhen X, Zhang J, Johnson GP, Friedman E (2001) D(4) dopamine receptor differentially regulates Akt/nuclear factor-kappa b and extracellular signal-regulated kinase pathways in D(4)MN9D cells. Mol Pharmacol 60:857–864
47. Choi EY, Jeong D, Park KW, Baik JH (1999) G protein-mediated mitogen-activated protein kinase activation by two dopamine D2 receptors. Biochem Biophys Res Commun 256:33–40
48. Jin M, Min C, Zheng M et al (2013) Multiple signaling routes involved in the regulation of adenylyl cyclase and extracellular regulated kinase by dopamine D(2) and D(3) receptors. Pharmacol Res 67:31–41
49. Guerrero C, Pesce L, Lecuona E, Ridge KM, Sznajder JI (2002) Dopamine activates ERKs in alveolar epithelial cells via Ras-PKC-dependent and Grb2/Sos-independent mechanisms. Am J Physiol Lung Cell Mol Physiol 282:L1099–L1107
50. Cai G, Zhen X, Uryu K, Friedman E (2000) Activation of extracellular signal-regulated protein kinases is associated with a sensitized locomotor response to D(2) dopamine receptor stimulation in unilateral 6-hydroxydopamine-lesioned rats. J Neurosci 20:1849–1857
51. Luttrell LM (2005) Composition and function of g protein-coupled receptor signalsomes controlling mitogen-activated protein kinase activity. J Mol Neurosci 26:253–264
52. Rozengurt E (2007) Mitogenic signaling pathways induced by G protein-coupled receptors. J Cell Physiol 213:589–602
53. Conway AM, Rakhit S, Pyne S, Pyne NJ (1999) Platelet-derived-growth-factor stimulation of the p42/p44 mitogen-activated protein kinase pathway in airway smooth muscle: role of pertussis-toxin-sensitive G-proteins, c-Src tyrosine kinases and phosphoinositide 3-kinase. Biochem J 337(Pt 2):171–177
54. Alderton F, Rakhit S, Kong KC et al (2001) Tethering of the platelet-derived growth factor beta receptor to G-protein-coupled receptors. A novel platform for integrative signaling by these receptor classes in mammalian cells. J Biol Chem 276:28578–28585
55. Pyne NJ, Pyne S (2011) Receptor tyrosine kinase-G-protein-coupled receptor signalling platforms: out of the shadow? Trends Pharmacol Sci 32:443–450
56. Pyne NJ, Waters C, Moughal NA, Sambi B, Connell M, Pyne S (2004) Experimental systems for studying the role of G-protein-coupled receptors in receptor tyrosine kinase signal transduction. Methods Enzymol 390:451–475
57. Freedman NJ, Kim LK, Murray JP et al (2002) Phosphorylation of the platelet-derived growth factor receptor-beta and epidermal growth factor receptor by G protein-coupled receptor kinase-2. Mechanisms for selectivity of desensitization. J Biol Chem 277:48261–48269
58. Wu JH, Goswami R, Kim LK, Miller WE, Peppel K, Freedman NJ (2005) The platelet-derived growth factor receptor-beta phosphorylates and activates G protein-coupled receptor kinase-2. A mechanism for feedback inhibition. J Biol Chem 280:31027–31035
59. Wu JH, Goswami R, Cai X et al (2006) Regulation of the platelet-derived growth factor

receptor-beta by G protein-coupled receptor kinase-5 in vascular smooth muscle cells involves the phosphatase Shp2. J Biol Chem 281:37758–37772

60. Cai X, Wu JH, Exum ST et al (2009) Reciprocal regulation of the platelet-derived growth factor receptor-beta and G protein-coupled receptor kinase 5 by cross-phosphorylation: effects on catalysis. Mol Pharmacol 75:626–636
61. Hupfeld CJ, Olefsky JM (2007) Regulation of receptor tyrosine kinase signaling by GRKs and beta-arrestins. Annu Rev Physiol 69:561–577
62. Chen CH, Cheng TH, Lin H et al (2006) Reactive oxygen species generation is involved in epidermal growth factor receptor transactivation through the transient oxidization of Src homology 2-containing tyrosine phosphatase in endothelin-1 signaling pathway in rat cardiac fibroblasts. Mol Pharmacol 69:1347–1355
63. Hu Y, Kang C, Philp RJ, Li B (2007) PKC delta phosphorylates p52ShcA at Ser29 to regulate ERK activation in response to H2O2. Cell Signal 19:410–418
64. Kim YK, Bae GU, Kang JK et al (2006) Cooperation of H2O2-mediated ERK activation with Smad pathway in TGF-beta1 induction of p21WAF1/Cip1. Cell Signal 18:236–243
65. Mbong N, Anand-Srivastava MB (2012) Hydrogen peroxide enhances the expression of Gialpha proteins in aortic vascular smooth cells: role of growth factor receptor transactivation. Am J Physiol Heart Circ Physiol 302: H1591–H1602
66. Tanimoto T, Lungu AO, Berk BC (2004) Sphingosine 1-phosphate transactivates the platelet-derived growth factor beta receptor and epidermal growth factor receptor in vascular smooth muscle cells. Circ Res 94: 1050–1058
67. Ushio-Fukai M, Hilenski L, Santanam N et al (2001) Cholesterol depletion inhibits epidermal growth factor receptor transactivation by angiotensin II in vascular smooth muscle cells: role of cholesterol-rich microdomains and focal adhesions in angiotensin II signaling. J Biol Chem 276:48269–48275
68. Means CK, Miyamoto S, Chun J, Brown JH (2008) S1P1 receptor localization confers selectivity for Gi-mediated cAMP and contractile responses. J Biol Chem 283:11954–11963
69. Long JS, Natarajan V, Tigyi G, Pyne S, Pyne NJ (2006) The functional PDGFbeta receptor-S1P1 receptor signaling complex is involved in regulating migration of mouse embryonic fibroblasts in response to platelet derived growth factor. Prostaglandins Other Lipid Mediat 80:74–80
70. Waters C, Sambi B, Kong KC et al (2003) Sphingosine 1-phosphate and platelet-derived growth factor (PDGF) act via PDGF beta receptor-sphingosine 1-phosphate receptor complexes in airway smooth muscle cells. J Biol Chem 278:6282–6290
71. Patel HH, Murray F, Insel PA (2008) G-protein-coupled receptor-signaling components in membrane raft and caveolae microdomains. Handb Exp Pharmacol 167–184
72. Bhatnagar A, Sheffler DJ, Kroeze WK, Compton-Toth B, Roth BL (2004) Caveolin-1 interacts with 5-HT2A serotonin receptors and profoundly modulates the signaling of selected Galphaq-coupled protein receptors. J Biol Chem 279:34614–34623
73. Fiorica-Howells E, Hen R, Gingrich J, Li Z, Gershon MD (2002) 5-HT(2A) receptors: location and functional analysis in intestines of wild-type and 5-HT(2A) knockout mice. Am J Physiol Gastrointest Liver Physiol 282: G877–G893
74. Gambara G, Billington RA, Debidda M et al (2008) NAADP-induced Ca(2+ signaling in response to endothelin is via the receptor subtype B and requires the integrity of lipid rafts/caveolae. J Cell Physiol 216:396–404
75. Kong MM, Hasbi A, Mattocks M, Fan T, O'Dowd BF, George SR (2007) Regulation of D1 dopamine receptor trafficking and signaling by caveolin-1. Mol Pharmacol 72:1157–1170
76. Oh YB, Gao S, Lim JM, Kim HT, Park BH, Kim SH (2011) Caveolae are essential for angiotensin II type 1 receptor-mediated ANP secretion. Peptides 32:1422–1430
77. Yamaguchi T, Murata Y, Fujiyoshi Y, Doi T (2003) Regulated interaction of endothelin B receptor with caveolin-1. Eur J Biochem 270: 1816–1827
78. Yu P, Yang Z, Jones JE et al (2004) D1 dopamine receptor signaling involves caveolin-2 in HEK-293 cells. Kidney Int 66:2167–2180
79. Barbeau A, Swift JL, Godin AG, De Koninck Y, Wiseman PW, Beaulieu JM (2013) Spatial intensity distribution analysis (SpIDA): a new tool for receptor tyrosine kinase activation and transactivation quantification. Methods Cell Biol 117:1–19
80. George AJ, Hannan RD, Thomas WG (2013) Unravelling the molecular complexity of GPCR-mediated EGFR transactivation using functional genomics approaches. FEBS J 280: 5258–5268

Chapter 13

Dopamine Receptors in Human Embryonic Stem Cell Differentiation

Glenn S. Belinsky, Mandakini B. Singh, Katerina D. Oikonomou, Michele L. McGovern, and Srdjan D. Antic

Abstract

Dopamine (DA) releasing axons are widely distributed in the adult central nervous system. They strongly influence\regulate signal integrations in the brain, including control of movement, functioning of limbic system, and cortical information processing. It is not widely known that DA axons invade the immature zones of the human fetal brain and participate in neuronal development long before they become engaged in information processing. This is interesting because both DA function and neuronal development are at the center of the modern schizophrenia research. More specifically, schizophrenia is now thought to be a consequence of abnormal processes that occur in the earliest stages of neuron differentiation. Due to multiple ethical and practical constrains, the structure and function of DA receptors in the human fetal brain is inaccessible to controlled scientific experiments. Recently developed human stem cell technology allows monitoring of molecular and cellular processes at every stage of in vitro neurodifferentiation: from a pluripotent stem cell to a differentiated neuron. This chapter describes several technical solutions for measuring human DA receptor mRNA, human DA receptor protein, and DA receptor-induced calcium release in human cells.

Key words Human, Stem cell, Dopaminergic, Neurodifferentiation, Postmitotic neurons, Neurites, Primers, PCR, Calcium-sensitive dye, Bulk-loading, Internal release of calcium

1 Introduction

Intense interest in the adult brain's dopaminergic system has been prompted by its involvement in reward-motivated behavior, Parkinson's disease, and mental diseases [1]. The function of the dopamine system in development is less well understood. Serious ethical and methodological obstacles preclude experiments with dopamine receptors during human embryonic development. Human pluripotent stem cells (hESCs and iPSCs), on the other hand, allow for examination of dopamine signaling at different stages of differentiation [2, 3]. An in vitro stem cell-based system of neuronal development will be useful for generating and testing hypotheses

Mario Tiberi (ed.), *Dopamine Receptor Technologies*, Neuromethods, vol. 96,
DOI 10.1007/978-1-4939-2196-6_13, © Springer Science+Business Media New York 2015

concerning which receptor subtypes are expressed and functional during the earliest stages of human neurodifferentiation, as pluripotent cellular lineages become committed to a neuronal lineage.

The existence of multiple DA receptors is indicative of the biological importance of the pathway and redundancy of receptor function. D1 and D5 DA receptors act via Gs-alpha to stimulate adenylyl cyclase, while D2–D4 receptors act via Gi-alpha, to inhibit cyclic AMP synthesis [4]. Dopamine receptor oligomerization, receptor-to-receptor interactions, or dopamine receptor interactions with scaffolding and signal-switching proteins also contribute to the complex regulation of dopamine receptor signaling [4–7]. Over 100 agonists and antagonists have been developed for these receptors, giving investigators the tools to probe DA receptor signal potentiation and attenuation in vitro prior to creating receptor knockout lines. However, investigators must carefully choose among the wide variety of endpoints for measurement, for example cell/colony morphology, percent yield of neuronal subtypes, or mRNA/protein expression. Researchers must also choose the precise stage of differentiation to measure those endpoints. With this in mind, this chapter presents protocols for measuring DA receptor expression, a method to verify coupling of signal transduction machinery, non-destructive methods to measure effects on gross morphology of live differentiating human stem cells, as well as the details for comparing yields of dopaminergic neurons. Some challenges associated with working with differentiating cultures are the length of time required to produce neurons (usually several weeks), the cellular heterogeneity of the cultures, and the propensity of neurons to migrate into clumps and form tangled masses of neurites. Other issues include the selection of primers for human dopamine receptors, practical problems with Western blot on stem cells, and fast delivery of dopaminergic drugs during multisite calcium imaging. Techniques to overcome these issues are discussed.

2 Materials

2.1 Stem Cell Culture

1. All Reagents are from Sigma (St. Louis, MO) unless otherwise noted. Stem cell media: 80 % Dulbecco's Modified Eagle Medium (DMEM)/F12 (Invitrogen, Carlsbad, CA, catalog# 11330), 20 % KnockOut Serum Replacement (KOSR, Invitrogen 10828), 1 mM glutamine (Invitrogen), 1× nonessential amino acids (NEAA, Invitrogen 11150), 4 ng/ml basic fibroblast growth factor (bFGF, Invitrogen 13256), and 7 nl/ml β-mercaptoethanol.
2. Six-well plates (Nunc, Rochester, NY) and 12 mm circular coverslips (Fisher 12545-80) coated with 0.1 % porcine gelatin.
3. Irradiated mouse embryo fibroblasts (MEF).

2.2 rtPCR

1. Trizol (Invitrogen), DNase (NEB, M0303S).
2. SSIII reverse transcriptase (Invitrogen), RNase OUT (Invitrogen), Oligo dT_{20} (Qiagen 79237 Germantown, Maryland), Random hexamers (Qiagen 79236) dNTP's (Promega U1511, Madison WI).
3. Primers (*see* **Note 1**):

   Dopamine receptors D1–D5

   D1 F 5′CAGTCCACGCCAAGAATTGCC, R 5′ATTGCACT CCTTGGAGATGGAGCC;

   D2 F 5′GCAGACCACCACCAACTACC, R 5′GGAGCTGTA GCGCGTATTGT;

   D3 F 5′TGGCTGCAGGAGCCGAAGT, R 5′GAGGGCAG GACACAGCAAAGGC;

   D4 F 5′CCCACCCCAGACTCCACC, R 5′GAACTCGGCGT TGAAGACAG;

   D5 F 5′GTCGCCGAGGTGGCCGGTTAC, R 5′GCTGGAGT CACAGTTCTCTGCAT;

   Tyrosine hydroxylase (TH)

   TH F 5′GGTTCCCAAGAAAAGTGTCAG, R 5′GGTGT AGACCTCCTTCCAG;

   Peptidyl-prolyl cis-trans isomerase A (PPIA)

   PPIA F 5′CCAGGCTCGTGCCGTTTTGC, R 5′GATGGA CTTGCCACCAGTGCCA;

   Hypoxanthine phosphoribosyltransferase 1 (HPRT)

   HPRT F 5′ GACTTTGCTTTCCTTGGTCA, R 5′ GGCTTT GTATTTTGCTTTTCC;

   β-actin (βACT)

   βACT F 5′CCTCGCCTTTGCCGATCC, R 5′ GATGCCGTG CTCGATGGGGT.
4. GoTaq Polymerase (Promega)

2.3 Westerns

1. Radioimmunoprecipitation assay (RIPA) buffer (Sigma R0278) containing 1 mM Phenylmethanesulfonyl fluoride (PMSF), 2.1 mM 4-(2-Aminoethyl)benzenesulfonyl fluoride (AEBSF), mM, 1.6 μM Aprotinin, 80 μM Bestatin, 28 μM E-64, 40 μM Leupeptin, 30 μM Pepstatin A, 2 mM Na3VO4, and 25 mM NaF.
2. Bicinchoninic acid (BCA) protein assay (Pierce, Rockford, IL).
3. 5× Loading buffer: 10 % SDS, 25 % βME, 50 % glycerol, 0.05 % bromophenol blue.
4. ColorPlus prestained marker proteins (NEB, Ipswich, MA)).
5. 15 or 12 lane 4–15 % TGX gradient mini gels (Bio-Rad, Hercules, CA).

6. Polyvinylidene fluoride (PVDF) membrane (Bio-Rad).
7. PBS + 5 % nonfat dry milk.
8. Rabbit anti-D1 (1:1,000 Abcam ab20066 Cambridge, MA), rabbit anti-D2 (1:300 Millipore AB5084P, Billerica, MA), Rabbit anti-D3 (1:1,000 LSBio, LS-C146243, Seattle, WA), mouse anti-D4 (1:1,000 Millipore MABN125), rabbit anti-D5 (1:300 Millipore AB9509), rabbit anti-TH (1:500 Pel-Freez P40101, Rogers, AR), rabbit anti-glyceraldehyde 3-phosphate dehydrogenase (GAPDH) (1:200 Santa Cruz Biotech. sc25778, Santa Cruz, CA) in PBS, 0.05 % Tween-20 and 0.5 % BSA.
9. Phosphate buffered saline (PBS) + 0.05 % Tween-20.
10. Anti-rabbit IgG-horseradish peroxidase (HRP) (1:2,000 Santa Cruz sc-2301) and anti-mouse IgG-HRP (1:2,000 Santa Cruz sc-2005) in PBS, 0.05 % Tween-20 and 0.1 % BSA.
11. Electrochemiluminescence (ECL) reagent (GE Healthcare, Pittsburgh).

2.4 Calcium-Sensitive Dye Loading

1. Oregon Green 488 Bapta-1 AM (OGB1-AM, Invitrogen, Cat. 06807).
2. 20 % Pluronic F-127 (Sigma Cat # P2443).
3. Dimethyl sulfoxide (DMSO).

2.5 Calcium Imaging: Equipment

1. BM-1 Bench Top Vibration Isolation Platform (Minus K Technology, Inc., Inglewood, CA).
2. Upright microscope with an epifluorescence module (Olympus BX51WI).
3. Recording chamber with feedback heater control (Warner Instruments, Hamden, CT).
4. Blue light-emitting diode (LED, 460 nm) for the light source (Luminus Devices, Inc., Billerica, MA). Custom-made computer-controlled LED power supply (Peter Lee, Essel R&D Inc, Toronto, Canada).
5. Charge-coupled device (CDD) camera (NeuroCCD-SMQ, RedShirtImaging, Decatur, GA).
6. Stimulus isolation unit (Isoflex, A.M.P.I., Jerusalem, Israel).
7. Micromanipulator MP-285 (Sutter Instrument, Novato, CA).
8. Micropipette Puller P-97 (Sutter Instrument, Novato, CA).

2.6 Live Colony Morphology

1. Inverted microscope with phase contrast optics.
2. Fine point markers (ultrafine point *Sharpie*, Oak Brook, Il).

2.7 Immunofluorescence

1. 4 % paraformaldehyde in PBS.
2. PBS + 0.2 % Triton TM X100 (Acros, Geel, Belgium).

3. 1:500 rabbit anti-TH (Pel-Freez, Rogers, AR), 1:1,000 mouse anti-TUJ1 (Sigma T5076), 1:1,000 mouse anti-NeuN (Clone 60, Millipore, Billerica, MA) in PBS + 10 % normal goat serum, 0.75 % BSA.
4. 1:500 Goat anti-mouse Alexa Fluor488 and goat anti-rabbit Alexa Fluor594 (Invitrogen) in PBS + 10 % goat serum.
5. Fluoromount-G (SouthernBiotech, Birmingham, AL).

3 Procedures

3.1 Stem Cell Culture

1. H9 cells are cocultured with mouse embryonic fibroblasts (MEFs) using standard methods [8]. Briefly, MEFs are grown in DMEM + 10% FBS + NEAA at 37 °C in 5 % carbon dioxide. Confluent MEFs are removed from the flask using 0.5 % trypsin, then irradiated with 8,000 rads, and then 1×10^6 cells seeded on a gelatin coated 6-well dish. For growth on coverslips, 0.9×10^6 irradiated MEFs are seeded on one 24-well plate containing one 12 mm glass coverslip per well.
2. Thaw stem cells and resuspend in 12 ml DMEM/F12. Centrifuge and aspirate off the DMEM/F12. Add 2–4 ml of stem cell media plus 3.3 μg/ml of Rock inhibitor, mix and seed 1–2 wells containing irradiated MEFs.
3. To passage stem cells, stem cell colonies are treated with 1 mg/ml collagenase in DMEM/F12 for 5 min. Colonies are rinsed twice with DMEM/F12, and then 2 ml of hESC growth media are added to each well. Colonies are scraped off the plate with a glass pipette. MEFs are rinsed once with DMEM/F12. Colonies are seeded on the rinsed MEFs at a 1:3 or 1:6 dilution.
4. Media should be changed daily and colonies should be checked for differentiating cells, which are first marked and later removed by aspiration. Colonies lacking well defined boarders or a uniform appearance across the colony are generally considered differentiated and must be removed.

3.2 Neuro-differentiation Protocol

hESCs were differentiated into neurons using a previously published protocol [9, 10]. The differentiation process consisted of five stages, starting with undifferentiated hESCs (Fig. 1a; Stage-1). hESC colonies were dissociated by collagenase, and stem cell aggregates (Embryoid Bodies, EBs) incubated for 3–4 days in hESC media without bFGF, on Ultralow adherence plates (Costar, Wilkes Barre, PA) (Stage-2). hESC aggregates were then seeded on dishes coated with 1:100 Geltrex, and allowed to expand for 4–8 days in NEP-basal medium (Stage-3) until neuroepithelial colonies appeared. NEP-basal medium consisted of DMEM/F12, 1 mg/mL BSA, 1× N2, 1× B27 supplements, and 1× penicillin/streptomycin/antimycotic. Colonies with neuroepithelial

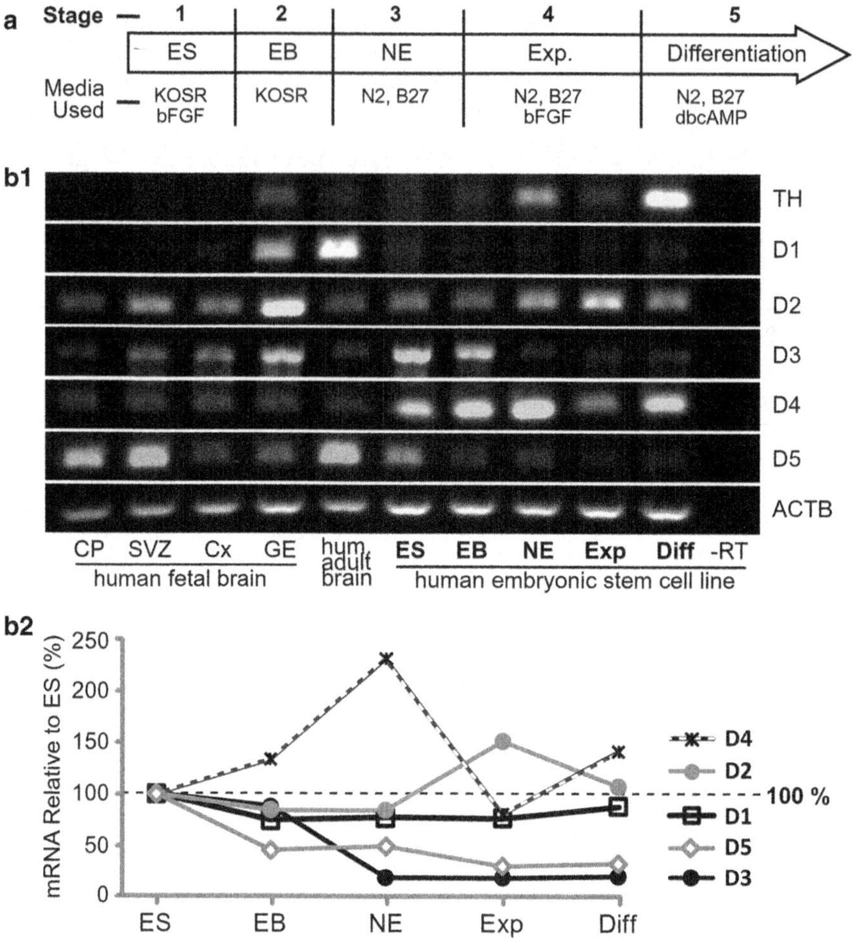

Fig. 1 Dopamine receptor mRNA levels during the dbcAMP-based differentiation protocol. (**a**) Time course of the differentiation protocol indicating media additives during each stage (Stages 1–5). (**b$_1$**) rtPCR of TH, five dopamine receptors and the housekeeping gene ACTB (β-actin). CP = cortical plate, SVZ = subventricular zone, Cx = cortex with all zones included, GE = ganglionic eminence, ES = Stage-1 undifferentiated H9 hES colonies, EB = Stage-2 embryoid bodies, NE = Stage-3 neuroepithelia, Exp = Stage-4 expansion of neuroepithelia, Dif = Stage-5 final neurodifferentiation. (**b$_2$**) Quantitation of hESC band intensities relative to Stage 1 (ES). Figure taken from Belinsky et al. [2]

morphology were removed by trituration and seeded on plates coated with 1:100 Geltrex. Cells were then grown in NEP-basal medium with 20 ng/mL bFGF for 7 days (*Expansion*, Stage-4). Cells were then maintained in NEP-basal medium with 1 mM Dibutyryl-cAMP (dbcAMP, Sigma) for 7 days, (*Differentiation*, Stage-5). For intermittent dopaminergic treatment, DA (Sigma) was added for 3–4 h each day during the expansion and differentiation stages using concentrations indicated in the figures [2].

3.3 RT-PCR of Dopamine Receptors, Tyrosine Hydroxylase, and Housekeeping Genes

1. Total RNA is purified with Trizol according to the manufacturer's recommendations. For expected yields (*see* **Note 3**).
2. RNA is treated with DNase according to the manufacturer's protocol (*see* **Note 1**).
3. cDNA is made according to the manufacturer's instructions.
4. PCR is done using the following protocol: 4 min 94 °C, (30 s 94 °C, 30 s 55 °C, 30 s 72 °C) 27–32 cycles, 10 min 72 °C. Extension time is increased to 45 s for amplicons above 400 bp. Housekeeping genes require the lower number of cycles to fall within the exponential phase of amplification.
5. Amplicons are subjected to electrophoresis and visualized with ethidium bromide.

3.4 Western Blots on Dopamine Receptors

1. Cells or tissue are lysed in cold RIPA buffer containing protease and phosphatase inhibitors. PMSF is added immediately before use (*see* **Note 2** for expected yields).
2. Adherent colonies or cells are washed quickly three times with cold PBS. After the last wash, as much liquid as possible is removed by aspiration out of the edge of the well held at an angle. For 6 wells, a total of 0.8 ml of lysis buffer is sufficient (~130 μl/well). The lysis buffer is pipetted into the center of the dish, and the flat end of a spatula is used to scrape the cells into one edge of the tilted dish. EBs require brief sonication on ice with a probe sonicator. Samples are stored at −80 °C.
3. Protein assays are performed using BSA as a standard.
4. Prior to loading, extracts are brought to 2 % sodium dodecyl sulfate (SDS), 5 % βME, 10 % glycerol, 0.01 % bromophenol blue, and heated to 95 °C for 5 min.
5. 15-lane or 12-lane gradient mini gels are loaded with 25–33 μg/lane respectively, then transferred to PVDF membrane according to the manufacturer's instructions.
6. Membranes are blocked for 1–3 h with 5 % nonfat dry milk in PBS.
7. Primary antibody incubations are carried out overnight in PBS + 0.05 % Tween-20 + 0.5 % BSA at 4 °C.
8. Membranes are washed 4 × 5 min with PBS + 0.05 % Tween-20, then secondary antibody incubation done in PBS + 0.05 % Tween-20 + 0.1 % BSA for 1.5 h.
9. After 4 × 5 min washes with PBS, bands are visualized using ECL chemiluminescent reagent.

The following sections describe several non-destructive methods of assaying effects of agonist/antagonist on live stem cell derived cultures:

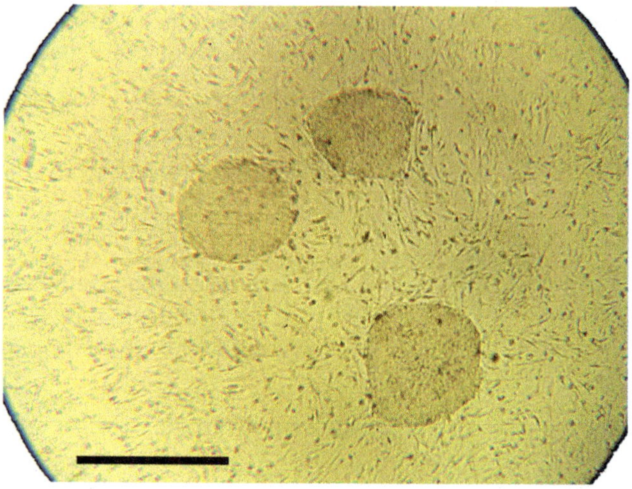

Fig. 2 Live hESC colonies in a culture well. Inverted microscope, phase contrast, 4×. Scale, 1 mm. hESC colonies are growing on a feeder layer of mouse embryonic fibroblasts (MEFs)

3.5 Live hES Colony

Morphology:

1. After experimental treatments of cultures, stem cell colonies are photographed at low power (4×) (Fig. 2). The files are coded so that the experiment can be scored blindly. The diameters of the colonies are measured and the means of treatment groups compared. Typically five serial fields containing multiple colonies are required. For counting numbers of differentiated colonies, the bottoms of plates can be marked and colonies quantitated (*see* **Note 4**).

3.6 Embryoid Body

Morphology:

We have noticed striking changes in embryoid body morphology after treatment with the D1 DA receptor antagonist SKF83566 (Fig. 3) [2]. To quantify effects such as this, EBs are photographed by swirling the dish, so that the EBs are massed in the center, prior to capturing images. The images of EBs are scored for morphology and different treatment groups compared. In this case, SKF83566 caused the EBs to have an irregular surface as opposed to the distinct smooth surface of the control EBs (Fig. 3) [2].

3.7 Neuroepithelial Stage

Morphology:

Many neuronal differentiation protocols require the adherence of EBs to plastic and subsequent formation of neuroepithelia in the resultant colony. The neuroepithelial morphology is apparent under phase contrast microscopy; thus, the percent of colonies with neuroepithelial morphology can be quickly quantitated in living preparations (Fig. 4a). The primitive neuroepithelial morphology, consisting of elongated pseudostratified cells that form

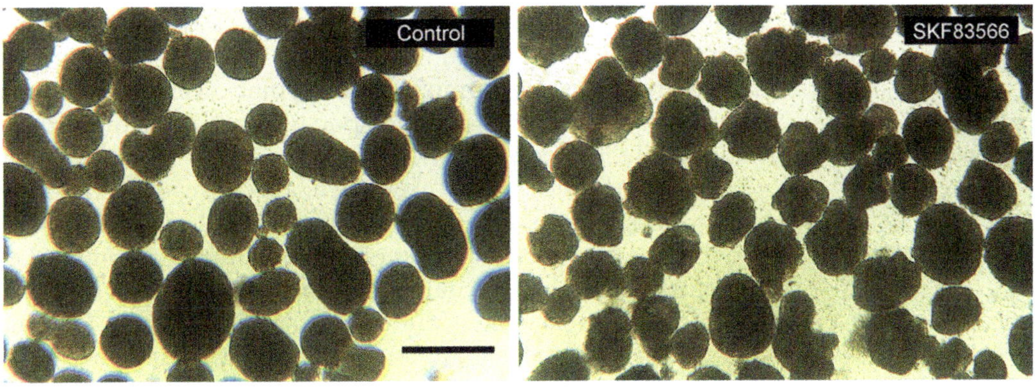

Fig. 3 Embryoid bodies (EBs) made from hESC colonies. *Left*: EBs in grown in the regular media. Inverted microscope, phase contrast, 4×. Scale, 0.5 mm. *Right*: Treatment with D1 DA receptor antagonist SKF83566 induces dramatic changes in EB morphology. Note the rugged and irregular edges of EBs treated with SKF83566, as opposed to smooth-edged and oval EBs grown in drug-free media

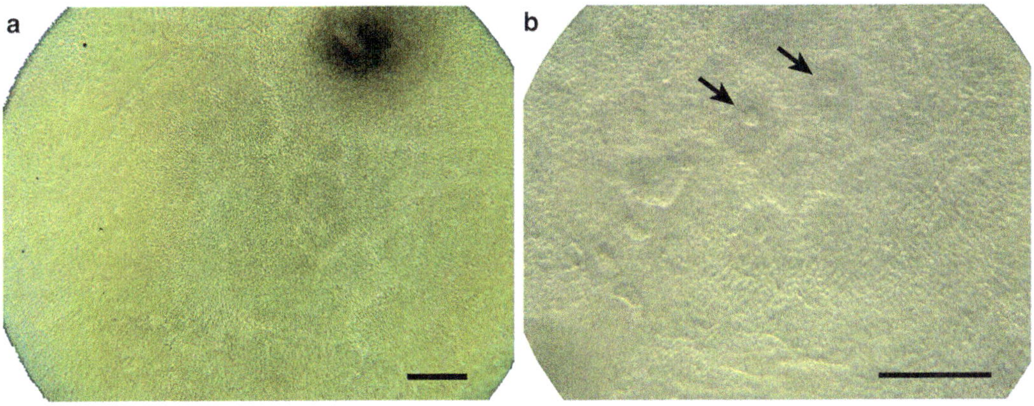

Fig. 4 Differentiation of hESC into neuronal epithelia—Stages 3 and 4. (**a**) Stage 3, neuroepithelia (NE). This is the result of adherence of EB's to the bottom of a well and subsequent formation of neural epithelium (neuroepithelia, NE). Scale, 500 μm. (**b**) An image taken 7 days after re-seeding of the neuroepithelial colonies. The formation of characteristic rosettes (*arrows*) is thought to be an equivalent of the neural tube formation in the earliest stages of brain development. Scale, 200 μm

characteristic ridge patterns in the colony, sometimes can be difficult to visualize at low power and generally requires 10× phase contrast (Fig. 4a).

3.8 Rosettes

After formation of primitive neuroepithelia (Fig. 4a), the subsequent stages usually include the formation of neuroepithelial rosettes (Fig. 4b). The appearance of rosettes is often the result of two steps: (1) manual selection of neuroepithelial colonies, (2) culturing of neurospheres and/or re-adherence (re-seeding of primitive neuroepithelia). Numbers of rosettes, as shown in Fig. 4b, can be quantitated on a per field basis using the appropriate number of serial fields for scoring (e.g., 20–30 visual fields per treatment group). Live scoring is a time and effort saving procedure. Live scoring generates data at multiple time points, which is extremely useful for understanding the temporal dynamics of the process under study. Live scoring also provides useful insights about the reliability and robustness of experimental data (repeated measurements on the same sample). Last, but not least, live scoring allows acquisition of two or more types of data from the same sample. For example, from the same culture well one can acquire several types of stereological parameters during in vitro development followed by molecular or biochemical tests at the end of the in vitro cycle.

3.9 Neurites

In cases where it is desirable to extract the culture for protein and nucleic acid, neurites can be quantitated on live neurons prior to any extractions that will destroy all morphological information.

1. Neuron precursors (neuroepithelia, Fig. 4a) are seeded on 6-well plates.
2. After neuronal differentiation has started (in our case 14 days), the entire well is examined for presence of neurites. For consistency, the 3–5 regions with the densest neurite growth are selected in each well. At 10× magnification (phase contrast), all the neurites are counted in the visual field. The six wells of a six-well plate per treatment group are typically needed.

3.10 Immunofluorescent Staining for Tyrosine Hydroxylase, β-Tubulin III, and NeuN

Staining for the dopaminergic marker tyrosine hydroxylase (TH) and the pan-neuronal markers β-Tubulin III or NeuN is used as an assay for determining if dopamine receptor ligands or other treatments change neuronal subtype specification. The procedure can also be used for DA receptor staining for immunofluorescence-validated antibodies.

1. Fix coverslips for 0.5–3 h in 4 % Paraformaldehyde (PFA) in PBS at room temperature.
2. Wash 2 × 4 min with PBS.
3. If storing coverslips, rinse off PFA with PBS twice, and store in PBS + azide at 4 °C for up to 3 days.
4. Permeabilize cells by incubating for 10 min in 0.2 % Triton in PBS.
5. Wash 3 × 4 min with PBS.
6. Block with PBS + 10 % normal goat serum + 0.75 % BSA for 30 min.
7. Incubate overnight at 4 °C.
8. Wash 3 × 4 min with PBS.
9. Incubate for 1 h with the secondary antibody in PBS + 10 % goat serum.

10. Wash 2 × 4 min with PBS.
11. Wash once in PBS with 1 μg/ml Hoechst 333258 7 min.
12. Wash 1 × 4 min with PBS.
13. Mount coverslips on slides with *Fluoromount G*. Do not let slips dry. *See* **Note 5** for a discussion of quantitating results. *See* **Note 6** for batch analysis of images. **Note 7** discusses how to count percentages of neurons.

3.11 Dye Loading for Calcium Imaging of ES Colonies

Temporal measurement of calcium levels in cells and colonies is an important method for demonstrating that ligands for a particular receptor produce a downstream response in the Ca^{2+} mediated signal transduction pathway. The following protocol describes a method for monitoring acute changes in calcium levels upon stimulation of a hESC or iPSC colony by a DA receptor agonist.

1. hESC (or iPSC) colonies are loaded with OGB1-AM dye. Preparation of OBG1-AM stock: The content of one original vial (50 μg) is dissolved in 2 μL of 20 % Pluronic F-127 and 8 μL of DMSO, vortexed for 30 min, and then 90 μL DMEM/F12 is added.
2. To load hESCs, 3.33 μL of stock is added to a well containing one glass coverslip in 0.5 ml of DMEM/F12 media, and incubated for 30 min at 37 °C.
3. Cells are washed twice with warm DMEM/F12 media, and transferred to a 35 mm dish containing growth media.
4. Colonies are allowed to recover for 30 min in the incubator.

3.12 Calcium Imaging

1. The dye-loaded colonies on glass coverslips were transferred to a heated recording chamber (Warner Instruments, Hamden, CT) positioned under an Olympus BX51WI microscope. The recording chamber was perfused with aerated saline (standard ACSF, 2 mM Ca^{2+}) preheated by a solution in-line heater (SH-27B, Warner Instrument). Both the platform and in-line heater were powered by a 2-channel *Temperature Controller* (TC-344B, Warner Instrument). Standard artificial cerebrospinal fluid (ACSF) contained (in mM) 125 NaCl, 26 $NaHCO_3$, 10 glucose, 2.3 KCl, 1.26 KH_2PO_4, 2 $CaCl_2$, and 2 $MgSO_4$. To exclude the possibility that calcium influx from the extracellular space is responsible for the observed calcium transients (Fig. 5), a "near zero calcium" ACSF (0.1 mM Ca^{2+}) was used in some experiments. Presence of slow onset and long-duration calcium transients in the "near zero calcium" solution suggests a calcium release from intracellular stores.
2. hESC colonies were illuminated with a computer-controlled LED (blue, 460 nm, Luminus Devices) via a 40× objective (NA = 0.8) and the resulting images were projected via the

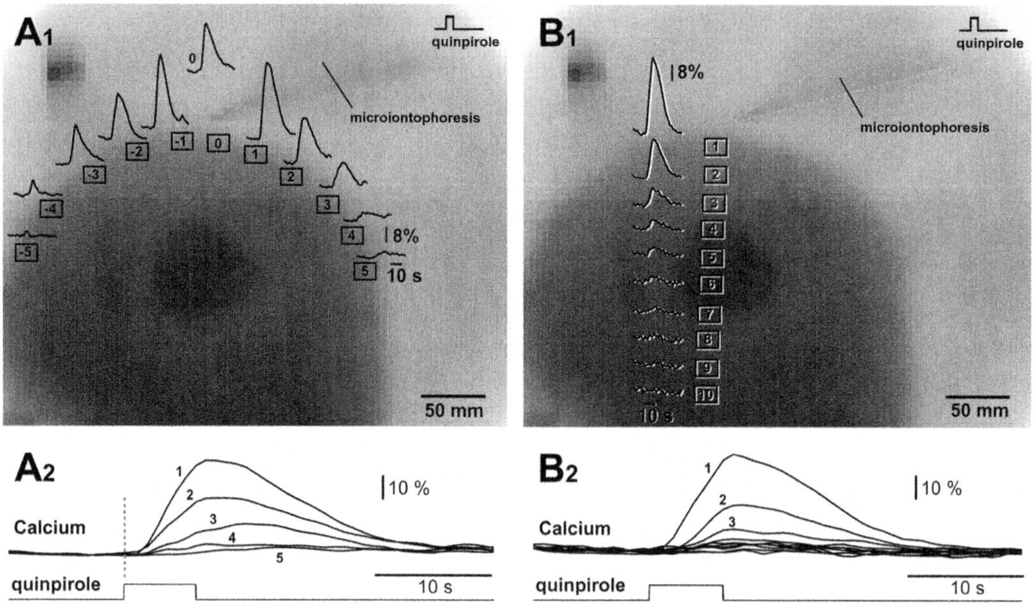

Fig. 5 D2 DA receptor agonist (quinpirole, QP) induces internal calcium release in hESC colonies. (**a_1**) A low resolution infrared microphotograph captured with the same camera used for fast calcium imaging (NeuroCCD-SMQ). The iontophoretic micropipette ("microiontophoresis") is positioned next to an OGB1-AM-loaded hESC colony. Note that the microelectrode tip is approximately 25 μm away from the outer edge of the colony. QP-induced calcium transients recorded simultaneously at 11 regions of interest (*boxes*) are shown on the same amplitude and time scale, and distributed according to their location. The regions of interest are chosen in a horizontal direction, along the edge facing the QP-filled pipette. (**a_2**) QP-induced calcium transients recorded simultaneously at 5 locations corresponding to boxes 1–5 in **a_1**. Each trace is a spatial average of 9 pixels. Bottom trace is the exact time course of the iontophoretic pulse. *Vertical dashed line* marks the onset of the drug ejection. Note the delay between the onset of pulse and infliction on trace 3 (~3 s). (**b_1**) Same as in **a_1**, except the contour of the iontophoretic electrode is outlined with a *dashed line*. The regions of interest are chosen in a vertical direction (radially) away from the electrode tip. (**b_2**) QP-induced calcium transients recorded simultaneously at 10 locations. Signals from the locations 4–10 are not labeled for clarity

same objective lens onto a fast CCD camera (80×80 pixel, NeuroCCD-SMQ, RedShirtImaging). Images were sampled at either 400 ms per full frame (2.5 Hz frame rate) or 800 ms interval (1.25 Hz frame rate) (*see* **Note 8**).

3. hESC colonies were stimulated by focal application of dopaminergic drugs (*see* **Note 9**), achieved using drug microiontophoresis as described in Zhou and Antic (2012) [11]. Briefly, dopaminergic drugs [SKF38393 (40 mM) and quinpirole (40 mM)] were dissolved in water and loaded into sharp micropipettes pulled out of borosilicate capillary glass (electrical resistance ~40 MΩ). Drug-filled micropipettes were attached to a microelectrode holder (E45P-F15NH, cat. # 64-1025, Warner Instruments) and positioned near the outer edge of a stem cell colony using a motorized micromanipulator (MP-285, Sutter Instrument). The tip of the drug-filled pipette

should not touch the surface of the colony (*see* **Note 10**). The nominal intensity of iontophoretic current required to successfully trigger internal calcium release was in the range of 15–30 nA. Duration of the iontophoretic pulse required to successfully trigger internal calcium release depends on the drug's potency and can be determined empirically. For example, with ATP (100 mM inside the micropipette), the duration of iontophoretic pulse could be as low as 0.5 s. Note that drug concentrations are in the millimolar (mM) range, because three factors (a–c) concurrently reduce the concentration of the drug reaching the colony: (a) the drug application pipettes are sharp (high resistance micropipettes normally used for intracellular recordings); (b) the pulse duration is short (in the order of a few seconds); and (c) the tips of drug application micropipettes are at some distance away from the surface of the colony (20–100 μm). With dopaminergic drugs (SKF38393 or quinpirole, 40 mM inside the micropipette) the effective duration of a drug pulse was in the range of 3–6 s. **Note 11** discusses identification and reduction of artifacts due to mechanical vibration.

4. Optical data were analyzed using *Neuroplex* software. To improve signal-to-noise ratio pixel outputs were spatially averaged (9–16 pixels per region of interest, ROI) and digitally filtered Gaussian low-pass cutoff 0.4 Hz. ROIs are represented by boxes in Fig. 5a_1, b_1. Signal amplitudes were reported as ΔF/F (%) (Fig. 5a_2, b_2). The resting light intensity "F" was not corrected for background light, because the background light was minimal due to low setting of the illumination intensity (LED power supply voltage set at 1.1 V) and only one layer of cells in culture.

4 Notes

1. Selection of Primers. The D5 DA receptor gene contains no splice junctions, so DNase is necessary to avoid amplifying genomic D5 receptor DNA. Because DA receptor and TH mRNA are not highly expressed, and homologous genes are present, primers were optimized for specificity as follows. In general, 1–4 potential primer pairs were tested and the pair producing the most robust band was used. Where possible, primer pairs were designed (in NCBI Primer-BLAST) with several criteria in mind. (1) To exclude amplification of genomic DNA, the amplicon should contain a splice site where an exon (>200 bp) exists, or one of the primers should encompass a splice site. (2) Primers with a tendency for hairpins, self-dimers, and heterodimers are avoided by checking on the manufacturers' websites (http://www.idtdna.com/analyzer/Applications/

OligoAnalyzer/, and http://www.sigma-genosys.com/calc/DNACalc.asp). (3) Primers are tested by in silico PCR for their ability to amplify genomic (for instance pseudogenes may be present in the genome that are very similar to the mRNA) using a Web based tool available at http://genome.ucsc.edu/cgi-bin/hgPcr). (4) In Silico PCR is done using NCBI Primer-BLAST with the appropriate mRNA sequence to verify that unwanted homologous receptor mRNA sequences are not amplified.

2. Yield of Total Protein. One issue with detecting DA receptors is that they may have low expression levels. Therefore, an optimal amount of protein should be loaded on a gel in order to see the bands. It is important to know what the protein yields to ensure enough protein for the Westerns. Below are typical yields of total protein level.

   From each six-well plate, the following yields were obtained when cells were subjected to our dopaminergic differentiation protocol [2]:

   MEFs → 0.6 mg

   Stage 1 hES colonies → 2.3 mg

   Stage 2 EBs → 2.6 mg

   Stage 3 Selected neuroepithelial colonies → 0.635 mg

   Stage 4 Expanding neuroepithelial colonies → 2.5 mg

   Stage 5 Differentiated neurons → 2.4 mg

3. Yield of Total RNA. The same issue (*see* **Note 2**) exists for detecting DA receptor message, as DA receptor messages may have low expression levels. It is important to know what the mRNA yields would be to ensure enough RNA for the qPCR. Given below are typical yields of total RNA obtained in our experiments.

   From each six-well plate, the following yields of total RNA were obtained when cells were subjected to our dopaminergic differentiation protocol [2]:

   MEFs → 30.5 μg

   Stage 1 hES colonies → 175 μg

   Stage 2 EBs → 147 μg

   Stage 3 Selected neuroepithelial colonies → 126 μg

   Stage 4 Expanding neuroepithelial colonies → 102 μg

   Stage 5 Differentiated neurons → 36 μg

4. Scanning the Colonies. For measuring changes in colony morphology, such as spontaneous differentiation (which may be a rare event), the entire plate is scored by marking the bottom of the plate with an ultrafine point marker (*Sharpie*), with separate

colors for each morphological type. Every colony in the well is examined. A marker is held with one hand with the tip positioned slightly above the objective and is used to quickly touch the bottom of the plastic dish. Colonies can also be circled in this manner. The bottoms of the plates can then be scanned using an image scanner (Fig. 6) and the plates returned to the incubator. The scanned images are scored manually or automatically using software such as ImageJ (NIH, www.rsbweb.nih.gov/ij/).

5. Scoring of Immunofluorescent Clusters. Although selected 10× or 20× fields can be found for photographing after immunofluorescent staining, we observe large heterogeneity from one field to another. In order to efficiently score the entire surface of a coverslip, and thus greatly reduce the coefficient of variation, clumps of TH positive neurons can be scored by moving back and forth along the entire length of the coverslip, and all clusters on a coverslip with ten or more positive cells counted. In our experience, ten coverslips are typically needed per treatment group to obtain ≥0.8 power. To verify beforehand that the cluster sizes in different treatment groups are not significantly different, serial photographs are captured of five or more clusters per coverslip, then size is compared by

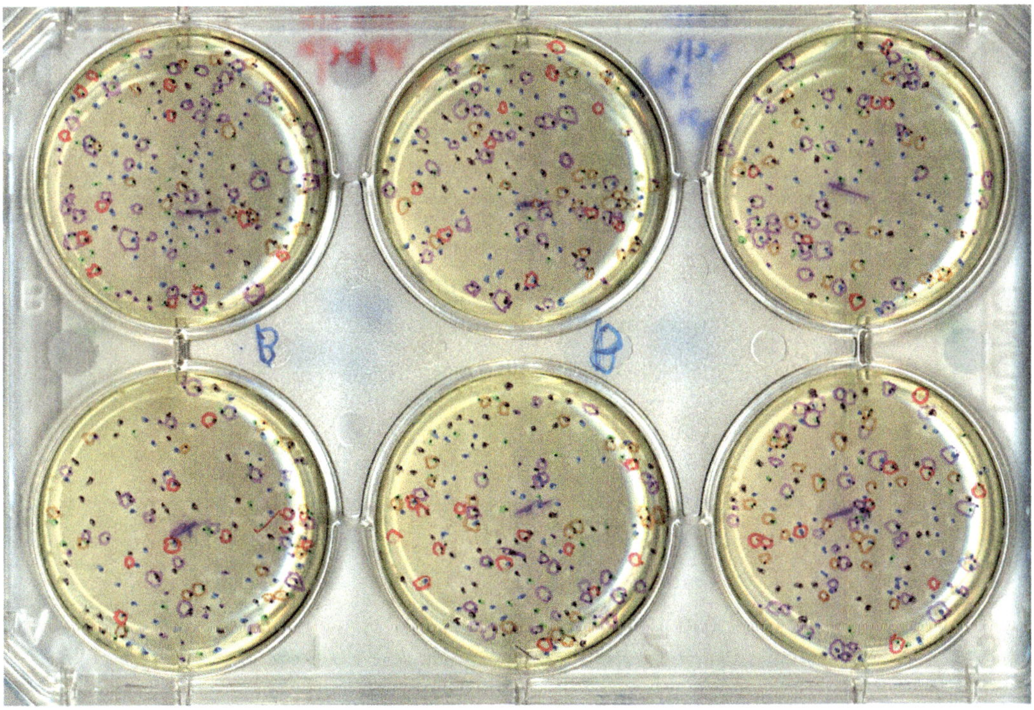

Fig. 6 Scoring live hESC colonies. Image of scanned plate (regular desk-top scanner), with differentiated stem cell colonies circled with a *colored marker* (ultrafine point, *Sharpie*)

measuring the diameter of clusters. We note that it is frequently impossible to count the number of positive cells in a field due to dense clumping, but it is clear that more than ten positive cells are present. Alternatively, integrated optical densities of signals in the appropriate channels can be quantitated and group means compared between different treatments. The latter can be accomplished by ImageJ, freely available from the NIH, or proprietary image analysis software. An ImageJ script for automatic batch processing of images to quantitate integrated optical densities is included in the next note.

6. *ImageJ Batch—Processing Script*. The following ImageJ script may be used to batch process all TIF images in a given folder in order to calculate integrated optical density. The min value (in this case 25) must be set by the user to gate out unwanted background signal.

// "BatchProcessFolders"

// This macro batch processes all the files in a folder and any subfolders in that folder. In this example, it runs the *Subtract Background* command on TIFF files then measures "Integrated optical density". For other kinds of processing, edit the processFile() function ("function processFile(path)") at the end of this macro:

```
requires("1.33 s");
dir = getDirectory("Choose a Directory ");
setBatchMode(true);
count = 0;
countFiles(dir);
n = 0;
processFiles(dir);
//print(count + " files processed");
function countFiles(dir) {
list = getFileList(dir);
for (i = 0; i < list.length; i++) {
if (endsWith(list[i], "/"))
countFiles("" + dir + list[i]);
else
count++;}}
function processFiles(dir) {
list = getFileList(dir);
for (i = 0; i < list.length; i++) {
if (endsWith(list[i], "/"))
```

```
processFiles("" + dir + list[i]);
else {
showProgress(n++, count);
path = dir + list[i];
processFile(path);
}}}
function processFile(path) {
if (endsWith(path, ".TIF")) {
open(path);
run("Window/Level...");
resetMinAndMax();
run("Set Scale...", "distance = 0 known = 0 pixel = 1 unit = pixel");
run("8-bit");
run("Set Measurements...", "area mean min integrated display redirect = None decimal = 3");
run("Window/Level...");
setMinAndMax(25, 255);
run("Apply LUT");
run("Measure");
run("Undo");
close();
}}
```

7. Scoring of Individual Cells. Although the TUJ1 antibody (directed against TUBB3) produces intense fluorescence ([10], their Fig. 2), the neurites stain stronger than the cell body and so individual cells are frequently unidentifiable in dense cellular regions. For quantitating % of neurons that are both positive for a pan-neuronal marker and the dopaminergic marker TH, the pan-neuronal marker NeuN is preferable as it stains the nuclei of mature neurons. Once again, due to large heterogeneity from one field to the next, care must be taken to score enough fields to obtain a power ≥0.8.
8. In order to avoid photodynamic (phototoxic) damage of dye-loaded hESCs it is necessary to keep the illumination intensity at minimum, while maintaining the quality of optical signals (signal-to-noise ratio). This can achieved by matching two experimental variables: [8a] Output of the LED power supply and [8b] Sampling interval of the CCD camera.

   [8a] Minimal Illumination Intensity. The LED power supply used in the present study was custom made (Dr. Peter Lee, Essel R&D Inc). Its main features are: (a) Trigger

input for synchronizing the LED pulse with the optical data acquisition episode; and (b) Adjustable LED voltage limit. Similar LED drivers are commercially available. One inexpensive driver (LEDD1B, ThorLabs, Newton, NJ) combines both the external triggering and adjustable output. The LEDD1B driver has already been used with a blue LED (Luminus Devices) in our laboratory; and seems perfectly suitable for calcium imaging applications at low light intensities. Maximum current output of the LEDD1B driver is only 1.2 A. This output is sufficient for driving Luminus Devices LEDs in low light experiments (Fig. 5). However, Luminus Devices LEDs can produce significantly brighter light when powered adequately (>>1.2 A).

[8b]Slow Sampling Rate. Increasing the sampling interval of the CCD camera (NeuroCCD-SMQ, RedShirtImaging) is very useful for optical imaging of internal calcium release [12, 13]. Internal calcium release is a relatively slow process (Fig. 5a_2); significantly slower than the action potential-induced calcium influx [11], for example. The two frame rates used in the present study (2.5 Hz and 1.25 Hz) 400 ms and 800 ms sampling intervals, respectively) are well suited for the slow dynamics of the calcium transient, but they are well below the frame rates this camera was designed for (2,000 Hz). Since rates lower than 40 Hz are not available from the scroll-down menu of the data acquisition software (Neuroplex, RedShirtImaging), it is necessary to enter the desired interval (e.g., 800 ms) manually. Forcing the camera to collect photons for longer periods of time alleviates the requirements for bright illumination. The relationship between the intensity of illumination and the optical signal quality was previously described by Homma and colleagues [14] (their Fig. 7). Briefly, the signal quality (signal-to-noise ratio) is proportional to the number of photons collected by the optical detector. The collection of photons can be increased by increasing the illumination intensity (which is harmful for stem cells), or increasing the sampling interval (which is the strategy used in the present study).

At a 800 ms sampling interval (1.25 Hz frame rate) the fluorescent image of hESC can easily be detected by the CCD camera even when the blue LED is hardly glowing in the back of the microscope. The minimal LED output was achieved by regulating the LED power supply. The calcium imaging technique described in this chapter is performed using a simple (standard) laboratory microscope with an epi-illumination module and single-photon excitation. For two-photon excitation of hESC-derived cultures see reference [15].

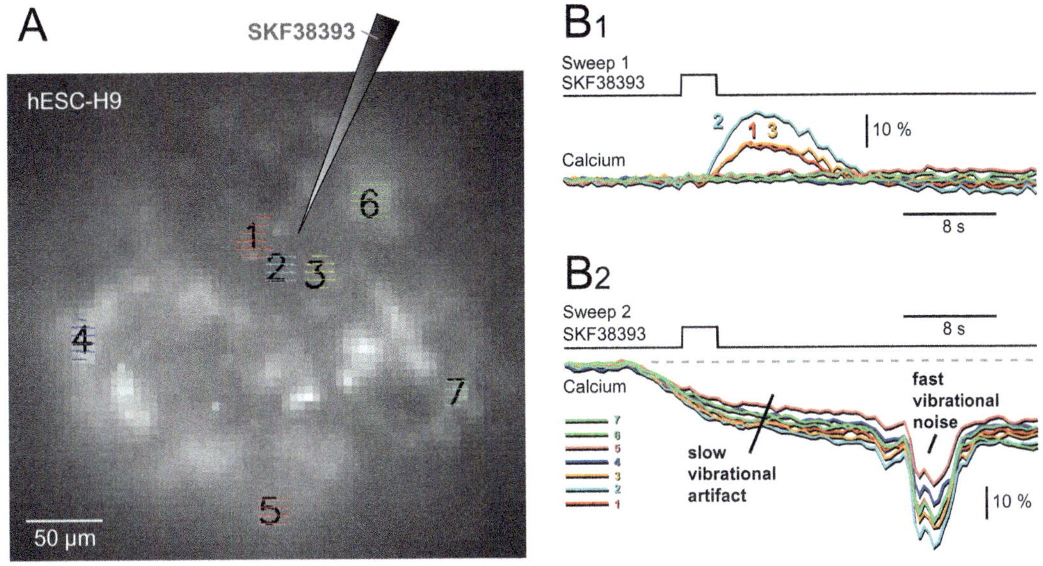

Fig. 7 Mechanical vibrations disrupt optical recordings. Low-resolution fluorescent image of a hESC-H9 colony bulk-loaded with a calcium-sensitive dye OGB1-AM and excited with blue light (460 nm). Drawing marks position of a glass micropipette filled with D1 DA receptor agonist SKF38393 (40 mM intrapipette concentration) and positioned 25 μm above the surface of the colony. *Colored horizontal dashes mark* individual detectors (pixels) selected for spatial averaging under each ROI (ROIs 1–7). (**b_1**) In response to a 3 s-long iontophoretic pulse of SKF38393 (*black trace*) stem cells near the tip of the drug pipette exhibited a delayed and slow-rising calcium transient (ROIs 1–3). At the same time, no changes in intracellular calcium were detected in locations 4–7. All traces 1–7 were recorded simultaneously. Each trace is the product of spatial averaging (10–15 pixels) and digital filtering (low-pass Gaussian, 0.4 Hz cutoff) (**b_2**) In the next experimental trial (Sweep 2) the same colony is exposed to the same stimulus (3 s of SKF38393) but calcium transients were not detected. The same regions of interest are used as in **b_1**. Note that a slow vibrational artifact is followed by a somewhat faster vibrational noise. Both artifacts are present in all detectors across the entire visual field, indicating that the colony moved in the visual filed during the 40 s optical recording episode, or an air bubble disturbed the media in front of the objective lens

9. Focal Application of Drugs has several major advantages over bath application. Focal application assures fast delivery and fast washout of the pharmacological agent. Repetitive applications can be carried out at the same stimulation site in order to study the sensitization of receptors or depletion of the intracellular calcium stores. Finally, focal application can be used to explore several application sites on the same colony, or to explore individual responses of several colonies in the same coverslip. Having multiple recordings on the same coverslip saves time, effort and resources for growing and differentiating stem cells.
10. Avoid Direct Electrical Stimulation. Keep a distance (20–100 μm) between the microiontophoresis electrode and the stem cell colony, because a physical contact during microiontophoresis produces direct electrical stimulation of stem cells, resulting in calcium release even when drugs are omitted from

the solvent (unpublished data, Antic lab). In the present study [2] the nominal intensity of iontophoretic current required to successfully trigger internal calcium release was in the range of 15–30 nA. Duration of the iontophoretic pulse required to successfully trigger internal calcium release depends on the drug's potency, and can be determined empirically. For example, with ATP [100 mM] inside the micropipette, the duration of iontophoretic pulse could be as low as 0.5 s. With dopaminergic drugs (SKF and QP) the effective duration of a drug pulse was in the range of 3–6 s.

11. Vibrational Noise. There are three main sources of mechanical vibrations that have a potential to severely affect optical recordings in hESC colonies.

    [1]. A poor vibration isolation table allows a transfer of mechanical noise from the floor and from the main table onto the microscope. To prevent this, it is important to properly adjust your air table (or mechanical isolation platform) according to manufacturer's specifications. The presence of mechanical noise should be quickly determined using infrared microscopy equipped with a video rate camera and a TV monitor. Focus on one microscopic object in the visual field (e.g., one distinctive cell) and project it on the TV monitor. At a highest microscope magnification the object on the TV monitor must be perfectly still.

    [2]. If coverslips are not properly secured to the bottom of the recording chamber, the jet coming from the perfusion line may push the coverslip forward, or to the side. Also, loose masses of cultured cells may partially detach from the culture and wobble in the perfusate jet.

    [3]. Bubbles of air trapped inside the perfusion system often introduce mechanical vibrations in optical recordings. This is especially apparent in optical imaging episodes lasting 10 or more seconds (Fig. 7).

    In order to eliminate vibrational noise from optical recordings use: [1] antivibration platforms, [2] metal anchors to weigh down and secure coverslips to the bottom of the recording chamber, [3] commercial or custom-made bubble-traps to prevent bubbles from entering the recording chamber.

    Vibrational artifacts can be easily identified in multisite optical records (Fig. 7). A majority of mechanical artifacts present themselves in all optical detectors indiscriminately (Fig. 7b_2), as opposed to biological signals that arise in distinct locations (Fig. 7b_1, ROIs 1–3, near the tip of the SKF38393-filled pipette).

    As a last line of defense from mechanical vibrations in optical recordings, the perfusion system can be stopped just before the optical trial begins and then restarted immediately after the end of the optical sweep. Keep the OFF time at minimum in order to maintain the partial pressure of O_2 and pH in the solution surrounding the living cells inside the recording chamber.

Acknowledgments

This study was supported by Connecticut Innovations grant 09-SCA-UCHC-13 to SDA and Institutional HCRAC grant to SDA. Calcium imaging experiments were performed in the *Stem Cell Physiology and Chemistry Core*, which is supported by Connecticut Innovations grant SCD-01–2009. LED epi-illumination system was kindly designed and built by Dr. Peter Lee, Essel R&D Inc, Toronto, Canada.

References

1. Tritsch NX, Sabatini BL (2012) Dopaminergic modulation of synaptic transmission in cortex and striatum. Neuron 76(1):33–50
2. Belinsky GS, Sirois CL, Rich MT, Short SM, Moore AR, Gilbert SE, Antic SD (2013) Dopamine receptors in human embryonic stem cell neurodifferentiation. Stem Cells Dev 22(10):1522–1540
3. Malmersjo S, Liste I, Dyachok O, Tengholm A, Arenas E, Uhlen P (2010) Ca2+ and cAMP signaling in human embryonic stem cell-derived dopamine neurons. Stem Cells Dev 19(9):1355–1364
4. Neve KA, Seamans JK, Trantham-Davidson H (2004) Dopamine receptor signaling. J Recept Signal Transduct Res 24(3):165–205
5. Pei L, Lee FJ, Moszczynska A, Vukusic B, Liu F (2004) Regulation of dopamine D1 receptor function by physical interaction with the NMDA receptors. J Neurosci 24(5):1149–1158
6. Fuxe K, Ferre S, Zoli M, Agnati LF (1998) Integrated events in central dopamine transmission as analyzed at multiple levels. Evidence for intramembrane adenosine A2A/dopamine D2 and adenosine A1/dopamine D1 receptor interactions in the basal ganglia. Brain Res Brain Res Rev 26(2–3):258–273
7. Perreault ML, Hasbi A, O'Dowd BF, George SR (2013) Heteromeric Dopamine Receptor Signaling Complexes: Emerging Neurobiology and Disease Relevance. Neuropsychopharmacology 39(1):156–168
8. Dravid G, Hammond H, Cheng L (2006) Culture of human embryonic stem cells on human and mouse feeder cells. Methods Mol Biol 331:91–104
9. Iacovitti L, Donaldson AE, Marshall CE, Suon S, Yang M (2007) A protocol for the differentiation of human embryonic stem cells into dopaminergic neurons using only chemically defined human additives: Studies in vitro and in vivo. Brain Res 1127(1): 19–25
10. Belinsky GS, Moore AR, Short SM, Rich MT, Antic SD (2011) Physiological Properties of Neurons Derived from Human Embryonic Stem Cells using a dbcAMP-based Protocol. Stem Cells Dev 20(10):1733–1746
11. Zhou WL, Antic SD (2012) Rapid dopaminergic and GABAergic modulation of calcium and voltage transients in dendrites of prefrontal cortex pyramidal neurons. J Physiol 590(Pt 16):3891–3911
12. Milojkovic BA, Zhou WL, Antic SD (2007) Voltage and Calcium Transients in Basal Dendrites of the Rat Prefrontal Cortex. J Physiol 585(2):447–468
13. Ross WN, Manita S (2012) Imaging calcium waves and sparks in central neurons. Cold Spring Harb Protoc 2012(10):1087–1091
14. Homma R, Baker BJ, Jin L, Garaschuk O, Konnerth A, Cohen LB, Zecevic D (2009) Wide-field and two-photon imaging of brain activity with voltage- and calcium-sensitive dyes. Philos Trans R Soc Lond B Biol Sci 364(1529):2453–2467
15. Fu W, Ruangkittisakul A, Mactavish D, Baker GB, Ballanyi K, Jhamandas JH (2013) Activity and metabolism-related Ca(2+) and mitochondrial dynamics in co-cultured human fetal cortical neurons and astrocytes. Neuroscience 250:520–535

Chapter 14

Calcium and Phospholipase Cβ Signaling Through Dopamine Receptors

Lani S. Chun, R. Benjamin Free, and David R. Sibley

Abstract

Dopamine receptors are highly validated therapeutic and scientific targets within the nervous system. Although dopamine receptor-mediated phospholipase Cβ activation has been well characterized, typical methods for investigating this are both time-consuming and inefficient, severely limiting the types of questions that can be addressed. Here we present a concise, controllable, efficacious, live cell-based signaling assay that exploits the D_1–D_2 dopamine receptor dimer's endogenous ability to activate phospholipase Cβ-mediated calcium mobilization. We present some background information to frame the subject of dopamine receptor-mediated calcium mobilization and describe a protocol for detecting calcium mobilization in cells heterologously co-expressing the D_1 and D_2 dopamine receptors. However, it is important to note that this method is highly adaptable to many other experimental paradigms and can work with the expression of other dopamine receptors and GPCRs. It illustrates a much needed in vitro method to observe calcium mobilization in real time that can be performed in different cell types, under many transfection conditions, and with various drug treatments.

Key words Calcium, Ca^{2+}, D_1–D_2 heteromer, Dopamine receptor, Fluo-8, G_q protein, Inositol 1,4,5-triphosphate, IP3, PLCβ

Abbreviations

AC	Adenylyl cyclase
D_1R	Dopamine receptor subtype 1
D_2R	Dopamine receptor subtype 2
D_3R	Dopamine receptor subtype 3
D_4R	Dopamine receptor subtype 4
D_5R	Dopamine receptor subtype 5
DA	Dopamine
DAR	Dopamine receptor
DMEM	Dulbecco's modified Eagle's medium
EBSS(−)	Calcium-free Earle's balanced salt solution
HBSS(−)	Calcium-free Hank's balanced salt solution
PLCβ	Phospholipase Cβ
RFU	Relative fluorescent units
SMB	Sodium metabisulfite

Mario Tiberi (ed.), *Dopamine Receptor Technologies*, Neuromethods, vol. 96,
DOI 10.1007/978-1-4939-2196-6_14, © Springer Science+Business Media New York 2015

1 Introduction

Dopamine (DA) targets its receptors to affect multiple signaling cascades. Five dopamine receptor (DAR) genes exist in mammals, each encoding a unique subtype (D_1R–D_5R), which are grouped by structure and function into the D1-like (D_1R, D_5R) and D2-like (D_2R, D_3R, D_4R) DAR families. Canonically, the D1-like receptors couple to the $G_{s/olf}$ proteins to activate adenylyl cyclase (AC)-mediated formation of cAMP, while the D2-like receptors couple to the $G_{i/o}$ proteins to inhibit AC [1, 2]. However, it has now been shown that the DARs can activate other signal transduction pathways that do not involve AC, and one of the primary "non-canonical" downstream targets of activated DARs is phospholipase Cβ (PLCβ)-mediated intracellular calcium signaling.

Development of the first D_1R-selective DAR agonists and antagonists [3–5]—all benzazepine class compounds—facilitated the realization that the D_1R could not only activate AC but could also stimulate inositol phosphate production and intracellular calcium mobilization. D_1R-mediated calcium mobilization was first discovered in renal cortical homogenates [6], and the link between the D_1R and $G_{q/11}$–PLCβ activation was subsequently confirmed in slices from rat and human neuronal striatum and cortex [7–13]. Production of additional D_1R-selective agonists and antagonists based on the benzazepine chemotype yielded some compounds with partial agonist/antagonist activity on D_1R-mediated AC function. Interestingly, the receptor affinity, signaling potency, and efficacy of partial versus full D_1R agonists were not necessarily proportional to the observed behavioral phenotypes, bolstering the theory that neuronal D_1R signaling was not limited to $G_{s/olf}$ protein activation [14].

In 2004, DAR-mediated calcium mobilization was linked to a heterodimer composed of the D_1R and D_2R (D_1–D_2 heteromer) [15]. Additional publications from this laboratory described D_1–D_2 heteromers in the nucleus accumbens and striatum of rat brain using confocal FRET techniques [16]. The D_1–D_2 heteromer was linked to G_q protein-mediated intracellular calcium release [15, 17], and co-internalization of the two receptors was seen following stimulation of the heteromer with DA or a D_1R-selective agonist with a D_2R-selective agonist [18, 19].

In addition, a benzazepine, SKF83959, was shown to selectively activate the D_1–D_2 heteromer-mediated calcium response without causing D_1R-mediated AC activation [17]. Antecedent publications had shown that SKF83959 inhibited D_1R-coupled cAMP formation [14, 20] and induced striatal intracellular calcium mobilization in rats and monkeys [21]. Interestingly, it did not cause epileptic seizures in rodents and monkeys, unlike typical D_1R agonists which stimulate cAMP production. Furthermore, SKF83959 caused typical D_1R agonist-like behaviors in rats [16] and had efficacy as an antiparkinsonian drug in MPTP-lesioned monkeys that did not respond to L-DOPA [22]. These data led to SKF383959 being designated a D_1–D_2 dimer-selective agonist. However, it was recently shown that in HEK293T cells transiently expressing the D_1R alone or D_1R+D_2R, SKF83959 did not stimulate a calcium response unless the $G_q\alpha$ subunit was also overexpressed [23], and behaviors induced by SKF83959 in unilaterally 6-OHDA-lesioned rats could be blocked by a D_1R antagonist, but not a D_2R antagonist [24]. The combined evidence surrounding D_1–D_2 dimer- and D_1R-mediated PLCβ activation makes it unclear which receptor state (D_1R monomer vs. D_1–D_2 heteromer) is present and applicable to SKF83959's actions in vivo.

The understanding of DAR-mediated PLCβ activation is further complicated by additional mechanisms which have been demonstrated to cause calcium mobilization, some of which include receptor cross talk/G_q protein priming [25–28] and D_2R–Gβγ signaling [29, 30]. The detection of intracellular calcium mobilization and PLCβ activation is important for understanding the role of the DARs in calcium signaling. Therefore, many methods for detecting and measuring both calcium mobilization and the more upstream PLCβ activation have been developed. Pulse-chase experiments in homogenized tissue samples using [^{3}H]-phosphatidylinositol or [^{3}H]-phosphatidylinositol 4,5-bisphosphate to measure PLCβ activity by quantifying the amount of the resulting metabolite, [^{3}H]-inositol 1,4,5-triphosphate, is widely used. However, it requires a relatively large amount of cells, and kinetic studies can be complicated, if not impossible, as PLCβ activation occurs rapidly. The same restrictions apply to [^{35}S]-GTPγS binding to measure G_q protein activation.

Here, we describe an assay that can be used in conjunction with various pharmacological agents as well as additional transfection conditions in live cells to make kinetic or single time-point measurements of intracellular calcium mobilization. The Fluo-8AM calcium dye is the latest in a series of calcium-sensing dyes, with improved sensitivity for better detection of cytosolic calcium. It is an efficient and easy way to measure intracellular calcium mobilization in live cell-based assays and can be scaled up for high-throughput experiments. We show how its especially sensitive nature in detecting calcium mobilization, as a result of DA-mediated PLCβ activation, is useful in understanding D_1–D_2 heteromer activation. This assay can also be used to further elucidate the various ways DARs can stimulate calcium mobilization, which, as of yet, is not fully understood. PLCβ activation through the DARs has been correlated to diseases such as schizophrenia [16] and depression [31, 32] and may play a role in aging-related disorders [22, 24, 33–35]. Understanding the full spectrum of DAR-mediated calcium signaling and consequent behavioral phenotypes may eventually be useful in understanding and treating such diseases.

2 Materials

2.1 Media and Buffers

1. Dulbecco's modified Eagle's medium (DMEM; GIBCO, Grand Island, NY) supplemented with 10 % fetal bovine serum, 10 μg/μL gentamicin, 1 mM sodium pyruvate, 10 U/mL penicillin, and 10 μg/mL streptomycin.
2. Dissociation buffer: calcium-free Earle's balanced salt solution (EBSS(−))—5.3 mM KCl, 26.2 mM $NaHCO_3$, 117.2 mM NaCl, 1.0 mM NaH_2PO_4–H_2O, and 5.6 mM D-glucose, pH 7.4; supplemented with 2.5 mM EDTA.
3. Assay buffer: calcium-free Hank's balanced salt solution (HBSS(−))—5.3 mM KCl, 0.44 mM KH_2PO_4, 4.2 mM $NaHCO_3$, 137.9 mM NaCl, 0.34 mM Na_2HPO_4, and 5.6 mM D-glucose; supplemented with 20 mM HEPES.

2.2 Tissue Culture and Transfection

1. HEK293T cells incubated at 37 °C, 5 % CO_2, and 90 % humidity.
2. 175 cm^2 cell culture flasks.
3. 150 cm^2 cell culture plates.
4. Transfection reagents (e.g., Clontech's CalPhos™ transfection kit, Clontech Laboratories, Inc., Mountain View, CA).
5. Expression vectors containing DARs such as rat D_1R or $D_{2L}R$ in the pCD-SRα vector [36–38]. *See* **Note 1** for additional constructs.

2.3 Assay Components

1. Fluo-8 calcium detection kit: keep light protected (e.g., Quest Fluo-8™ calcium dye kit, AAT Bioquest, Sunnyvale, CA; *see* **Note 2**).
2. Agonist and antagonist compounds, including DA. Stock solutions are 10–100 mM, depending on compound solubility, kept in the presence of 0.2 mM sodium metabisulfite (SMB), which prevents compound oxidation.
3. 384-well, optical, clear bottom, cell culture, black-walled plates (e.g., Thermo Scientific Nunc 142761; *see* **Note 3**).
4. 384-well, clear compound plate (e.g., Corning 3702).

2.4 Equipment

1. Kinetic plate reader capable producing ~490 nm excitation wavelength and reading emission at ~520 nm (e.g., FDSS/μCell kinetic plate reader, Hamamatsu, Bridgewater, NJ).
2. Multidrop® Combi reagent dispenser (Thermo Scientific, Hudson, NH; optional).
3. 40 μm sterile cell strainer (Fisher Scientific, Pittsburgh, PA; optional).

3 Methods

3.1 Using Fluo-8AM to Detect Intracellular Calcium Mobilization

The Fluo-8AM calcium dye is cell membrane permeable but undergoes cleavage via cell-endogenous esterases, resulting in an inability to exit the cell once it has been internalized. Once cytosolic calcium is present, calcium ions bind to the Fluo-8 dye, causing a 200-fold increase in emission at 514 nm when excited at 490 nm (Figs. 1 and 2). The assay requires no washes and can be incubated at room temperature, making it a very easy, potentially high-throughput assay that can be used in many different applications in many different cell types. Furthermore, this assay can be changed to different assay plate sizes ranging from 96- to 3,456-well applications.

In the assay protocol presented in this chapter, we use cells transiently co-transfected with the D_1R and D_2R in a 384-well format treated with DA to stimulate the receptors (Table 1). However, this assay is amenable to the addition of other protein constructs, different receptors, and various reagents that can inhibit or stimulate various parts of the DAR signaling cascade. We have had success in adapting this protocol to D_1R alone co-transfected with G_{15}

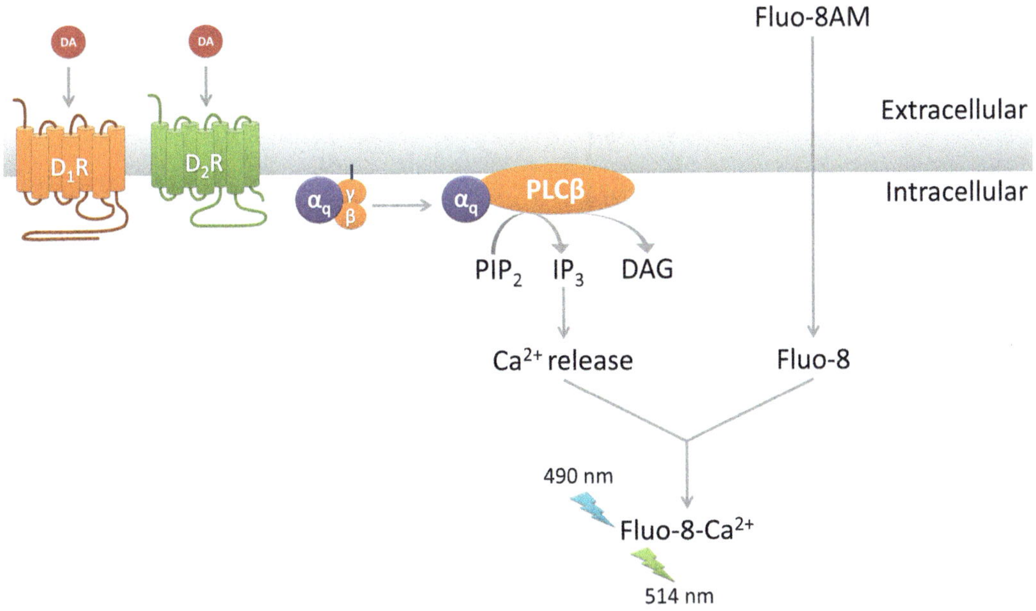

Fig. 1 Mechanism of Fluo-8 calcium live cell-based assay. Cells are incubated with the Fluo-8AM dye in the presence of a quencher that inhibits the fluorescence of extracellular Fluo-8. Fluo-8AM becomes passively internalized, and endogenous esterases cleave off a lipophilic moiety, resulting in a negatively charged Fluo-8 that is trapped in the cell. Once the D_1R and D_2R are co-activated, the G_q protein activates the PLCβ-mediated inositol 1,4,5-triphosphate production (IP_3), resulting in the release of ionic calcium from intracellular stores into the cytosol. This calcium binds to the Fluo-8 dye, changing its conformation so that it fluoresces at 514 nm when excited with a 490 nm light

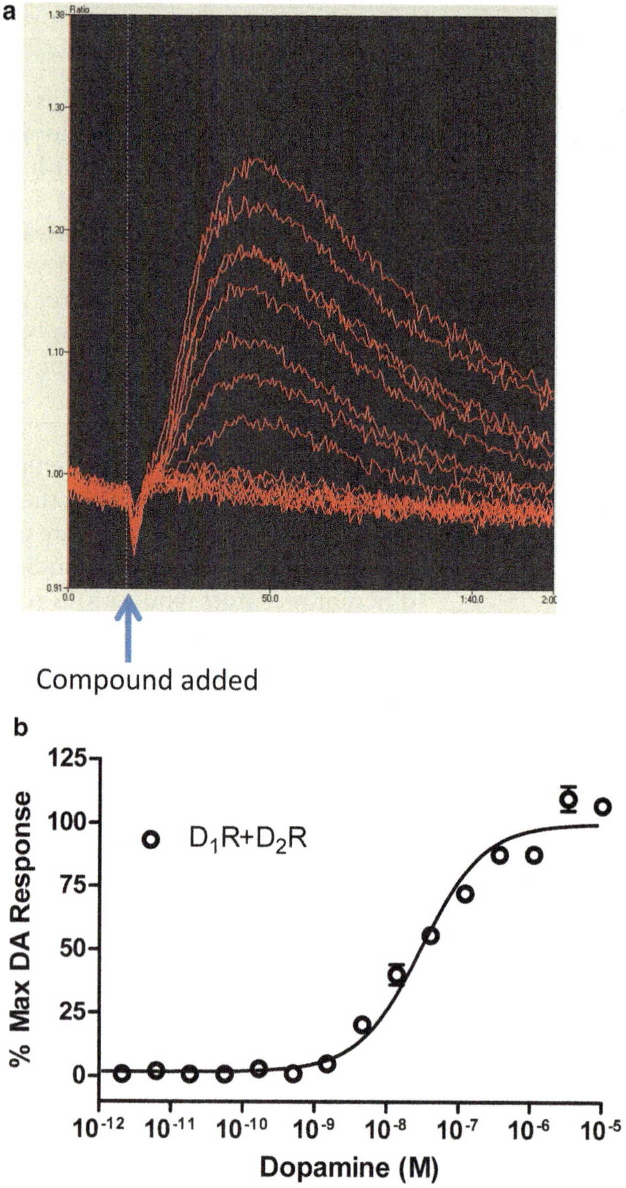

Fig. 2 Calcium assay data output in cells co-transfected with the D_1R and D_2R. (**a**) The FDSS μCell is set to read every 0.6 s for 200 intervals, resulting in a 2 min kinetic read. Increasing concentrations of DA stimulate increasing levels of calcium mobilization in cells co-expressing the D_1R and D_2R. Shown is an example of a set of kinetic reads from a single column on the 384-well plate. Each *curve* represents a single well in the column that has been treated with DA concentrations ranging from 0.02 nM to 100 μM. (**b**) The maximal response in each kinetic read is graphed to form a concentration response curve and normalized to set a maximal dopamine response to 100 % (EC_{50} = 0.03 μM). The graph shown is an example of a typical curve with *each point* representing the average of three wells ± SEM

as well as D_2R alone co-transfected with the chimeric G protein, G_{qi5}. Furthermore, although not described here, the assay can be run in multiple modes including one to test antagonists. For antagonist mode testing, the antagonist compound is added to the cells first, followed by an EC_{80} concentration of agonist, and inhibition of this agonist stimulation is observed (*see* **Note 4**).

In addition, this is a very malleable assay that can be read on many different machines. Although we describe a protocol for an output read on the FDSS μCell system and associated software, we have successfully used the FlexStation-3 Benchtop Multi-Mode Microplate Reader (Molecular Devices, Sunnyvale, CA) and the FDSS-7000 plate reader with no appreciable difference in signal strength. Another plate reader commonly used to read calcium assays is the FLIPR (Molecular Devices, Sunnyvale, CA).

3.2 Preparation of Transiently Transfected Cells Expressing the D_1R and D_2R

1. Chemically lift HEK293T cells via incubation with EBSS(−), and pellet by centrifugation at 1,000 rpm for 10 min.
2. Resuspend cells in supplemented DMEM and seed in 150 cm^2 plates at 10×10^6 cells/plate (*see* **Note 5**). Incubate overnight at 37 °C, 5 % CO_2, and 90 % humidity.
3. After 16–20 h, transfect cells using the CalPhos transfection kit according to the manufacturer's recommendations (*see* **Note 6**), and incubate overnight at 37 °C, 5 % CO_2, and 90 % humidity.
4. After 16–20 h, lift cells as described in step 1 and resuspend cells to make a final concentration of 1,500,000 cells/mL.
5. Dispense 20 μL cell suspension into each well (30,000 cells/well) in a 384-well, black, optical plate. Incubate overnight at 37 °C, 5 % CO_2, and 90 % humidity.
   (a) Consistency will be improved by using automated pipetting solutions for high well number plates, such as electronic pipettes or multiplate dispensers.
   (b) If using Multidrop® to dispense cells into the optical plate, make sure to filter cells through a 40 μm sterile cell strainer prior to counting cell density to improve fluorescence signal and prevent clogging in the multidrop due to cell debris.
   (c) Wash Multidrop® tubing by aspirating 1 mL 70 % ethanol, then at least 2 mL dH_2O. Aspirate cell suspension until you see it dispensed from the pipette cassette. Then start automated dispensation of 20 μL cell suspension into each well. To clean multidrop after cell dispensation, run water first, and then ethanol through as described above, making sure to completely empty tubing before shutting down the machine. Running ethanol immediately after cells through the instrument can cause cells to dehydrate and stick to the dispenser.

Table 1
Flow chart of steps for cell preparation and the calcium assay. The flowchart assumes that the format is for 384-well plates using calcium phosphate transfection procedures and Fluo-8 dye-based calcium detection

Step	Protocol section	Approx. time	Description	See additional
Plate HEK293T cells	3.2 Steps 1 and 2	20–30 min	10×10^6 cells per 150 cm^2 plate	Note 6
Incubate	3.2 Step 2	16–20 h	37 °C, 5 % CO_2, 90 % humidity	
Calcium phosphate transfection	3.2 Step 3	10–20 min	Up to 25 μg DNA per 150 cm^2 plate	Notes 1 and 6
Incubate	3.2 Step 3	16–20 h	37 °C, 5 % CO_2, 90 % humidity	
Plate cells in optical plate	3.2 Steps 4 and 5	20–30 min	20 μL, 30,000 cells/well	Note 3
Incubate	3.2 Step 5	16–20 h	37 °C, 5 % CO_2, 90 % humidity	
Prepare compound plate	3.3 Step 1	~5–10 min per compound	Serial 1:3 dilution	Note 7
Add dye mix to cells	3.3 Steps 2 and 3	10 min	25 μL/well	Notes 2, 8, and 9
Incubate	3.3 Step 3	30–60 min	Room temperature, light protected	
Read	3.4 Steps 1–6	5 min	Excite 490 nm Emit 514 nm 0.6 s interval 201 intervals	Note 10
a. Baseline read b. Add compound c. Post-add read		~14 s ~106 s	23 intervals 5 μL/well	

3.3 Calcium Assay Preparation

1. Prepare compound for assay in 384-well compound plates by diluting compound(s) in HBSS(−) to 100–1,000 μM, depending on compound solubility. While other dilution buffers may work, calcium-containing buffers should be avoided to limit background emission. Serially dilute 1:3 in HBSS(−) supplemented with 0.2 mM SMB. The compounds will be diluted 1:10 when added to the cells to give the final concentration, so one may need to experimentally determine the top concentration

sufficiently high to make a full concentration response curve for a particular compound.

(a) Make a compound plate with 2–3 columns per serial compound dilution in a 384-well assay plate. It is common to include controls such as the buffer alone, DMSO blank, an and a DA control dose–response curve in the plates (*see* **Note 7**). A starting point is to make a per-well volume of at least 40 μL to ensure robotic pipettes are able to aspirate up liquid.

2. While every manufacturer will have different concentrations of dye and quantity of quenchers, for the AAT Bioquest Kit, make the dye mix by adding 20 μL reconstituted component A (dye) to 9.5 mL 1× component B (Pluronic® F127 + quencher) for each 384-well plate (*see* **Note 8**). Vortex dye mix vigorously.
3. Dispense 25 μL dye mix to each well in the cell plate (*see* **Note 9**), and incubate at room temperature for 30–60 min in light-protected conditions.

3.4 Running the Calcium Assay Protocol

1. Plate reader should be set to excite at 490 nm and read emission at or around 514 nm.
2. Onboard pipetting should be set so that pipettes aspirate 5 μL from compound plate, and dispense 5 μL into the assay plate (which already contains 45 μL of dye and cells) at the programmed time. This results in a tenfold dilution of the compound to the final concentration on the cells.
3. Load pipette tips on the FDSS μCell.
4. Install 540 nm filter and 530 nm light components in μCell.
5. Put the cell plate and compound plate into corresponding platforms in the μCell.
6. Set the protocol to take a kinetic reading every 0.6 s for at least 201 intervals, and dispense 5 μL compound per well at approximately the 24th read interval. This allows you to obtain a background read for the first 20 s or so to record a baseline. Then upon addition of an agonist, record for an additional few minutes to ensure that each concentration has time to reach its peak effect as seen in Fig. 2 (*see* **Note 10**).

4 Anticipated Results

There are a myriad of ways to export data from the many different plate readers available. Most manufacturers have onboard software that is capable of exporting data analyzed by a number of different methods. For the examples chosen here, we export data as the maximum signal, recorded in relative fluorescence units (RFU), minus the minimum (baseline) signal for each well's kinetic curve (Fig. 2a). For some assays it may be convenient to normalize the E_{max} response seen with DA to 100 %, such as shown in Fig. 2b.

It is often necessary to set the extracted data to use the baseline (minimum) at a particular time point prior to the addition of compound. This is due to what we call an "add artifact" that sometimes occurs during pipetting of the compound from the compound plate to the cell plate. This may be due to the presence of the pipette tip in the well or the mixture of solutions or some other unknown factor. Regardless, it often results in a small dip in the kinetic fluorescence read below baseline (Fig. 2a), before rising again. Samples treated with buffer alone also often display an add artifact followed by a rise back to baseline. For this reason it is important to note that the software must take the minimum baseline reading before compound addition begins, or it could incorrectly set the bottom of the add artifact as the baseline.

As observed in Fig. 2a, the recorded calcium signal for DARs has a very fast onset, is transient, and rapidly returns to baseline in less than 10 min. For this reason it is necessary to measure the signal kinetically in real time. Furthermore, it should also be noted that when using transiently transfected cells, some variations in potency and efficacy are observed due to the variations inherent in receptor expression levels. Signal-to-noise ratios in our hands can range from 1.5- to 2-fold for transiently transfected D_1R–D_2R cells to upward of fivefold if using stably transfected cell lines.

5 Notes

1. The D_1–D_2 heteromer is able to activate the G_q protein without additional manipulations. However, monomeric DARs are not easily linked to calcium mobilization in heterologous cell systems. Despite this, we have found ways to artificially link D_1R and D_2R activation to calcium mobilization, and the same methods may be used for the other DAR subtypes and G proteins. To link monomeric D_2R to a calcium response, we have used cells co-expressing the D_2R with a chimeric G_{qi5} protein (G_q protein with the last five residues replaced with the last five residues of the G_i protein) which links D_2R stimulation to PLCβ activation instead of AC inhibition [39]. Alternatively a promiscuous G protein expression vector such as the G_{15} protein can be used to link the D_1R and/or the D_2R to a calcium response. The G_{15} protein is able to couple to receptors that are normally linked to the $G_{i/o}$, G_s, and G_q proteins, activating the PLCβ-mediated calcium mobilization.
2. Dilute components and refreeze aliquots for future use as directed by vendor.

3. The use of poly-D-lysine-coated plates are not recommended for readers that read from the bottom of the plate.
4. We have found that an EC_{80} concentration of the agonist is preferred to maximize the signal window for an antagonist to block, limiting errors associated with a smaller stimulation window. The EC_{80} of a dose–response curve is the concentration at which 80 % of the maximal response is reached. Assuming that you are using a standard curve fit equation with a hill slope of 1, and have normalized the maximum and minimum curve values to 100 and 0, respectively, the EC_{80} can be easily calculated from the equation $y = \text{minimum} + \dfrac{\text{maximum} - \text{minimum}}{1 + 10^{\log EC_{50} - x}}$, where y = % maximum and x = log(concentration), which can be condensed to $\log EC_{80} = \log EC_{50} + 0.6$.
5. Need two 150 cm^2 plates to make one 384-well plate of cells with 30,000 cells/well.
6. We have transfected 10 μg of the D_1R and 5 μg of the D_2R per 150 cm^2 cell plate, which gives sufficient receptor expression for the calcium assay to work (1–4 pmol receptor binding per mg protein).
7. Caution: sometimes the solvent will give a weak false-positive result, giving the appearance of a partial agonist/antagonist. Run solvent as control.
8. Component B should be diluted in HBSS(−) to make a 1× solution if a buffer is not provided by the kit. Pluronic® F127 increases the solubility of acetomethyl (AM) esters. The purpose of the quencher is to bind to excess Fluo-8AM dye that has not been internalized by the cell. This minimizes background fluorescence caused by the extracellular dye.
9. If using a Multidrop® for dye dispensation, note the tubing may become stained, so it may be preferable to have a separate set of tubing for dye dispensing.
10. Plate readers with onboard electronics can vary greatly in terms of the volume they can pipette, the frequency at which they can take readings, the sensitivity settings for the camera, the wavelengths they are capable or reading, and the user interface. These are general guidelines, extracted from our use of the FDSS/μCell, but may have to be modified (in some cases significantly) depending on the plate reader.

Acknowledgments

This research was supported by the NINDS/NIH intramural research program and the NIH Graduate Partnership Program in conjunction with Johns Hopkins University.

References

1. Missale C, Nash SR, Robinson SW et al (1998) Dopamine receptors: from structure to function. Physiol Rev 78:189–225
2. Sibley DR, Monsma FJ Jr (1992) Molecular biology of dopamine receptors. Trends Pharmacol Sci 13:61–69
3. Setler PE, Sarau HM, Zirkle CL et al (1978) The central effects of a novel dopamine agonist. Eur J Pharmacol 50:419–430
4. Hyttel J (1983) SCH 23390—the first selective dopamine D-1 antagonist. Eur J Pharmacol 91:153–154
5. O'Boyle KM, Gaitanopoulos DE, Brenner M et al (1989) Agonist and antagonist properties of benzazepine and thienopyridine derivatives at the D1 dopamine receptor. Neuropharmacology 28:401–405
6. Felder CC, Jose PA, Axelrod J (1989) The dopamine-1 agonist, SKF 82526, stimulates phospholipase-C activity independent of adenylate cyclase. J Pharmacol Exp Ther 248: 171–175
7. Mahan LC, Burch RM, Monsma FJ et al (1990) Expression of striatal D1 dopamine receptors coupled to inositol phosphate production and Ca2+ mobilization in Xenopus oocytes. Proc Natl Acad Sci 87:2196–2200
8. Undie AS, Friedman E (1990) Stimulation of a dopamine D1 receptor enhances inositol phosphates formation in rat brain. J Pharmacol Exp Ther 253:987–992
9. Undie AS, Friedman E (1992) Selective dopaminergic mechanism of dopamine and SKF38393 stimulation of inositol phosphate formation in rat brain. Eur J Pharmacol 226: 297–302
10. Undie AS, Friedman E (1994) Inhibition of dopamine agonist-induced phosphoinositide hydrolysis by concomitant stimulation of cyclic AMP formation in brain slices. J Neurochem 63:222–230
11. Undie AS, Weinstock J, Sarau HM et al (1994) Evidence for a distinct D1-like dopamine receptor that couples to activation of phosphoinositide metabolism in brain. J Neurochem 62:2045–2048
12. Jope RS, Song L, Powers R (1994) [3H]PtdIns hydrolysis in postmortem human brain membranes is mediated by the G-proteins Gq/11 and phospholipase C-beta. Biochem J 304(Pt 2):655–659
13. Pacheco MA, Jope RS (1997) Comparison of [3H]phosphatidylinositol and [3H]phosphatidylinositol 4,5-bisphosphate hydrolysis in postmortem human brain membranes and characterization of stimulation by dopamine D1 receptors. J Neurochem 69:639–644
14. Arnt J, Hyttel J, Sánchez C (1992) Partial and full dopamine D1 receptor agonists in mice and rats: relation between behavioural effects and stimulation of adenylate cyclase activity in vitro. Eur J Pharmacol 213:259–267
15. Lee SP, So CH, Rashid AJ et al (2004) Dopamine D1 and D2 receptor co-activation generates a novel phospholipase C-mediated calcium signal. J Biol Chem 279: 35671–35678
16. Perreault ML, Hasbi A, Alijaniaram M et al (2010) The dopamine D1-D2 receptor heteromer localizes in dynorphin/enkephalin neurons. J Biol Chem 285:36625–36634
17. Rashid AJ, So CH, Kong MMC et al (2007) D1–D2 dopamine receptor heterooligomers with unique pharmacology are coupled to rapid activation of Gq/11 in the striatum. Proc Natl Acad Sci 104:654–659
18. So CH, Varghese G, Curley KJ et al (2005) D1 and D2 dopamine receptors form heterooligomers and cointernalize after selective activation of either receptor. Mol Pharmacol 68: 568–578
19. O'Dowd BF, Ji X, Alijaniaram M et al (2005) Dopamine receptor oligomerization visualized in living cells. J Biol Chem 280:37225–37235
20. Cools AR, Lubbers L, van Oosten RV et al (2002) SKF 83959 is an antagonist of dopamine D1-like receptors in the prefrontal cortex and nucleus accumbens: a key to its antiparkinsonian effect in animals? Neuropharmacology 42:237–245
21. Panchalingam S, Undie AS (2001) SKF83959 exhibits biochemical agonism by stimulating [35S]GTP[gamma]S binding and phosphoinositide hydrolysis in rat and monkey brain. Neuropharmacology 40:826–837
22. Andringa G, Stoof JC, Cools AR (1999) Subchronic administration of the dopamine D1 antagonist SKF 83959 in bilaterally MPTP-treated rhesus monkeys: stable therapeutic effects and wearing-off dyskinesia. Psychopharmacology (Berl) 146:328–334
23. Chun LS, Free RB, Doyle TB et al (2013) D1-D2 dopamine receptor synergy promotes calcium signaling via multiple mechanisms. Mol Pharmacol 84:190–200
24. Zhen X, Goswami S, Friedman E (2005) The role of the phosphatidylinositol-linked D1 dopamine receptor in the pharmacology of SKF83959. Pharmacol Biochem Behav 80: 597–601

25. Calderon DP, Leverkova N, Peinado A (2005) Gq/11-induced and spontaneous waves of coordinated network activation in developing frontal cortex. J Neurosci 25:1737–1749
26. Dai R, Ali MK, Lezcano N et al (2008) A crucial role for cAMP and protein kinase A in D1 dopamine receptor regulated intracellular calcium transients. Neurosignals 16:112–123
27. Otani S, Auclair N, Desce JM et al (1999) Dopamine receptors and groups I and II mGluRs cooperate for long-term depression induction in rat prefrontal cortex through converging postsynaptic activation of MAP kinases. J Neurosci 19:9788–9802
28. Zhang L, Bai J, Undie AS et al (2005) D1 dopamine receptor regulation of the levels of the cell-cycle-controlling proteins, cyclin D, P27 and Raf-1, in cerebral cortical precursor cells is mediated through cAMP-independent pathways. Cereb Cortex 15:74–84
29. Camps M, Hou C, Sidiropoulos D et al (1992) Stimulation of phospholipase C by guanine-nucleotide-binding protein βγ subunits. Eur J Biochem 206:821–831
30. Hernandez-Lopez S, Tkatch T, Perez-Garci E et al (2000) D2 dopamine receptors in striatal medium spiny neurons reduce L-type Ca2+ currents and excitability via a novel PLC[beta]1-IP3-calcineurin-signaling cascade. J Neurosci 20:8987–8995
31. Liu J, Wang W, Wang F et al (2009) Phosphatidylinositol-linked novel D(1) dopamine receptor facilitates long-term depression in rat hippocampal CA1 synapses. Neuropharmacology 57:164–171
32. Pei L, Li S, Wang M et al (2010) Uncoupling the dopamine D1-D2 receptor complex exerts antidepressant-like effects. Nat Med 16:1393–1395
33. Undie AS, Friedman E (1992) Aging-induced decrease in dopaminergic-stimulated phosphoinositide metabolism in rat brain. Neurobiol Aging 13:505–511
34. Undie AS, Friedman E (1993) Diet restriction prevents aging-induced deficits in brain phosphoinositide metabolism. J Gerontol 48: B62–B67
35. Undie AS, Wang HY, Friedman E (1995) Decreased phospholipase C-beta immunoreactivity, phosphoinositide metabolism, and protein kinase C activation in senescent F-344 rat brain. Neurobiol Aging 16:19–28
36. Monsma FJ, Mahan LC, McVittie LD et al (1990) Molecular cloning and expression of a D1 dopamine receptor linked to adenylyl cyclase activation. Proc Natl Acad Sci 87:6723–6727
37. Takebe Y, Seiki M, Fujisawa J et al (1988) SR alpha promoter: an efficient and versatile mammalian cDNA expression system composed of the simian virus 40 early promoter and the R-U5 segment of human T-cell leukemia virus type 1 long terminal repeat. Mol Cell Biol 8:466–472
38. Zhang LJ, Lachowicz JE, Sibley DR (1994) The D2S and D2L dopamine receptor isoforms are differentially regulated in Chinese hamster ovary cells. Mol Pharmacol 45:878–889
39. Coward P, Chan SD, Wada HG et al (1999) Chimeric G proteins allow a high-throughput signaling assay of Gi-coupled receptors. Anal Biochem 270:242–248

Chapter 15

Intracellular Trafficking Assays for Dopamine D2-Like Receptors

Chengchun Min, Mei Zheng, and Kyeong-Man Kim

Abstract

G protein-coupled receptors (GPCRs) follow two endocytic pathways: homologous pathway, which is GPCR kinase (GRK) mediated and agonist induced, and heterologous pathway mediated by second messenger-dependent protein kinases. Dopamine D_2 receptor undergoes GRK-mediated and, to a lesser extent, protein kinase C (PKC)-mediated endocytosis. Dopamine D_3 receptor almost totally undergoes PKC-mediated endocytosis. D_3 receptor also uniquely undergoes pharmacological sequestration, which involves conformational changes of receptor proteins that abrogates hydrophilic agonist binding. Pharmacological sequestration does not involve an actual movement of receptor to the cytosol. Internalization of GPCRs can be determined using ligand binding, flow cytometry, ELISA, fluorescence, or cell surface biotinylation assays. Here, we review convenient and commonly used methodologies involved in the internalization of dopamine receptors focusing on D_2 and D_3 receptor.

Key words Receptor internalization, Ligand binding, GRK, PKC, Clathrin, Caveolae, Fluorescence-activated cell sorting, ELISA

1 Introduction

Following agonistic stimulation, G protein-coupled receptors (GPCRs) undergo conformational changes that allow binding of G proteins, leading to the activation of various signaling pathways and initiation of intracellular trafficking, such as receptor endocytosis (internalization or sequestration). Clathrin-mediated and caveolae-dependent pathways are the best-characterized major internalization routes [1, 2].

After agonist stimulation, the receptor is phosphorylated by GPCR kinases (GRKs) [3], enhancing the binding of β-arrestins, which connects to adaptors such as AP-2 and clathrin [4–6]. The functional roles of homologous (agonist-induced, GRK-mediated) receptor endocytosis could be diverse and still remain unclear. The roles of receptor-mediated endocytosis were initially recognized as the desensitization of GPCR signal transduction because this

Mario Tiberi (ed.), *Dopamine Receptor Technologies*, Neuromethods, vol. 96,
DOI 10.1007/978-1-4939-2196-6_15, © Springer Science+Business Media New York 2015

process involves removal of binding sites for hydrophilic agonists from the cell surface [7–9]. This hypothesis agrees with the findings that some receptors become internalized and are transported to late endosomes and lysosome for degradation. However, other studies showed that internalized receptors recycle back to the plasma membrane through early endosomes, which eventually results in resensitization of the desensitized receptors [10, 11]. Compared with homologous regulatory processes of GPCRs, molecular mechanisms of heterologous internalization mediated by second messenger-dependent kinases are poorly understood.

Dopamine D_2 receptor (D_2R) mainly undergoes agonist-induced internalization [12] even though it still undergoes heterologous (protein kinase C (PKC)-mediated) internalization [13–15]. Agonist-induced internalization seems to mediate resensitization; however, more complicated mechanisms might be involved in a microdomain-dependent manner [16]. PKC-mediated internalization of D_2R seems to mediate receptor desensitization [14, 15]. Dopamine D_3 receptor (D_3R) rarely undergoes agonist-induced internalization but undergoes robust PKC-mediated internalization, desensitization, and downregulation [12, 13, 17]. The relationship between PKC-mediated desensitization and internalization of D_3R is unclear. D_3R also undergoes a characteristic form of sequestration, called pharmacological sequestration, which mediates receptor desensitization [18].

Various experimental approaches have been utilized to measure GPCR internalization or sequestration [19]. In the ligand binding method, either radiolabeled hydrophilic ligands or a combination of radiolabeled hydrophobic ligands combined with hydrophilic ligands (to compete with surface binding of hydrophobic radioligands) can be utilized [12, 20]. Receptor internalization is also commonly determined using receptors tagged at the N-terminus with epitopes specifically recognized by antibodies (usually the hemagglutinin (HA) or FLAG epitope) [12, 19]. Sometimes, receptor internalization can be measured by fluorescence [17] or cell surface biotinylation [21].

Here, we describe principles and assay procedures, which are commonly used, and radioligand methods and approaches using epitope-tagged receptors at the N-terminus. In addition, pharmacological sequestration, which is a unique mode of receptor sequestration observed in some GPCRs including dopamine D_3R [18, 22], is described.

2 Materials

2.1 Ligand Binding Method

The method requires the following: MEM (for HEK-293 cells), DMEM (for COS-7 cells), [^{3}H]-sulpiride and [^{3}H]-spiperone (specific activity higher than 70 Ci/mmol is recommended, needs to be stored frozen), 24-well plates, poly-L-lysine (PLL; 25 μg/mL, 4 °C), ascorbic acid (100 mg in 25 mL MEM, needs to be prepared fresh), dopamine (need to be prepared fresh), sodium metabisulfite (prepare and freeze stock solution), haloperidol (prepare and freeze stock solution; haloperidol is dissolved in 0.1 N HCl, up to 3 mg/mL; haloperidol·HCl is soluble in H_2O), and 1 % sodium dodecyl sulfate (SDS).

2.2 Fluorescence-Activated Cell Sorting

The following are required: MEM (for HEK-293 cells), DMEM (for COS-7 cells), 6-well plates, ascorbic acid, dopamine, anti-HA or anti-FLAG antibody, anti-mouse fluorescein isothiocyanate (FITC)-conjugated IgG (Fc specific), phosphate buffered saline (PBS)/EDTA (5 mM), and formaldehyde 8 % pH 8.5 (needs to be prepared fresh).

2.3 ELISA Method

Enzyme-linked immunosorbent assay (ELISA) requires the following: MEM (for HEK-293 cells), DMEM (for COS-7 cells), 1 % bovine serum albumin (BSA) in PBS, anti-FLAG or anti-HA antibody, sodium azide, horseradish peroxidase (HRP)-conjugated anti-mouse IgG, 0.05 M phosphate-citrate buffer (0.05 M sodium phosphate, 0.025 M citric acid), *o*-phenylenediamine dihydrochloride (OPD) (0.4 mg/mL dissolved in 50 mL of 0.05 M phosphate-citrate buffer), 30 % H_2O_2, and 3 N HCl or 3 M H_2SO_4.

2.4 Pharmacological Sequestration Method

The following are required: MEM (for HEK-293 cells), DMEM (for COS-7 cells), [^{3}H]-sulpiride and [^{3}H]-spiperone (specific activity higher than 70 Ci/mmol is recommended, needs stored frozen), and low pH buffer (150 mM NaCl and 50 mM acetic acid, pH 2.0).

3 Methods

Here, we describe methods that involve transfection into heterologous cells. To employ appropriate assay methods, it is important to understand the fundamental principles involved in the trafficking of receptors. If trafficking involves movement of receptors from the plasma membrane to the cytosol, the radioligand method or approaches using epitope-tagged receptor can be used (Fig. 1). If trafficking involves conformational changes of receptor such that it cannot bind to hydrophilic ligands, rather than actual movement to cytosolic region, the ligand binding method can be used but the epitope-tagged method cannot.

Theoretically, any cell type can be used for receptor internalization assays. Here, we describe assay protocols in HEK-293 cells and COS-7 cells, which are relatively convenient cells to use for the delivery of cDNAs. COS-7 cells express relatively lower endogenous GRK2 or levels of β-arrestins compared with HEK-293 cells [12, 19]. Thus, COS-7 cells are an excellent experimental system

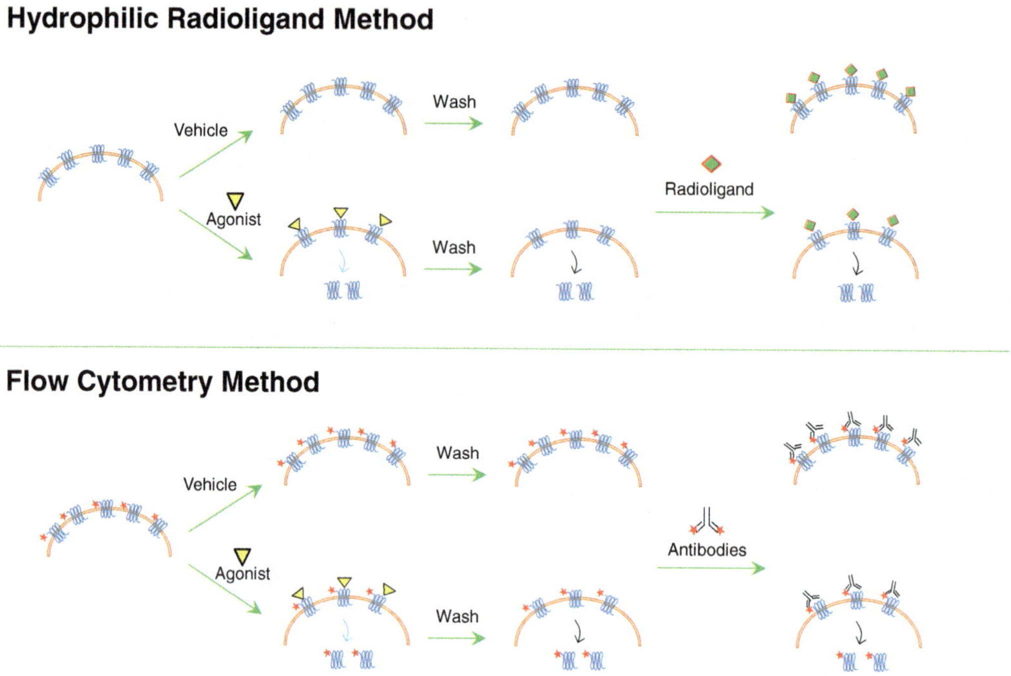

Fig. 1 Diagrams describing the principles of determining receptor internalization. The *upper panel* shows the basic concept to measure receptor internalization using the hydrophilic radioligand method. The process involves induction of receptor internalization through treatment with agonist for a desired period of time, washing, and labeling with hydrophilic radioligand. The ratio internalization is calculated by (vehicle-treated – agonist-treated)/(vehicle-treated). The *lower panel* shows how to measure receptor internalization by utilizing epitope-tagged receptors. The process involves induction of receptor internalization through treatment with agonist for a desired period of time, washing, and labeling with antibodies. The ratio of internalization is calculated as in the *upper panel*

to test the effects of various cellular components, such as GRKs or β-arrestins, in promoting receptor internalization. On the other hand, HEK-293 cells can better assess inhibition of internalization and are better to prepare knockdown cells for GRK or β-arrestin [11, 18, 19] (*see* **Note 1**).

3.1 Ligand Binding

Hydrophilic radiolabeled ligand ([^{3}H]-sulpiride) can be used to determine cell surface receptor density of D_2R for agonist-induced and PMA (phorbol 12-myristate 13-acetate)-induced endocytosis. The radioligand method using [^{3}H]-sulpiride is not appropriate to assess agonist-induced endocytosis of D_3R because it is extremely difficult to wash away pretreated agonists, such as dopamine or quinpirole, from cells expressing D_3R. This method cannot be used for assessing D_4R internalization because of its low affinity for D_4R.

1. Transfect cells with proper transfection methods. It is important to adjust receptor expression levels so they are similar (*see* **Note 2**).

2. Treat each well of a 24-well plate with 250 μL of 0.1 mg/mL poly-L-lysine (PLL) (Sigma) dissolved in sterile water for 30 min. The PLL solution is removed and the wells are allowed to dry (*see* **Note 3**).
3. After 24 h of transfection, the cells are treated with trypsin. Culture medium (14 mL; enough to fill 24 wells) is dispensed in a 100 mm diameter culture dish. Aliquots of 0.5 mL are added to wells (*see* **Note 4**). For a row of six wells, three wells are used for total binding and the other three are used for nonspecific binding.
4. Next day, serum-free MEM or DMEM containing 20 mM HEPES is warmed to 37 °C. Medium is removed from the wells and replaced with 200 μL of warm serum-free medium followed by 200 μL warm agonist-containing serum-free medium (*see* **Note 5**).
5. Each plate is incubated at 37 °C in a CO_2 incubator for the desired time (*see* **Note 6**).
6. At the designated time, the plate is removed and placed on ice. Each well is washed three times with 250 μL ice-cold serum-free medium (5 min incubation on ice per wash). After the final wash, each plate is positioned on a slant place on ice and the medium is completely removed from each well (*see* **Note 7**).
7. Each well then receives 250 μL ice-cold serum-free medium containing [^{3}H]-sulpiride (2.2 nM for D_2R and 7.2 nM for D_3R). To measure nonspecific binding, the same concentration of [^{3}H]-sulpiride and 10 μM haloperidol is added. The plate is incubated for 150 min at 4 °C (*see* **Note 8**).
8. Each well is washed three times with ice-cold serum-free medium and 200 μL 1 % SDS is added. The plate is shaken at room temperature 2 h or overnight.
9. The resulting cell lysate from each well is transferred to an individual scintillation vial, 2.5 mL of scintillation fluid is added, and counts are obtained using a liquid scintillation counter.

Examples of internalization results of D_2 receptor from HEK-293 cells and COS-7 cells are shown in Fig. 2.

3.2 Fluorescence-Activated Cell Sorting (Flow Cytometry)

Detection of GPCR endocytosis by FACS requires a GPCR-tagged epitope at the N-terminus. Since the N-terminus disappears from the cell surface when GPCR internalizes, FACS can be used to quantify removal of a GPCR from the plasma membrane.

1. Cells are transfected with HA- or FLAG-tagged receptor at the N-terminus. The next morning, each well of 6-well plate is treated with PLL. The PLL solution is removed and the wells are allowed to dry. Cells are treated with trypsin, followed by

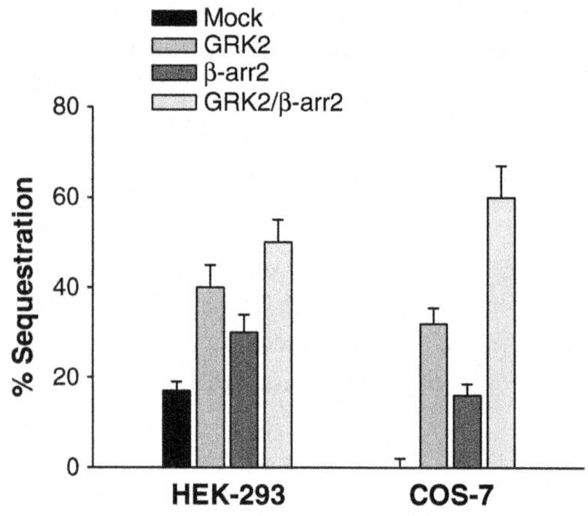

Fig. 2 Comparison of internalization of dopamine D_2 receptor in HEK-293 cells and COS-7 cells. Cells were transfected with dopamine D_2R in pCMV5 with or without GRK2 and/or β-arrestin2. Receptor internalization was determined by the radioligand method ([^{3}H]-sulpiride). In HEK-293 cells, some fraction of D_2R was internalized in response to agonistic stimulation. In COS-7 cells, virtually no internalization of D_2R occurred in the absence of exogenous GRK2 or β-arrestin2

culture medium to stop trypsin action. Aliquots are added to wells (*see* **Note 9**).

2. Next day, cells in each well are washed with 2 mL of serum-free medium.
3. One milliliter of serum-free medium is dispensed in each well and incubated 30 min at 37 °C.
4. One milliliter of 2× stimulation buffer is then added to each well (*see* **Note 5**) and the plate is put on ice to stop further receptor trafficking.
5. The medium is removed and each well is washed three times with 2 mL of ice-cold serum-free medium.
6. Each well receives a 1:500 dilution of anti-HA or anti-FLAG antibody in serum-free medium and is incubated on ice for 50 min.
7. Each well is then washed three times with 2 mL cold serum-free medium. Each well receives an aliquot of a 1:250 dilution of anti-mouse FITC-conjugated IgG (Fc specific) in serum-free medium.
8. Each plate is kept on ice in the dark for 50 min and then wells are washed three times with cold serum-free medium.
9. Four hundred microliters of cold PBS/EDTA is added to each well and incubated on ice for 10 min to detach the cells from the wells.

10. The cells in each well are resuspended and then transferred to Falcon 2052 tubes containing 100 μL 8 % formaldehyde taking care to shield the samples from light.
11. Mean cell surface fluorescence and number of fluorescence-positive cells are determined by FACS.

3.3 ELISA

This section details the ELISA assay of receptor internalization using the D_2R as an example. Analysis by ELISA is based on the presence of an N-terminal epitope tag on the receptor, such as HA or FLAG, which is no longer recognized by the cognate antibody once the receptors are internalized.

1. Cells are transfected with receptor that is tagged at the N-terminus with HA or FLAG (*see* **Note 9**).
2. Next day, wells of the 24-well plate are treated with 250 μL of a 0.1 mg/mL PLL solution per well for 30 min and dried for 2 h.
3. The cells are dispensed in wells of a 24-well plate and left overnight.
4. The cells are washed with 400 μL of warm, serum-free medium and then stimulated for 0–60 min with agonists at 37 °C (*see* **Note 5**).
5. Each plate is placed on ice to stop the reaction, and the cells are washed with 1 mL ice-cold PBS three times on ice (10 min between washes).
6. Each well was treated with 1 % BSA in PBS (cold) on ice for 30 min.
7. After removing BSA solution, each well was treated with 250 μL 1:1,000 anti-HA/anti-FLAG antibody in 1 % BSA/PBS (0.1 % sodium azide) for 1 h at 4 °C.
8. The cells are washed three times with ice-cold PBS (10 min between washes).
9. The medium is aspirated and the cells are fixed in 0.25 mL/well of 4 % formaldehyde prepared fresh in PBS. The fixation is done for 15 min on ice. The remainder of the assay is performed at room temperature.
10. The cells are washed three times with ice-cold PBS prior to being treated with 1 % BSA in ice-cold PBS for 30 min (*see* **Note 10**).
11. After removing BSA solution, each well is incubated with 250 μL of a 1:1,000 dilution of horseradish peroxidase-conjugated anti-mouse antibody in 1 % BSA in PBS for 1 h at room temperature.
12. Cells are washed three times with PBS at room temperature (10 min between washes).

13. For the next step, OPD is dissolved in phosphate-citrate buffer (0.05 M), and 200 μL of the fresh substrate solution (containing 40 μL of fresh 30 % hydrogen peroxide per 100 mL of OPD solution) is added to each well.
14. The plates are incubated until the color of the solution in the well changes and the reaction is then stopped by adding 50 μL of 3 N HCl.
15. The supernatants are individually dispensed in wells of a 96-well plate and read using an ELISA reader at 492 nm. The background reading obtained from mock-transfected or non-transfected control cells is subtracted to calculate the percentage of internalization.

3.4 Pharmacological Sequestration

Typical receptor trafficking, which is mediated by GRK and β-arrestins, involves clear receptor endocytosis in which receptors move from cell surface to cytosol. For this kind of endocytosis, either radioligand binding method or epitope-tagged receptor approaches (flow cytometry or ELISA) can be utilized. In contrast, D_3R undergoes a unique mode of sequestration (pharmacological sequestration), which involves conformational changes accompanied by shift of receptors toward more hydrophobic domains within the plasma membrane without translocation into other intracellular compartments (Fig. 3). Thus, the ligand binding method, but not flow cytometry or ELISA, can be used to determine pharmacological sequestration (Fig. 4).

1. Cells were transfected with D_2R or D_3R.
2. The following day, wells of a 24-well plate are coated with PLL and cells are seeded at a density of 1.5×10^5 cells/well.

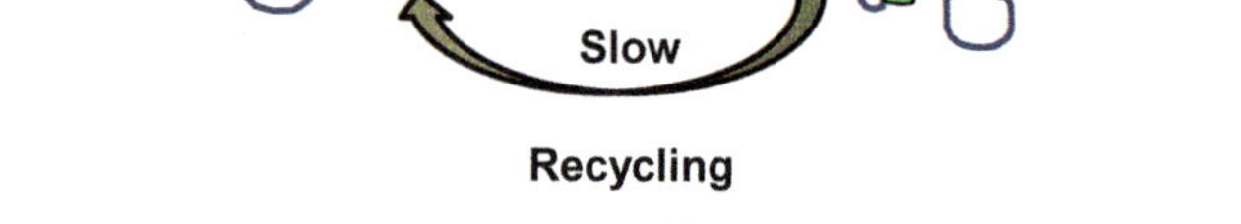

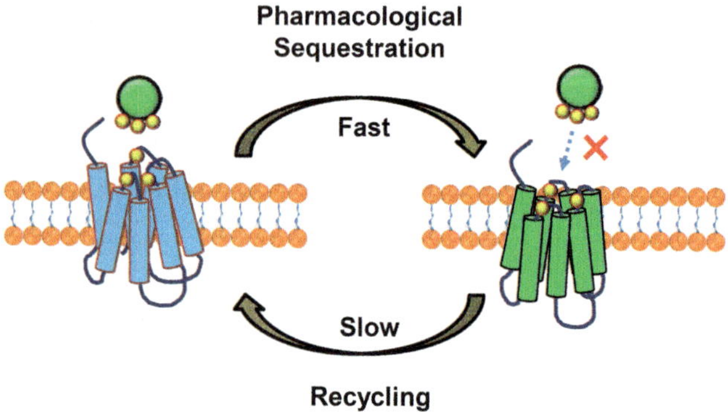

Fig. 3 Diagram showing the principle of pharmacological sequestration of receptors. In response to agonistic stimulation, some receptors undergo conformational changes through which receptors cannot bind hydrophilic ligands. Stimulated receptors do not move to cytosol but translocate to more hydrophobic environment within the plasma membrane

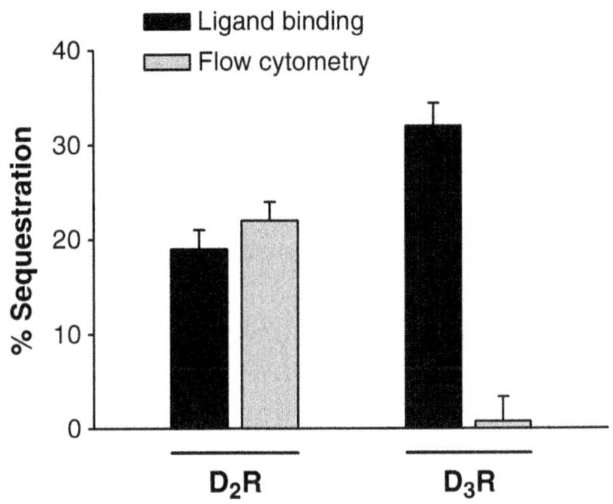

Fig. 4 Comparison of D_2 and D_3 receptor sequestration by radioligand method and flow cytometry. In the radioligand method, cells expressing D_2R were treated with 10 μM of dopamine for 1 h. In contrast, cells expressing D_3R were treated with 1 μM dopamine for 5 min. Cells were washed with low pH buffer and were treated with 2.2 nM (D_2R) or 7.2 nM (D_3R) [^{3}H]-sulpiride. In flow cytometry, cells were treated with 10 μM dopamine for 1 h

3. The following day, each well is rinsed once prior to preincubation with 0.5 mL of pre-warmed, serum-free medium containing 20 mM HEPES, pH 7.4, for 15 min at 37 °C.
4. Cells are then stimulated with agonists for 5 min (*see* **Note 4**) and the reaction is stopped by placing each plate on ice.
5. Each well is washed three times with ice-cold, low pH buffer (150 mM NaCl, 50 mM acetic acid, pH 2.0) with 5 min between washes (*see* **Note 11**).
6. Cells are washed two times with ice-cold 20 mM, pH 7.4, HEPES buffer (5 min between washes) and incubated with 250 μL of [^{3}H]-sulpiride at 4 °C for 150 min in the absence or presence of 10 μM haloperidol.
7. The cells are washed and lysed with 1 % SDS, and the remaining radioactivity is counted using a liquid scintillation counter.

4 Notes

1. In our experience, HEK-293 cells yield more variable results even when the same cells were used continuously. More consistent results can be obtained from COS-7 cells.
2. Total receptor expression levels can be determined using [^{3}H]-spiperone, which labels receptors located intracellularly as well as surface localized. Receptor expression levels are recommended between 1 and 2 pmol/mg proteins.

3. PLL treatment is strongly recommended for HEK-293 cells. PLL can be reused several times.
4. If cell numbers are not sufficient, the cell suspension can be prepared to 10 mL, which will be sufficient for 16 wells.
5. Agonists or PMA should be prepared at concentrations twice those of the final concentrations. For example, if 10 μM dopamine is the final concentration of agonist, prepare 20 μM dopamine in 10 μM sodium metabisulfite or 200 μM ascorbic acid, and then add 200 μL serum-free media containing 20 μM dopamine per well.
6. Cells are stimulated with agonists usually between 0 and 60 min. It is recommended to have time course experiments to test the effects of some treatments on receptor internalization.
7. Do not use suction pump when removing media from the wells in all washing steps. Instead, use a micropipette to remove media very carefully.
8. A 30 min incubation results in similar outcomes.
9. It should be confirmed that adding epitope tags does not affect ligand binding, signaling, and trafficking properties of receptor. Also confirm the expression levels using the ligand binding method.
10. You may skip this step if it does not affect the results under your experimental conditions.
11. It was confirmed that ligand binding properties of D_2R, which has similar ligand binding properties with D_3R, were not changed under these experimental conditions.

Acknowledgment

This study was financially supported by the Ministry of Knowledge Economy (MKE) and Korea Institute for Advancement of Technology (KIAT) through the Inter-ER Cooperation Project (R0002019).

References

1. Doherty GJ, McMahon HT (2009) Mechanisms of endocytosis. Annu Rev Biochem 78:857–902
2. Hansen CG, Nichols BJ (2009) Molecular mechanisms of clathrin-independent endocytosis. J Cell Sci 122(Pt 11):1713–1721
3. Pitcher JA, Freedman NJ, Lefkowitz RJ (1998) G protein-coupled receptor kinases. Annu Rev Biochem 67:653–692
4. Ferguson SS et al (1996) Role of beta-arrestin in mediating agonist-promoted G protein-coupled receptor internalization. Science 271(5247):363–366
5. Goodman OB Jr et al (1996) Beta-arrestin acts as a clathrin adaptor in endocytosis of the beta2-adrenergic receptor. Nature 383(6599): 447–450

6. Laporte SA et al (1999) The beta2-adrenergic receptor/betaarrestin complex recruits the clathrin adaptor AP-2 during endocytosis. Proc Natl Acad Sci U S A 96(7):3712–3717
7. Hausdorff WP, Caron MG, Lefkowitz RJ (1990) Turning off the signal: desensitization of beta-adrenergic receptor function. FASEB J 4(11):2881–2889
8. Gurevich VV, Benovic JL (1997) Mechanism of phosphorylation-recognition by visual arrestin and the transition of arrestin into a high affinity binding state. Mol Pharmacol 51(1):161–169
9. Sibley DR, Strasser RH, Benovic JL, Daniel K, Lefkowitz RJ (1986) Phosphorylation/dephosphorylation of the beta-adrenergic receptor regulates its functional coupling to adenylate cyclase and subcellular distribution. Proc Natl Acad Sci U S A 83(24): 9408–9412
10. Yu SS, Lefkowitz RJ, Hausdorff WP (1993) Beta-adrenergic receptor sequestration. A potential mechanism of receptor resensitization. J Biol Chem 268(1):337–341
11. Cho D et al (2010) Agonist-induced endocytosis and receptor phosphorylation mediate resensitization of dopamine D(2) receptors. Mol Endocrinol 24(3):574–586
12. Kim KM et al (2001) Differential regulation of the dopamine D2 and D3 receptors by G protein-coupled receptor kinases and beta-arrestins. J Biol Chem 276(40):37409–37414
13. Cho EY et al (2007) Roles of protein kinase C and actin-binding protein 280 in the regulation of intracellular trafficking of dopamine D3 receptor. Mol Endocrinol 21(9):2242–2254
14. Namkung Y, Sibley DR (2004) Protein kinase C mediates phosphorylation, desensitization, and trafficking of the D2 dopamine receptor. J Biol Chem 279(47):49533–49541
15. Cho DI et al (2013) ARF6 and GASP-1 are post-endocytic sorting proteins selectively involved in the intracellular trafficking of dopamine D(2) receptors mediated by GRK and PKC in transfected cells. Br J Pharmacol 168(6):1355–1374
16. Cho DI et al (2012) The N-terminal region of the dopamine D2 receptor, a rhodopsin-like GPCR, regulates correct integration into the plasma membrane and endocytic routes. Br J Pharmacol 166(2):659–675
17. Thompson D, Whistler JL (2011) Dopamine D(3) receptors are down-regulated following heterologous endocytosis by a specific interaction with G protein-coupled receptor-associated sorting protein-1. J Biol Chem 286(2):1598–1608
18. Min C, Zheng M, Zhang X, Caron MG, Kim KM (2013) Novel roles for beta-arrestins in the regulation of pharmacological sequestration to predict agonist-induced desensitization of dopamine D3 receptors. Br J Pharmacol 170(5):1112–1129
19. Kang DS, Tian X, Benovic JL (2013) beta-Arrestins and G protein-coupled receptor trafficking. Methods Enzymol 521:91–108
20. Itokawa M et al (1996) Sequestration of the short and long isoforms of dopamine D2 receptors expressed in Chinese hamster ovary cells. Mol Pharmacol 49(3):560–566
21. Vickery RG, von Zastrow M (1999) Distinct dynamin-dependent and -independent mechanisms target structurally homologous dopamine receptors to different endocytic membranes. J Cell Biol 144(1):31–43
22. Mostafapour S, Kobilka BK, von Zastrow M (1996) Pharmacological sequestration of a chimeric beta 3/beta 2 adrenergic receptor occurs without a corresponding amount of receptor internalization. Recept Signal Transduct 6(3–4):151–163

Chapter 16

Study of Crosstalk Between Dopamine Receptors and Ion Channels

Ping Su, Albert H.C. Wong, and Fang Liu

Abstract

Dopamine receptors and ion channels are important membrane proteins, which play critical roles in regulating the function of the central nervous system. Protein-protein interactions between dopamine receptors and ion channels have been widely investigated and are relevant to many neuropsychiatric disorders, including schizophrenia, Parkinson's disease, stroke, drug addiction, and Alzheimer's disease. These interactions are mediated by dynamic associations of protein domains with temporal and spatial specificity. Peptides that mimic the amino-acid sequence of the interacting domains are an effective tool for disrupting protein-protein interactions. These peptides can selectively modulate particular signaling pathways, distinct from the primary signaling cascades initiated by ligand binding to either receptor alone. The trans-activator of transcription (TAT) domain of the HIV virus can be added to these small peptides to facilitate entry across the cell membrane. Here, we summarize the methods to identify the interactions between dopamine receptors and ion channels, determine the interacting sites between these proteins, and synthesize TAT-tagged interfering peptides. These peptides can be useful as an experimental tool and may also serve as a potential treatment for neuropsychiatric diseases involving dopamine receptors and ion channels. These methods may also be applied to the investigation of other protein (receptor)-protein (receptor) interactions.

Key words Dopamine receptors, Ion channels, TAT-tagged peptide, Protein-protein interaction, Co-immunoprecipitation, GST pull-down, In vitro binding assay, DNA cloning

1 Introduction

Dopamine is an important neurotransmitter in the mammalian central nervous system. It regulates a variety of functions, including cognition [1], reward [2], motor activity [3, 4], motivation [1], and the endocrine system [5] notably prolactin [6–9]. Abnormal dopamine signaling is implicated in the pathophysiology of many neural disorders, such as Parkinson's disease [10–14], schizophrenia [15–20], depression [21], and drug addiction [22–27]. Dopamine signaling is mediated by five transmembrane G protein-coupled receptors (D1R, D2R, D3R, D4R, and D5R). Protein-protein interactions involving heterotrimeric G proteins and ion

Mario Tiberi (ed.), *Dopamine Receptor Technologies*, Neuromethods, vol. 96,
DOI 10.1007/978-1-4939-2196-6_16, © Springer Science+Business Media New York 2015

channels regulate dopamine receptor signaling [28–37]. Ion channels are membrane proteins that are responsible for establishing the resting membrane potential, shaping action potentials, and modulating neuronal excitability.

Interactions between dopamine receptors and ion channels have been subject to extensive investigation, and these interactions are important for receptor trafficking, receptor localization, and downstream signaling [38, 39]. The dopamine D1R/glutamate N-methyl-D-aspartate receptor (NMDAR) interaction is involved in long-term potentiation (LTP), NMDAR-mediated currents [40], NMDAR-mediated excitotoxicity [41], and D1R surface expression, as well as D1R-mediated cAMP accumulation [42]. The D2R interaction with the NR2B subunit of the NMDAR is involved in cocaine addiction [43]. The D2R also interacts with the α-amino-3-hydroxy-5-methyl-4-isoxazolepropionic acid (AMPA) receptor subunit GluR2 to modulate AMPA receptor-mediated excitotoxicity [44]. The dopamine D2R-transient receptor potential channel 1 (TRPC1) interaction enhances cell surface expression of TRPC1 [45], while the D5R interaction with the gamma-aminobutyric acid (GABA) A receptor γ2 subunit can regulate $GABA_AR$-mediated currents and miniature inhibitory postsynaptic currents (mIPSC) [46]. D2R and D4R can form stable complexes with the G protein-coupled inwardly rectifying potassium (Kir3) channels and regulate downstream signaling [47]. Therefore, the study of interactions between dopamine receptors and ion channels is relevant to understanding neural disorders and to the development of new treatments.

The regulation of physiological functions through dopamine receptors and ion channels is involved in many kinds of signaling pathways that regulate numerous physiological functions. The traditional pharmacological approach to modifying receptor function is to activate or block with agonists or antagonists [48]. While many effective drugs are based on this strategy [49–53], it also can produce problematic side effects when only one of several downstream pathways is the intended target. An alternative approach is to identify specific receptor-protein interactions mediating particular functions, and to selectively target those, rather than blocking or stimulating the receptor itself [54, 55]. A critical starting point for this new strategy is to identify the interacting site between the receptor and its protein partner. Once the binding site has been identified, then peptides corresponding to the amino-acid sequence of the site can be used to disrupt the interaction to investigate the pathophysiological role of the interaction. Thus, the generated peptide can be used as either an experimental tool for the investigation of the function of protein-protein interactions or as a potential treatment for disease. In this chapter, we will describe the methods we have used to investigate the interactions between dopamine receptors and ligand-gated ion channels. We believe these methods can also be used to investigate protein-protein interactions involving other neurotransmitter receptors.

1.1 Immunohistochemistry and Immunofluorescence

The first step in studying receptor-ion channel interactions is to confirm the presence of the protein complex. Co-localization and co-immunoprecipitation are the most commonly used methods to define the interactions between dopamine receptors and ion channels. The co-localization of dopamine receptors and ion channels within a particular brain structure, and within the same cells, is a prerequisite for the formation of a protein complex. This method has been used to demonstrate the interactions between dopamine receptors and ion channels, such as the $GABA_AR\gamma2$-D5R interaction [46]. Standard immunohistochemical (IHC) techniques can be used to visualize the localization of the receptor and ion channel of interest [56]. Such techniques require specific antibodies that bind to either dopamine receptors or ion channels, which are detected by fluorescence-conjugated secondary antibodies binding to the primary antibody. Different fluorophores can be chosen for each secondary antibody, permitting each protein to be visualized as a different color under fluorescent microscopy. The different color images can then be overlaid to determine the degree of co-localization. Co-localization of dopamine receptors and ion channels at the cell surface membrane is required for their functional interaction.

1.2 Co-immunoprecipitation

Although the co-localization of dopamine receptors and ion channels by IHC can suggest a potential interaction between these two proteins, direct evidence for such protein-protein interactions must be obtained through different experimental approaches. The traditional method to assess protein-protein interactions is co-immunoprecipitation, which is based on the premise that a protein complex containing both the receptor and the ion channel can be precipitated with antibodies against either the receptor or the channel. Co-immunoprecipitation has been the primary biochemical technique used to demonstrate the D5R-$GABA_AR\gamma2$ [46], D1R-NMDAR [41], D2R-GluR2 [44], D2R and D4R interact with Kir3 [47], and D2R-TRPC1 interactions [45].

While co-immunoprecipitation can demonstrate the formation of a complex between two or more proteins, this approach cannot resolve whether the interactions are direct or indirect. For example, two proteins may form a complex by both binding to a third intermediary protein, in which case the observed interaction is indirect. To distinguish direct and indirect protein-protein interactions, fluorescence and bioluminescence resonance energy transfer (FRET and BRET), in vitro blot overly/overlay assay, and yeast two-hybrid systems can be used.

1.3 FRET and BRET

FRET/BRET is based on the principle that energy transfer between two chromophores requires a certain physical proximity. Thus, these methods can demonstrate that two molecules have come close enough to make it improbable that another intermediary molecule separates them [57]. For assessing protein-protein interactions using FRET or BRET, one of the putative interacting proteins is labeled with a donor, such as cyan fluorescent protein (CFP), and the other with an acceptor, such as yellow fluorescent protein (YFP). When there is interaction between the two proteins, the donor and acceptor are in sufficient proximity (1–10 nm) for the acceptor emission to increase because of the intermolecular energy transfer from the donor. One limitation of FRET is the requirement for external illumination to initiate the fluorescence transfer, which can lead to background noise due to direct excitation of the acceptor. BRET can address this limitation by relying on a bioluminescent luciferase rather than CFP to produce an initial photon emission compatible with YFP. Both FRET and BRET can be used to visualize the interaction over time, but are limited by several other considerations. The first is that both proteins to be studied using BRET or FRET must be modified with fluorophores that may also affect other aspects of protein function. This is especially problematic if the fluorophore interferes with ligand binding, intracellular trafficking, localization, or signal transduction. Another conceptual limitation is that FRET and BRET can be used only to test the interactions between particular pairs of proteins that have been strongly supported by previous research. The requirement to construct the chimeric protein containing the fluorophore prevents this approach from being useful for screening experiments aimed at identifying previously unknown or unsuspected protein-protein interactions. For such screens, protein labeling assays are more suitable.

1.4 Protein Labeling Methods

Protein labeling methods for protein-protein interaction screens begin with enzymatic biotinylation of a protein of interest, resulting in biotinylation of specific lysine residues by a bacterial biotin ligase [58, 59]. This is often accomplished by genetically linking the protein of interest at its N-terminus, at its C-terminus, or at an internal loop to a 15-amino-acid peptide, termed AviTag or Acceptor Peptide (AP). The tagged protein is then incubated with biotin ligase (BirA) in the presence of biotin and ATP. For the detection of labeled proteins, anti-biotin antibodies or avidin-/streptavidin-tagged detection strategies can be used, and this method can also be used to purify proteins. The purified proteins can be analyzed by two-dimensional (2D) electrophoresis or mass spectrometry to identify the proteins interacting with the target. BirA can react specifically with its target peptide inside mammalian and bacterial cells and at the cell surface, while other cellular proteins are not modified. As this technique can only detect the proteins interacting with the tagged proteins, indirect interactions cannot be detected. On the other hand, this approach requires minimal disruption of cellular physiology since only the tagged protein requires modification. The most important advantage of this technique is the potential to discover all the proteins interacting with the target protein, making it especially useful for generating new hypotheses.

1.5 Glutathione S-Transferase (GST) Pull-Down Assay

One of the critical steps in characterizing protein-protein interactions is to determine the specific binding site. For this purpose, several techniques are often used: glutathione *S*-transferase (GST) pull-down assays, in vitro binding assays, and yeast two-hybrid systems. For GST pull-down assays, different fragments of each protein interacting partner are fused to a GST tag. The principle is the same as for co-immunoprecipitation, except that the GST tag can be visualized more easily with an anti-GST antibody as compared to an antibody against membrane receptors/ion channels. Protein fragments may not be bound by antibodies used for co-immunoprecipitation of the intact whole protein. The GST-tagged protein fragment is used to "pull down" the interacting protein from solubilized tissue/cell extracts, and the GST-tagged protein fragment and the interacting protein complexes are separated by sodium dodecyl sulfate polyacrylamide gel electrophoresis (SDS-PAGE) and transferred to nitrocellulose filter membranes or polyvinylidene fluoride (PVDF) membranes. The membrane blots can be probed with the anti-GST antibody and antibodies against the other intact interacting partner. As with conventional co-immunoprecipitation, indirect protein-protein interactions cannot be excluded, and the GST fusion can create or destroy native interactions by modifying the structure of the tagged protein. The use of protein fragments may disrupt the binding motif essential for protein-protein interaction. Thus, overlapping protein fragments should be used to identify the interacting sites.

1.6 Yeast Two-Hybrid System

The yeast two-hybrid system is a technique used to detect protein-protein interactions and protein-DNA interactions by testing for physical interactions (such as binding) between two proteins or a single protein and a DNA molecule, respectively. This method was used to discover the interaction between D2R and TRPC1 ion channels [45]. The yeast two-hybrid system exploits the principle that binding of a transcription factor to an upstream activating sequence can activate a downstream reporter gene. A genetically engineered strain of yeast is often used, in which the biosynthesis of certain nutrients (usually amino acids or nucleic acids) is dependent on the transcription of a reporter gene that is under the control of a special transcription factor. If the reporter gene is not expressed, the yeast will fail to survive since they lack a critical nutrient. The transcription factor required for the yeast survival is divided into the binding domain and the activating domain, each attached to one of the putative interacting proteins (or interacting protein regions). The binding domain is responsible for binding to the upstream activating sequence, while the activating domain promotes transcription activation. The interaction between the two proteins under study brings the transcription factor fragments together and permits transcription to occur. If these two domains are not brought together by the protein-protein interaction, transcription will not occur.

The mutant yeast strain can be made to incorporate foreign DNA in the form of plasmids to produce fusion proteins. The protein fused to the binding domain may be referred to as the bait and is typically a known protein used to identify new binding partners. The protein fused to the activating domain may be referred to as the prey and can be either a single known protein or a library of known or unknown proteins. For this technique, when using a library, each cell should be transfected with no more than a single plasmid so that each cell expresses only one member from the protein library. Yeast two-hybrid systems can detect direct protein-protein interactions, can probe interacting domains between proteins, and can screen for new interacting proteins. However, this method requires generating mutant yeast strains, which may not represent the typical physiological environment in a mammalian cell. In addition, this method easily generates false-positive results due to its high sensitivity.

1.7 In Vitro Binding Assay

For in vitro binding assays, different fragments of either dopamine receptors or ion channels are fused to a GST tag and then incubated with [^{35}S]-methionine-labeled probes (designed according to the fragment of the other protein) as well as cDNA of fragment of the other protein. The samples are then separated by SDS-PAGE and [^{35}S]-methionine-labeled proteins visualized by autoradiography using X-ray films. Since both GST-fusion protein and the [^{35}S]-labeled probe are synthesized in vitro, no other proteins exist in the reaction system except the two interacting proteins. Therefore, this method can confirm direct interactions between these two kinds of proteins. Moreover, because GST-fusion proteins are used in this method, researchers can also map the detailed interacting domains within interacting proteins.

In summary, various techniques can be used to detect the interactions between dopamine receptors and ion channels and to find the protein domains responsible for these interactions. Once interactions have been defined and the discrete protein-protein interacting domains have been characterized, it is necessary to find an effective way to disrupt the interactions to evaluate biological functions and determine the functional importance of these protein-protein interactions. The interacting site is fused to a TAT fragment, the 11-amino-acid protein transduction domain from human immunodeficiency virus (HIV). The TAT fragment sequence allows the attached peptide to cross cell membranes and thus delivers the peptide into the intracellular space. The use of TAT-tagged peptides that mimic the amino-acid sequence of interacting domains between proteins has emerged as an effective method to disrupt protein-protein interactions, without influencing the other functions of each individual protein. What follows is a presentation of the methods used to identify the interacting sites in order to synthesize TAT-tagged peptides.

2 Materials

2.1 Immunofluorescence Staining

1. Phosphate-buffered saline (PBS): NaCl 8.0 g, $Na_2HPO_4{\cdot}12H_2O$ 2.08 g, KCl 0.2 g, and KH_2PO_4 0.2 g dissolved in 1 L ultrapure water at pH 7.4.
2. 4 % (w/v) paraformaldehyde (PFA) prepared in 1×PBS (pH 7.4). PFA powder is often used in this step, so just weigh 4 g PFA powder, and dissolve in 100 ml PBS. Once prepared in PBS, the solution should be filtered with normal filter paper or filter membranes. PFA is toxic, which should be made in the hood, and researchers should wear proper personal protective equipment to make the solution.
3. Bovine serum albumin (BSA) (Sigma-Aldrich).
4. Primary antibody against protein(s) of interest, which should be stored according to the manufacturer's instructions.
5. Fluorescence-conjugated secondary antibodies against species of primary antibodies, which is light sensitive.
6. Mounting medium: 30 % glycerol (v/v) prepared in 1×PBS (pH 7.4).
7. Glass microscope slides (Fisher Scientific, 1.0 mm), cover slips (Fisher Scientific, 0.13–0.17 mm), and confocal microscope.
8. Permeabilization buffer: 0.3 % Triton X-100 prepared in 1×PBS (pH 7.4), which should be made fresh before use.
9. Blocking buffer and antibody dilution buffer: 0.3 % Triton X-100 and 1 % BSA prepared in 1×PBS (pH 7.4), which can be stored at −20 °C for up to 1 year.

2.2 Immunoprecipitation and Western Blotting

1. Lysis buffer: 150 mM NaCl, 2 mM EDTA, 50 mM Tris-HCl (pH 7.4), 0.5 % sodium deoxycholate, 1 % NP-40, 1 % Triton X-100, and 0.1 % SDS. Add before use: 1 mM (working concentration) phenylmethanesulfonyl fluoride (PMSF) and protease inhibitor cocktail (Sigma-Aldrich) (*see* **Note 1**).
2. Tissue homogenizer.
3. Protein A/G plus agarose beads (Santa Cruz Biotechnology).

4. Refrigerated centrifuge.
5. Rocking/rotating platform.
6. 2× SDS sample buffer (Bio-Rad): 65.8 mM Tris-HCl (pH 6.8), 2.1 % (w/v) SDS, 26.3 % (w/v) glycerol, and 0.01 % (w/v) bromophenol blue. This is used as loading buffer in samples for SDS-PAGE and Western blotting.
7. SDS-PAGE and membrane transfer equipment.
8. Nitrocellulose membrane or PVDF membranes. Nitrocellulose membrane can be soaked directly into transfer buffer before use, while PVDF membranes should be activated with methyl alcohol before soaked into transfer buffer. Meanwhile, methyl alcohol should be used carefully as it is toxic.
9. Blocking buffer: 5 % nonfat dry milk and 0.1 % Tween 20 prepared in 1×PBS (pH 7.4) or 1× Tris-buffered saline (TBS) (pH 7.4) (10 mM Tris (hydroxymethyl) aminomethane (Tris), 150 mM NaCl).
10. Primary antibody buffer: 1 % BSA, 0.02 % NaN_3 (extremely toxic, wear personal protective equipment), and 0.1 % Tween 20 prepared in 1×PBS or 1×TBS (pH 7.4).
11. Wash buffer: 0.1 % Tween 20 prepared in 1×PBS or 1×TBS (pH 7.4).
12. Primary antibodies against protein(s) of interest.
13. Horseradish peroxidase-conjugated secondary antibody against the primary antibody species.
14. Enhanced chemiluminescence (ECL) reagents (make fresh before use according to the manufacturer's instructions) and X-ray film (light sensitive, and should be used only in dark room).

2.3 Cloning of GST-Fusion Protein cDNA Constructs

1. pGEX-4T-3 cloning vector.
2. cDNA of protein(s) of interest.
3. Restriction enzymes of interest.
4. Taq polymerase with buffer.
5. dNTP.
6. Custom-made primers used for polymerase chain reaction (PCR) of genes of interest.
7. Agarose gel: 1–3 % agarose dissolved in Tris-acetate-EDTA (TAE) buffer, heat with microwave, and add 0.1 % ethidium bromide (EB) before use. EB is a hazardous chemical. Wear personal protective equipment to avoid skin touch. EB is a highly sensitive visualizing DNA reagent classically used in DNA electrophoresis. Alternatively, safer chemicals such as SYBR Green available from Invitrogen can also be used (*see* **Note 2** for safely carrying EB procedures).

8. 100× BSA (10 mg/ml, New England BioLabs, #B9001): 100 time working concentration.
9. T4 DNA ligase with buffer (New England BioLabs, #M0202).
10. GenElute Plasmid Miniprep Kit (Sigma-Aldrich).
11. Luria broth (LB) medium: 25 g LB powder dissolved in 1 L ultrapure water (add 15 g agar for bacterial growing plates, and autoclave for 45 min (121 °C at 205.8 KPa)). Add 100 μg/ml ampicillin in sterilized LB before use.
12. 37 °C bacterial incubator.
13. 37 °C bacterial shaker.
14. Competent cells: DH5α and BL21 (purchased from Invitrogen).
15. 50× TAE buffer (should be prepared in a fume hood): Tris-HCl 242 g, acetic acid 57.1 ml (may hurt eyes, wear personal protective equipment), and 0.5 M EDTA (pH 8.0) 100 ml are added into ultrapure water to a final volume of 1 L.

2.4 GST Pull-Down Assay

1. Isopropyl β-D-1-thiogalactopyranoside (IPTG): IPTG powder dissolved in ultrapure water to a final concentration as 1 M (aliquot and stock at −20 °C).
2. Triton X-100 (Sigma-Aldrich).
3. Glutathione-Sepharose 4B beads (GE Healthcare).
4. Elution buffer: 50 mM Tris-HCl and 10 mM reduced glutathione (pH 8.0). It is possible to dispense in 1–10 ml aliquots and store at −20 °C until needed. Avoid more than five freeze/thaw cycles.
5. 2-Mercaptoethanol (BME) (toxic, should be used in a fume hood, and wear personal protective equipment).

2.5 In Vitro Binding Assay

1. T7 Quick Master Mix (Promega).
2. [^{35}S]-Methionine.
3. Luciferase control DNA.
4. BioMax (Kodak) films.
5. Filter paper (VWR, used for coarse precipitates).
6. Gel dryer.
7. Vacuum pump.

3 Methods

3.1 Immunofluorescence Staining

1. Staining cells:
   (a) Preparation of cells: put small cover slips into 24-well plates, and coat with 0.1 mg/ml poly-D-lysine solution

prepared in sterile ultrapure water overnight. The next day, wash twice with sterile ultrapure water (put 1 ml into each well every time). Plate cells at the proper density ($1.5–2\times10^5$/ml) into wells containing cover slips, and incubate in 5 % CO_2 cell incubator at 37 °C.

(b) After treatments (e.g., drug exposure, transfection), wash cells twice with 1 ml PBS per well, 5 min each.

(c) Fixation: fix cells with 1 ml 4 % PFA per well at room temperature for 20 min.

(d) Suck out PFA solution, and wash three times with 1 ml PBS per well, 5 min each.

(e) Permeabilization: add 1 ml 0.3 % Triton X-100 solution prepared in PBS into every well, and permeabilize cells for 15 min.

(f) Suck out the solution, add 1 ml 0.3 % Triton X-100 and 1 % BSA solution prepared in PBS into every well, and block cells at room temperature for 1 h.

(g) Suck out the solution and add 300–500 μl (300 μl is the smallest volume to cover the cells well enough) primary antibody diluted in blocking solution into every well. Incubate at 4 °C overnight.

(h) The next day, suck out the solution, and wash cells three times with 1 ml PBS per well, 5 min each.

(i) Suck out the solution, add 300–500 μl (300 μl is the smallest volume to cover the cells well enough) fluorescence-conjugated secondary antibody diluted in blocking solution into every well, and incubate at room temperature for 2 h. Here and in subsequent steps, the cover slips should be protected from light.

(j) Suck out the solution, and wash cells six times with 1 ml PBS per well, 5 min each.

(k) Put a drop of mounting medium on glass slides (do not put too much, just make it big enough for the cover slips), and put the cover slips onto the mounting medium facing down.

(l) Dry slides at 4 °C overnight. Examine slides next day under confocal fluorescence microscope. Alternatively, slides can be stored at 4 °C in dark for up to 1 month. See Fig. 1 for a prototypical example of an immunofluorescence staining experiment [46].

2. Staining frozen slices (in wells):

(a) Cut frozen slices of tissues of interest (8–30 μm thickness is recommended to obtain reliable results).

(b) Put slices into 12- or 24-well plates with 1 ml PBS per well for 30 min.

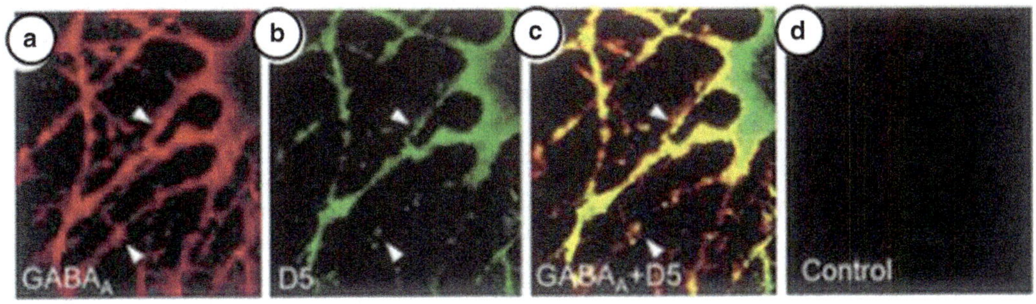

Fig. 1 Co-localization of GABA$_A$ and dopamine D5 receptors in cultured hippocampal neurons. (**a**, **b**) Confocal optical sections of the distribution of GABA$_A$ receptors (**a**) and D5 receptors (**b**) in cultured hippocampal neurons were obtained by double immunostaining with bd-17 and D5 antibodies. Dopamine D5 receptors were revealed by FITC-conjugated secondary antibodies (*green*) and GABA$_A$ receptors by Cy3. (**c**) Superimposition of confocal images in (**a**) and (**b**). *Arrowheads* highlight areas of punctate GABA$_A$ and D5 receptor clustering. No staining was observed when primary antibodies were eliminated (**d**) or when antibodies were preabsorbed with antigen (data not shown). The figure is taken with permission from the journal Nature [46]

(c) Suck out the solution and permeabilize slices with 1 ml 0.3 % Triton X-100 prepared in PBS per well for 15 min.

(d) Suck out the solution and incubate slices with 1 ml per well of blocking solution (0.3 % Triton X-100 and 1 % BSA prepared in PBS) at room temperature for 1 h with gentle shaking.

(e) Suck out the solution, add 300–500 μl of primary antibody diluted in blocking solution into every well, and incubate at 4 °C overnight (at least 300 μl of antibody solution is needed to cover well enough the slices).

(f) The next day, wash slices three times with 1 ml PBS per well, 5 min each.

(g) Suck out the solution and add 300–500 μl of secondary fluorescence-conjugated antibody diluted in blocking solution into every well (at least 300 μl of antibody solution is needed to cover well enough the slices). Incubate at room temperature for 2 h. Here and in subsequent steps, the cover slips should be protected from light.

(h) Wash three times with 1 ml PBS per well, 5 min each.

(i) Drop mounting medium onto the slices (do not drop too much, just big enough for one slice), transfer the slices extremely carefully from PBS to the melting mounting with Chinese writing brush, and put cover slip on carefully, avoiding bubbles.

(j) Dry slices at 4 °C overnight. Examine slices under confocal fluorescence microscope. Alternatively, the slices on cover slips can be stored at 4 °C in dark for up to 1 month.

3.2 DNA Cloning

1. Design and synthesize primer pairs for target gene fragments (without TAT).
2. Obtain cDNA encoding the proteins of interest.
3. PCR amplification of target gene fragments:
   (a) PCR setup:

10× Taq polymerase buffer	5 μl
dNTP mix (2.5 mM each)	4 μl
Template	10–100 ng
Forward primer (10 μM)	0.625 μl
Reversed primer (10 μM)	0.625 μl
Taq DNA polymerase	0.25 μl
Add ultrapure water to	50 μl

   Template: the cDNA encoding proteins of interest without TAT tag.

   (b) PCR condition:

94 °C	3 min	
94 °C 58–65 °C 72 °C	30 s 30–60 s 60–90 s (1 kb/min)	30 cycles
72 °C	10 min	
4 °C	Pause	

   (c) Add 6× DNA loading buffer (Thermo Scientific, #R6011) into PCR product, load the mixture into 1–3 % DNA agarose gel, and run the gel. Typically, we used the 6× DNA loading buffer, but any other concentration should also be fine. (As agarose gels are prepared with EB, a hazardous material, wear personal protective equipment and handle gels safely.)

   (d) Check the results with UV light, and cut the right size bands (*see* **Notes 3** and **4**).

   (e) Extract the DNA with DNA extraction kit, and store at -20 °C until use (up to 6 months). Typically, the purified DNA is used immediately or within 1 week.

4. Digestions of PCR product and cloning vector using two restriction enzymes:
   (a) Restriction digestion setup:

10× NEB buffer	5 μl
Vector or PCR products	1 μg
100× BSA	0.5 μl
Enzyme I (NEB, USA)	1 μl
Enzyme II (NEB, USA)	1 μl
Add ultrapure water to	50 μl

(b) Incubate the mixture at 37 °C for 3–4 h (*see* **Note 5** for the use of restriction enzymes and BSA). Here, restriction enzymes have been selected to obtain a digested DNA with cohesive ends.

(c) Add 6× DNA loading buffer to digestion reaction, load the mixture onto 1–3 % DNA agarose gel, and run the gel (dephosphorylation of the cloning vector can be also done prior to adding DNA loading buffer).

(d) Check the results with UV light, and cut the right size bands.

(e) Extract the DNA with DNA extraction kit, and store at −20 °C until use (up to 6 months). Typically, the purified DNA is used immediately or within 1 week. However, for this step, long-term storage of the DNA extract is not recommended since the digested double-stranded DNA has cohesive ends.

5. Ligation of the gene fragments with vector:

(a) Ligation reaction setup (*see* **Notes 6** and **7**):

10× T4 DNA ligase buffer	2 μl
Vector/insert	1:3–10
T4 DNA ligase (NEB)	1 μl
Add ultrapure water to	20 μl

(b) Incubate at 4 °C overnight.

6. DNA transformation of DH5α competent cells:

(a) Grab heat shock competent cells from −80 °C and let thaw on ice (10–20 min) (*see* **Note 8** for the use of different competent cells).

(b) Add the ligation mixture into 100 μl of competent cells, gently mix, and let sit on ice for 30 min.

(c) Heat shock the mixture at 42 °C for 90–120 s, put on ice immediately, and let sit for 2 min.

(d) Add 800 μl LB medium without ampicillin, and incubate in shaker at 37 °C with gentle agitation (around 120 rpm) for 40 min.

(e) Centrifuge the mixture at 1,000 × *g* at room temperature for 2 min and discard 700–800 μl of supernatant. Resuspend the competent cells with supernatant leftover, transfer and spread evenly onto LB-ampicillin plates, wait until there is no liquid flowing on the plates, and put into 37 °C incubator (LB-ampicillin plates do not need to be prewarmed at 37 °C before use; plates can be left at room temperature 10–20 min before use). Put plates in bacterial incubator to grow colonies overnight (12–16 h). Wait until the colonies have reached a big enough size to be picked up.

7. DNA miniprep and sequencing:
   (a) When the bacterial colonies have reached the right size, pick single colony with a 10 μl pipette tip, and put the tip into a 10 ml bacterial culture tube prefilled with 5 ml LB medium with ampicillin.
   (b) Place the tube into bacterial shaker and incubate until the liquid becomes cloudy (it usually takes an overnight incubation).
   (c) Do the DNA miniprep with GenElute Plasmid Miniprep Kit (Sigma).
   (d) Send the DNA miniprep samples to sequencing facility to confirm integrity of DNA sequences following PCR (some confirmation steps can be carried out before sending the samples, such as double restriction digestion and PCR as described in Sects. 3.3 and 3.4).

3.3 In Vitro Binding Assay

1. Probe synthesis:
   (a) Reaction setup:
   T7 Quick Master Mix (Promega; use according to the manufacturer's instructions): 40 μl cDNA (negative control, without plasmid DNA; positive control, using luciferase control DNA 1 μg): 0.5–1.0 μg (approximate 1 μl)
   [^{35}S]-Methionine (1,000 Ci/mM at 10 mCi/ml): 2 μl (if specific activity of the radiolabeled probe is low, the amount of μCi [^{35}S]-methionine can be increased)
   ddH_2O 7 μl
   Total 50 μl
   (b) Incubate at 30 °C for 1.5 h (reaction temperature and time can be modified to optimize specific activity of radiolabeled probe).
   (c) Run gel to check the probe:
   The radiolabeled products can be stored at −20 °C for up to 2 months or at −70 °C for up to 6 months. Handling of radioactive material should be done according to Institutional Radiation Compliance Safety Office.

2. In vitro binding:
   (a) PBS (1 ml) + GST-fusion protein (20 μg) + Glutathione-Sepharose 4B beads (10 μl) + probe synthesized in Sect. 3.3, step 1 (10 μl, stored at −20 °C).
   (b) Incubate at room temperature for 1 h.
   (c) Wash twice with 0.1 % Triton X-100 prepared in 1× PBS.
   (d) Add 15 μl of 2× SDS loading buffer.
   (e) Boil samples (negative control, GST protein; positive control, 1 μl of probe) at 100 °C for 5 min and load on gels (*see* **Note 9**).
   (f) Run samples on SDS-PAGE for approximately 1 h.
3. Gel handling and exposure (after running the gel):
   (a) At the end of the run, put the gel into plastic box container in a fume hood. Soak the gel in enhancer solution. The volume of enhancer solution used depends on the size of the gel and the box container. Make sure the gel is soaked completely. Incubate for 20–40 min. Discard enhancer into a bottle for radioactive liquid waste.
   (b) Wash gel with ultrapure water three times, 20 min each. The volume of water should be 1.5–2 times of the volume of enhancer solution used in Sect. 3.3, step 3(a).
   (c) Cut filter paper to fit the size of gel.
   (d) Carefully place gel on a cut filter paper. Avoid bubble in the middle of the gel when placing.
   (e) Put the gel on gel dryer. Pour some dry ice around the vacuum pump. Turn on power of gel dryer and vacuum pump.
   (f) Let dry for 2 h.
   (g) Dried gels are subjected to autoradiography using BioMax (Kodak) film (a 7-day exposure is recommended at room temperature, and the exposure time can be modified depending on the signal).

Figure 2 shows a representative example of results obtained with in vitro binding assays [41].

3.4 Co-immunoprecipitation Assay

1. Preparation of tissue extract:
   (a) Put animal tissue into 1.5 ml or 10 ml centrifuge tubes (acutely dissected tissue is best but frozen tissue kept at −80 °C can also be used). Add lysis buffer into the tubes (1 ml lysis buffer per 100 mg tissue). Homogenize at 8,000–10,000 rpm three times for 15 s each (*see* **Note 10** for selection of lysis and washing buffers to use in co-immunoprecipitation).
   (b) Shake at 4 °C for 1 h using rocking/rotating platform.
   (c) Centrifuge for 20 min at 4 °C, 14,000 × g.

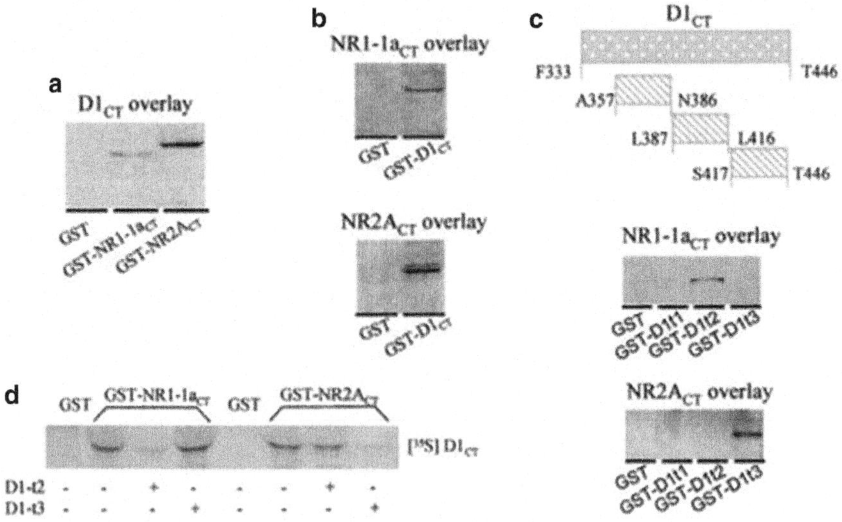

Fig. 2 Association of D1 and NMDA receptors in vitro. (**a**) Blot overlay assay depicting the direct binding of [^{35}S]-D1CT to the GST-NR1-1aCT and GST-NR2ACT. (**b**) Direct binding of the [^{35}S]-NR1-1aCT (*top*) and [^{35}S]-NR2ACT (*bottom*) to GST-D1CT. GST was used as control. (**c**) *Top*: schematic representation of the generated D1-t1, D1-t2, and D1-t3 mini-genes. Amino-acid sequence D1-t2 was critical for direct binding to NR1-1a subunit (*middle*); D1-t3 was critical for direct binding to NR2A subunit (*bottom*). (**d**) In vitro binding assay showing the blockade of direct binding of NR1-1a to D1CT upon the addition of D1-t2, but not D1-t3 peptide (*left*), and blockade of direct binding of NR2A to D1CT upon the addition of D1-t3, but not D1-t2 peptide (*right*). The figure is taken with permission from the journal Cell [41]

(d) Collect the supernatant, which contains the solubilized tissue extract.

(e) Measure the protein concentration with BCA protein assay (Pierce).

2. Co-immunoprecipitation setup:

(a) Prepare protein A/G plus agarose beads (Santa Cruz Biotechnology) as follows. Wash 25 μl of protein A/G plus agarose beads with 1 ml cold PBS in a 1.5 ml Eppendorf tube for 5 min using a rocking/rotating platform. Centrifuge at 1,000×g in a microfuge for 1 min at 4 °C. Discard supernatant and repeat washing twice.

(b) Add beads to 500–1,000 μg of solubilized tissue extract, and add tissue lysis buffer to the mixture to a final volume of 500–1,000 μl in a 1.5 ml Eppendorf tube.

(c) Mix the beads and the extract mixture at 4 °C for 30 min using rocking/rotating platform.

(d) Centrifuge at 1,000×g for 5 min at 4 °C to pellet the beads. Transfer the supernatant to a fresh Eppendorf tube (*see* **Note 11**).

(e) Add 2–4 μg primary antibodies or related IgG (as a negative control) to each sample tube (*see* **Note 12** for selection of primary antibodies).

(f) Mix the clarified supernatant and the primary antibody by rotation at 4 °C for 3 h to allow binding of primary antibodies to the protein of interest.

(g) Add 25 μl of prewashed protein A/G plus agarose beads.

(h) Rotate at 4 °C overnight to allow binding of beads to the primary antibody.

(i) Centrifuge at 1,000×g for 5 min to pellet the beads and discard the supernatant.

(j) Wash the beads with 1 ml cold tissue lysis buffer at 4 °C. Spin and remove supernatant. Repeat two more times.

(k) Following the third wash, discard supernatant. Add 25 μl of 2× SDS sample buffer to beads, and heat at 100 °C for 5 min. Figure 3 shows a representative example of a co-immunoprecipitation experiment [44].

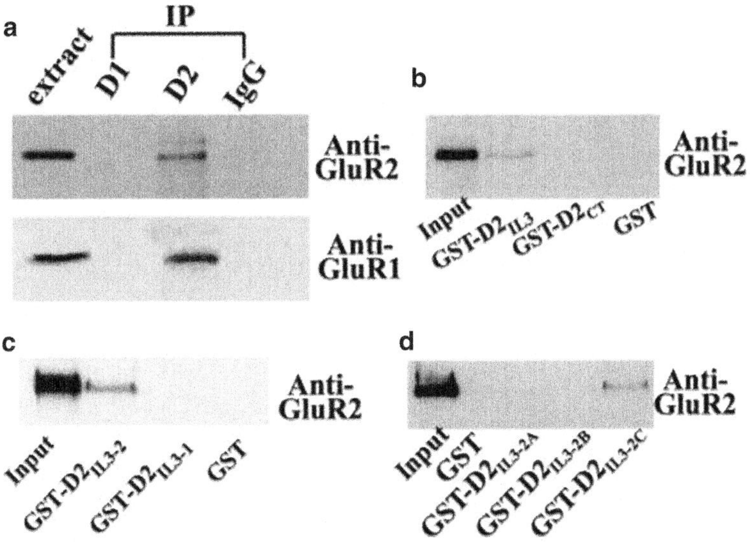

Fig. 3 D2 receptors exhibit a biochemical interaction with GluR2. (**a**) Co-immunoprecipitation of GluR2 subunit from solubilized rat hippocampal tissue with D2, but not D1, receptor antibody. (**b**) Western blots of GluR2 after affinity precipitation from solubilized rat hippocampal tissue by GST-D2 (IL3) but not by GST-D2 (CT) or GST alone. (**c**) Identification of the D2 receptor region involved in the D2-GluR2 interaction. GST-fusion proteins encoding regions within the D2 third intracellular loop were used to affinity purify the GluR2 subunit. Only GST-D2 (IL3-2) was able to affinity purify the GluR2 subunit. (**d**) Additional delineation of the D2 receptor third intracellular loop involved in the D2-GluR2 interaction. Western blot analysis reveals that GST-D2 (IL3-2C), but not GST-D2 (IL3-2A), GST-D2 (IL3-2B), or GST alone, was able to pull down the GluR2 subunit from solubilized rat hippocampus. Furthermore, the D1 receptor does not co-immunoprecipitate the GluR1 or GluR2 AMPA receptor subunits. *IP* immunoprecipitation. The figure is taken with permission from the Journal of Neuroscience [44]

3.5 Wet Transfer Western Blotting

1. Prepare an 8 or 10 % SDS-polyacrylamide gel (*see* **Note 9**).
2. Load co-immunoprecipitation samples into individual wells of the gel. Run at a constant voltage of 80 V until the dye front reaches the interface between stacking gel and separating gel. Then, increase voltage to 120 V and let the dye front run to the bottom of separating gel.
3. Transfer the resolved proteins from gel to a nitrocellulose (or PVDF) membrane using constant current of 400 mA for 2 h at 4 °C (*see* **Note 13**).
4. After transfer, incubate the membrane with 3 % BSA or 5 % nonfat dry milk blocking buffer at room temperature for 1 h.
5. Following the blocking step, gently wash the membrane with 0.05 % Tween 20 prepared in PBS and incubate the membrane with the primary antibody solution (*see* **Note 14**).

3.6 GST Pull-Down Assay

1. Expression of GST-fusion protein:
   (a) Transform BL21 competent cells with plasmids. Grow bacteria in 15 ml LB with antibiotics (100 μg/ml) in a 50 ml tube. Incubate bacteria at 37 °C (250 rpm) in a shaking incubator overnight (*see* **Note 15** for selection of antibiotics).
   (b) Take out tubes and pour 25 ml of bacterial culture into 225 ml LB with antibiotics (100 μg/ml).
   (c) Grow bacteria at 37 °C in a shaking incubator (250 rpm) for 1.5–2 h.
   (d) Quantify bacterial growth at an optical density of 600 nm (OD600). OD600 should be between 0.6 and 1.2. If OD600 has not reached the optimal value, continue growing the bacteria and quantify the OD600 every 30 min until OD600 reaches the optimal value.
   (e) Add IPTG to a final concentration of 0.5 mM, and incubate at 30 °C or 28 °C (to reduce degradation of unstable proteins), 250 rpm for 3 h (*see* **Note 16**).
   (f) Pour bacterial culture into 250 ml centrifuge bottle and spin at 1,000 × g for 20 min at 4 °C.
   (g) Discard the supernatant.
   (h) Add 5 ml of bacterial lysis buffer (1 % Triton X-100 and protease inhibitor prepared in PBS), resuspend pellet, and transfer to a 15 ml centrifuge tube. Sonicate the resuspended pellet for 40 s three times to break the bacterial wall.
   (i) Place tubes in a shaker at 4 °C and solubilize bacterial lysate for 1 h.
   (j) Centrifuge at 12,000 × g at 4 °C for 20 min, and keep the supernatant.

(k) Add 150 μl Glutathione-Sepharose 4B beads to a 1.5 ml Eppendorf tube and wash with 1 ml of 0.1 % Triton X-100 prepared in PBS. Place the tube on a rocking/rotating platform at 4 °C and mix for 5 min. Centrifuge at 1,000 × g for 1 min at 4 °C.

(l) Discard the supernatant and keep the beads.

(m) Mix supernatant (isolated in step 1(j)) to beads in a new 15 ml centrifuge tube. Place the tube on a rocking/rotating platform and mix the supernatant and beads for 2 h at 4 °C.

(n) Centrifuge samples at 1,000 × *g* for 1–2 min at 4 °C. Discard the supernatant, and transfer beads into a 1.5 ml Eppendorf tube.

(o) Add 1 ml 0.1 % Triton X-100 prepared in PBS into beads, and place on the shaker to rotate at 4 °C for 5 min. Centrifuge at 4 °C, 1,000 × *g* for 1–2 min, and discard the supernatant. Repeat for three times.

(p) Discard the supernatant as much as possible.

(q) Add 200 μl elution buffer into beads. Resuspend beads with the pipette, put on a rocking/rotating platform, and mix at room temperature for 20 min.

(r) Centrifuge 10,000 × *g* for 20 min at room temperature. Keep the supernatant, and store at −80 °C.

2. Preparation of tissue extract:

(a) Put animal tissue into 1.5 ml or 10 ml centrifuge tubes (acutely dissected tissue is best but frozen tissue kept at −80 °C can also be used). Add lysis buffer into the tubes (1 ml lysis buffer per 100 mg tissue). Homogenize at 8,000–10,000 rpm for three times, 15 s each. Put the tube on a rocking/rotating platform at 4 °C and solubilize the lysate for 1 h.

(b) Centrifuge at 10,000 × g for 10 min at 4 °C. Transfer the supernatant (solubilized tissue extract) to a new 1.5 ml or 10 ml centrifuge tube.

(c) Measure the protein concentration of tissue extract and GST-fusion protein samples with BCA protein assay kit.

3. GST pull-down:

(a) Prepare 30 μl of Glutathione-Sepharose 4B beads for GST pull-down as described in Sect. 3.6, step 1(k) and (l).

(b) Add GST-fusion protein (50–100 μg) and solubilized tissue extract (500–1,000 μg) to washed Glutathione-Sepharose 4B beads. Complete to a final volume of 500 μl with a solution of 0.1 % Triton X-100/PBS/protease inhibitor cocktail (*see* **Note 17**).

(c) Incubate at 4 °C overnight using rocking/rotating platform.

(d) Centrifuge samples at 1,000 × *g* for 1 min at 4 °C, and discard the supernatant. Add 1 ml of 0.1 % Triton X-100/PBS, and put on a rocking/rotating platform at 4 °C for 5 min. Repeat the washing step three times (*see* **Note 18**).

(e) Following the last wash, discard supernatant leaving ~30 μl in tube and add 30 μl sample buffer (95 % 2× SDS sample buffer + 5 % 2-mercaptoethanol (BME)). Heat tubes at 100 °C for 5 min. Centrifuge the samples at 1,000 × *g* for 1 min at room temperature. Subject samples to SDS-PAGE and Western blotting.

4 Notes

1. 100 mM PMSF solution prepared in isopropyl alcohol can be stocked at −20 °C long-term usage, and it needs to be vortexed to get crystal clear solutions before use. Wear gloves when using PMSF, an extremely toxic substance that can be absorbed through the skin.
2. EB handling:

EB safely operating procedure:

(a) Set up a designated area for work with ethidium bromide, and label it with the following wording: "Ethidium Bromide in use. Mutagen and irritant."

(b) Line the work area with a disposable plastic-backed absorbent pad.

(c) Keep containers closed as much as possible.

(d) If weighing dry powders and the balance cannot be located in a fume hood or BSC, tare a container and then add the material to the container in a hood and seal the container before returning to the balance to weigh the powder.

(e) Change gloves regularly (at least every 2 h) and wash hands at the time of the glove change. Check the work area for contamination using a UV light (EB will fluoresce a reddish brown). If decontamination is needed, try the methods below after wiping up excess liquid with paper towels.

(f) Wipe the contaminated area or equipment with fresh towels and a soap/water solution multiple times. You can also wipe with towels soaked in ethanol. Check for any remaining contamination using UV light.

(g) Take fresh paper towels soaked in ethanol and place them over the contaminated surface. Sprinkle activated charcoal on the ethanol-saturated towel in contact with contaminated surface. Wipe up ethanol/charcoal mixture with additional towels and place all cleanup materials into a plastic bag. Check for any remaining contamination with UV light and repeat if needed.

(h) Use a solution of 4.2 g of sodium nitrite ($NaNO_2$), 20 ml of 50 % hypophosphorous acid solution (H_3PO_2), and 300 ml of water to decontaminate. Check the area again with the UV light until all EB has been removed, and then rinse with water. It should be noted that hypophosphorous acid is a DEA listed chemical and may require additional authorization for purchase.

All decontamination materials must be disposed of as hazardous waste.

Personal protective equipment:

(a) Standard nitrile laboratory gloves and a fully buttoned lab coat with sleeves extending to the wrists should be worn when handling EB (powder and solutions).

(b) If splashes may occur, wear goggles and a face shield. Otherwise, wear standard laboratory safety glasses.

(c) When using a UV light to visualize EB contamination, wear UV-blocking eyewear or work behind a UV-shielding glass. (Most standard safety glasses will block UV, but employees should check the approval of their safety glasses.)

(d) In cases where the arms or torso may be exposed to liquid suspensions or dry particles, wear Tyvek sleeves and/or gowns (or other airtight nonwoven textile).

Disposal EB waste:

(a) For electrophoresis gels, trace amounts of EB in gels should not pose a hazard. Higher concentrations, e.g., when the color of the gel is dark pink or red, should not be placed in laboratory trash: less than 0.1 % EB (place in laboratory trash) and more than or equal to 0.1 % (place in biohazard box for incineration).

(b) Gloves, test tubes, paper towels, etc., that are grossly contaminated with EB should be placed in medical waste for incineration. Consider deactivating in bleach before disposal if the items are significantly contaminated.

(c) In some laboratories, EB solutions may be used to stain the gels. EB solutions should be disposed as below:

Aqueous solutions containing <10 μg/ml EB can be released to the drain; aqueous solutions containing >10 μg/ml EB should be filtered or deactivated using charcoal filtration or chemical neutralization; and solutions containing heavy metals, organics, cyanides, or sulfides should be disposed as hazardous waste.

3. After PCR of DNA, the time for running the gel is not fixed and depends on the size of PCR product. It is necessary to check regularly with UV light until the bands are sufficiently separated. The concentration of the gel and type of DNA ladder used should be adjusted based on the PCR product size. For smaller PCR products, higher concentration of agarose gel should be used, which reduces band migration in the gel.
4. Before sequencing the plasmid constructs, digestion and PCR can be used to assess if plasmid vector and insert size are right. With diagnostic restriction digests, only two bands should be detected in the gel, corresponding to the vector and the insert fragments. With diagnostic PCR, there should be only one band for the insert fragment. For shorter insert fragments, it is better to use PCR, whereas for longer fragments, both digestion and PCR should be conducted.
5. It is better to choose enzymes that use the same buffer, and a double-digestion system from NEB can be used. Furthermore, all the restriction enzymes used in DNA subcloning should be stored at −20 °C. BSA is necessary in most restriction digestions, but the experimenter will need deciding whether or not to use it according to the manufacturer's instructions. 37 °C is normally used for restriction digestions, but temperature will be fixed as per manufacturer's instructions.
6. The vector/insert ratio is essential for a successful ligation reaction. Usually, the ratio is between 1:3 and 1:10. Different ratios for one pair of vector and insert can be tested. For smaller fragments (<200 bp), the experimenter can initially choose a ratio close to 1:10, while selecting a ratio close to 1:3 for longer fragments (>1 kb). Several tries may be needed for successful ligations.
7. T4 ligase buffer is frozen in the −20 °C, and upon thawing a salt precipitate is formed. Dissolve the salt precipitate by vortexing vigorously or by warming buffer at 37 °C.
8. Two kinds of competent cell strains are often used in transformation. For recombinant DNA cloning, bacterial strains such as DH5α and TOP10 can be employed, while for expression of fusion proteins, the *E. coli* strain BL21 is usually chosen.
9. The concentration of acrylamide chosen for the gels depends on the molecular weight of proteins in samples. Use higher concentration if protein under study has a low molecular weight and vice versa for higher molecular weight proteins. The selection of 10-well or 15-well combs depends on the amount of your samples to be loaded on gels.

10. Since both dopamine receptors and ion channels are membrane proteins, the type of lysis buffer used for isolation of tissue protein, co-immunoprecipitation, and GST pull-down assays is critical. Selection of lysis and washing buffers should depend on the binding affinity between interacting proteins. Triton X-100 and SDS can be initially considered. Alternatively, commercially available membrane protein isolation kits may also be utilized.
11. Co-immunoprecipitation experiments using IgG as negative control will allow experimenter to determine the extent of nonspecific immunoprecipitated bands. If nonspecific immunoprecipitated bands are observed, a preclearing step is recommended to reduce the amount of nonspecific immunoprecipitation. Prior to adding antibody, incubate solubilized tissue extract with protein A/G plus agarose beads at 4 °C for 1 h. Centrifuge and transfer the supernatant to a new tube. Add new prewashed protein A/G plus agarose beads and antibodies to the tube, and incubate at 4 °C overnight.
12. The specificity of co-immunoprecipitation assays depends mostly on antibodies used for immunoprecipitation. Thus, proper selection of antibodies is essential for a successful experiment. If the selected antibody targets an epitope that mediates the protein-protein interaction, it will fail to pull down interacting proteins. The basic principle is that antibodies used for immunoprecipitation should not be the same species as those used for Western blotting detection. Sometimes, many attempts are required to optimize a protocol that works best for one co-immunoprecipitation assay.
13. Larger-size proteins move slower than smaller ones on SDS-PAGE and wet transfer processes. Thus, larger-size proteins may take longer to transfer. However, the experimenter can change the icebox during transfer to keep the transfer buffer cold to improve protein transfer.
14. For Western blotting, highly specific antibodies are critical to obtain reliable results.
15. For GST pull-down assays, two antibiotics are commonly used for culturing bacteria: ampicillin and kanamycin. The experimenter should choose antibiotics according to the antibiotic resistance marker of plasmids used to transform bacteria. For example, expression plasmids encoding GST-tagged proteins typically carry an ampicillin resistance marker.
16. Time and temperature for GST-fusion protein expression can be modified depending on the OD600 value, protein stability, and protein expression levels. As for the protein stability, generally speaking, the larger-size proteins are more prone to degradation. Thus, lower temperature will be used for these proteins to inhibit protease activity of bacteria.

17. A final volume of 500 μl is typically used in our GST pull-down assays. If final volume is brought up to 1 ml, the experimenter will need using more GST-fusion protein and solubilized tissue extract. This may improve detection sensitivity on protein-protein interactions on Western blots.
18. Wash solution recipes for GST pull-down assays depend on the specificity and binding affinity of the interactions. The experimenter can try various concentrations of Triton X-100 from low to high. Sometimes, nonspecific bands can be detected with the GST alone, in which case a new transformation of BL21 with GST vector can be done to generate a new purified batch of GST alone for negative control.

Acknowledgments

The projects presented in this chapter were supported by operating grants from the Canadian Institutes of Health Research (previously MRC Canada), the NIDA, the Ontario Mental Health Foundation, the Canadian Psychiatric Research Foundation, and the Heart and Stroke Foundation of Canada.

References

1. Holmes A, Lachowicz JE, Sibley DR (2004) Phenotypic analysis of dopamine receptor knockout mice; recent insights into the functional specificity of dopamine receptor subtypes. Neuropharmacology 47:1117–1134
2. Wise RA (1994) Cocaine reward and cocaine craving: the role of dopamine in perspective. NIDA Res Monogr 145:191–206
3. Clark D, White FJ (1987) D1 dopamine receptor—the search for a function: a critical evaluation of the D1/D2 dopamine receptor classification and its functional implications. Synapse 1:347–388
4. Svensson A, Carlsson ML, Carlsson A (1995) Crucial role of the accumbens nucleus in the neurotransmitter interactions regulating motor control in mice. J Neural Transm Gen Sect 101:127–148
5. Banihashemi B, Albert PR (2002) Dopamine-D2S receptor inhibition of calcium influx, adenylyl cyclase, and mitogen-activated protein kinase in pituitary cells: distinct Galpha and Gbetagamma requirements. Mol Endocrinol 16:2393–2404
6. Jackson DM, Westlind-Danielsson A (1994) Dopamine receptors: molecular biology, biochemistry and behavioural aspects. Pharmacol Ther 64:291–370
7. Caronti B, Calderaro C, Passarelli F, Palladini G, Pontieri FE (1998) Dopamine receptor mRNAs in the rat lymphocytes. Life Sci 62:1919–1925
8. Kikuchi de Beltran K, Koshikawa N, Miwa Y, Kobayashi M (1994) Dorsal striatal mechanisms involved in the dopamine D2 receptor-mediated potentiation of apomorphine-induced jaw movements. Eur J Pharmacol 252:99–104
9. Waddington JL, O'Boyle KM (1987) The d-1 dopamine receptor and the search for its functional role: from neurochemistry to behaviour. Rev Neurosci 1:157–184
10. Nanko S, Hattori M, Ueki A, Ikeda K (1993) Dopamine D3 and D4 receptor gene polymorphisms and Parkinson's disease. Lancet 342:250
11. Barbeau A (1968) Dopamine and dopamine metabolites in Parkinson's disease—a review. Proc Aust Assoc Neurol 5:95–100
12. Triarhou LC (2002) Introduction. Dopamine and Parkinson's disease. Adv Exp Med Biol 517:1–14
13. Chetrit J et al (2013) Inhibiting subthalamic D5 receptor constitutive activity alleviates abnormal electrical activity and reverses motor impairment in a rat model of Parkinson's disease. J Neurosci 33:14840–14849

14. Berthet A et al (2012) L-DOPA impairs proteasome activity in parkinsonism through D1 dopamine receptor. J Neurosci 32:681–691
15. Kane JM, Freeman HL (1994) Towards more effective antipsychotic treatment. Br J Psychiatry Suppl:22–31
16. Matthysse S (1974) Dopamine and the pharmacology of schizophrenia: the state of the evidence. J Psychiatr Res 11:107–113
17. Reynolds GP (1989) Beyond the dopamine hypothesis. The neurochemical pathology of schizophrenia. Br J Psychiatry 155:305–316
18. Sigmundson HK (1994) Pharmacotherapy of schizophrenia: a review. Can J Psychiatry 39: S70–S75
19. Seeman P (2006) Targeting the dopamine D2 receptor in schizophrenia. Expert Opin Ther Targets 10:515–531
20. Carlsson A (1995) Neurocircuitries and neurotransmitter interactions in schizophrenia. Int Clin Psychopharmacol 10 Suppl 3:21–28
21. Nestler EJ, Carlezon WA Jr (2006) The mesolimbic dopamine reward circuit in depression. Biol Psychiatry 59:1151–1159
22. Newman AH et al (2012) Medication discovery for addiction: translating the dopamine D3 receptor hypothesis. Biochem Pharmacol 84: 882–890
23. Kiyatkin EA (1994) Dopamine mechanisms of cocaine addiction. Int J Neurosci 78:75–101
24. Lobo MK, Nestler EJ (2011) The striatal balancing act in drug addiction: distinct roles of direct and indirect pathway medium spiny neurons. Front Neuroanat 5:41
25. Missale C, Nash SR, Robinson SW, Jaber M, Caron MG (1998) Dopamine receptors: from structure to function. Physiol Rev 78: 189–225
26. Girault JA, Greengard P (2004) The neurobiology of dopamine signaling. Arch Neurol 61:641–644
27. Santini E, Valjent E, Fisone G (2008) Parkinson's disease: levodopa-induced dyskinesia and signal transduction. FEBS J 275:1392–1399
28. Kim OJ, Ariano MA, Lazzarini RA, Levine MS, Sibley DR (2002) Neurofilament-M interacts with the D1 dopamine receptor to regulate cell surface expression and desensitization. J Neurosci 22:5920–5930
29. Zeng C et al (2005) Interaction of angiotensin II type 1 and D5 dopamine receptors in renal proximal tubule cells. Hypertension 45: 804–810
30. Lan H, Teeter MM, Gurevich VV, Neve KA (2009) An intracellular loop 2 amino acid residue determines differential binding of arrestin to the dopamine D2 and D3 receptors. Mol Pharmacol 75:19–26
31. Lan H, Liu Y, Bell MI, Gurevich VV, Neve KA (2009) A dopamine D2 receptor mutant capable of G protein-mediated signaling but deficient in arrestin binding. Mol Pharmacol 75: 113–123
32. Liu Y, Buck DC, Neve KA (2008) Novel interaction of the dopamine D2 receptor and the Ca2+ binding protein S100B: role in D2 receptor function. Mol Pharmacol 74:371–378
33. Liu Y, Buck DC, Macey TA, Lan H, Neve KA (2007) Evidence that calmodulin binding to the dopamine D2 receptor enhances receptor signaling. J Recept Signal Transduct Res 27:47–65
34. Macey TA, Liu Y, Gurevich VV, Neve KA (2005) Dopamine D1 receptor interaction with arrestin3 in neostriatal neurons. J Neurochem 93:128–134
35. Torvinen M, Kozell LB, Neve KA, Agnati LF, Fuxe K (2004) Biochemical identification of the dopamine D2 receptor domains interacting with the adenosine A2A receptor. J Mol Neurosci 24:173–180
36. Wang M, Lee FJ, Liu F (2008) Dopamine receptor interacting proteins (DRIPs) of dopamine D1-like receptors in the central nervous system. Mol Cells 25:149–157
37. Lee FJ et al (2007) Dopamine transporter cell surface localization facilitated by a direct interaction with the dopamine D2 receptor. EMBO J 26:2127–2136
38. Sedaghat K, Tiberi M (2011) Cytoplasmic tail of D1 dopaminergic receptor differentially regulates desensitization and phosphorylation by G protein-coupled receptor kinase 2 and 3. Cell Signal 23:180–192
39. Kuzhikandathil EV, Oxford GS (2002) Classic D1 dopamine receptor antagonist R-(+)-7-chloro-8-hydroxy-3-methyl-1-phenyl-2,3,4,5-tetrahydro-1H-3-benzazepine hydrochloride (SCH23390) directly inhibits G protein-coupled inwardly rectifying potassium channels. Mol Pharmacol 62:119–126
40. Nai Q et al (2010) Uncoupling the D1-N-methyl-D-aspartate (NMDA) receptor complex promotes NMDA-dependent long-term potentiation and working memory. Biol Psychiatry 67:246–254
41. Lee FJ et al (2002) Dual regulation of NMDA receptor functions by direct protein-protein interactions with the dopamine D1 receptor. Cell 111:219–230
42. Pei L, Lee FJ, Moszczynska A, Vukusic B, Liu F (2004) Regulation of dopamine D1 receptor function by physical interaction with the NMDA receptors. J Neurosci 24:1149–1158

43. Liu XY et al (2006) Modulation of D2R-NR2B interactions in response to cocaine. Neuron 52:897–909
44. Zou S et al (2005) Protein-protein coupling/uncoupling enables dopamine D2 receptor regulation of AMPA receptor-mediated excitotoxicity. J Neurosci 25:4385–4395
45. Hannan MA, Kabbani N, Paspalas CD, Levenson R (2008) Interaction with dopamine D2 receptor enhances expression of transient receptor potential channel 1 at the cell surface. Biochim Biophys Acta 1778:974–982
46. Liu F et al (2000) Direct protein-protein coupling enables cross-talk between dopamine D5 and gamma-aminobutyric acid A receptors. Nature 403:274–280
47. Lavine N et al (2002) G protein-coupled receptors form stable complexes with inwardly rectifying potassium channels and adenylyl cyclase. J Biol Chem 277:46010–46019
48. Dyck B et al (2011) PAOPA, a potent analogue of Pro-Leu-glycinamide and allosteric modulator of the dopamine D2 receptor, prevents NMDA receptor antagonist (MK-801)-induced deficits in social interaction in the rat: implications for the treatment of negative symptoms in schizophrenia. Schizophr Res 125:88–92
49. Seeman P (2002) Atypical antipsychotics: mechanism of action. Can J Psychiatry 47:27–38
50. Fischer PA (1995) Treatment strategies in Parkinson's disease after a quarter century experiences with L-DOPA therapy. J Neural Transm Suppl 46:381–389
51. Kleber HD (1992) Treatment of cocaine abuse: pharmacotherapy. Ciba Found Symp 166:195–200, discussion 200-196
52. Grunder G, Carlsson A, Wong DF (2003) Mechanism of new antipsychotic medications: occupancy is not just antagonism. Arch Gen Psychiatry 60:974–977
53. Simpson EH et al (2011) Pharmacologic rescue of motivational deficit in an animal model of the negative symptoms of schizophrenia. Biol Psychiatry 69:928–935
54. Perreault ML, Hasbi A, O'Dowd BF, George SR (2014) Heteromeric dopamine receptor signaling complexes: emerging neurobiology and disease relevance. Neuropsychopharmacology 39:156–168
55. Agnati LF, Tarakanov AO, Ferre S, Fuxe K, Guidolin D (2005) Receptor-receptor interactions, receptor mosaics, and basic principles of molecular network organization: possible implications for drug development. J Mol Neurosci 26:193–208
56. Ciliax BJ et al (2000) Dopamine D(5) receptor immunolocalization in rat and monkey brain. Synapse 37:125–145
57. Verma V, Hasbi A, O'Dowd BF, George SR (2010) Dopamine D1-D2 receptor Heteromer-mediated calcium release is desensitized by D1 receptor occupancy with or without signal activation: dual functional regulation by G protein-coupled receptor kinase 2. J Biol Chem 285: 35092–35103
58. Roux KJ, Kim DI, Raida M, Burke B (2012) A promiscuous biotin ligase fusion protein identifies proximal and interacting proteins in mammalian cells. J Cell Biol 196:801–810
59. Rhee HW et al (2013) Proteomic mapping of mitochondria in living cells via spatially restricted enzymatic tagging. Science 339:1328–1331

Part V

Behavioral Analysis of Dopamine Function

Chapter 17

Study of Dopamine Receptor and Dopamine Transporter Networks in Mice

Victor Gorgievski, Eleni T. Tzavara, and Bruno Giros

Abstract

Dopamine (DA) dysregulation is a core feature in Parkinson's disease and in addictive disorders. DA has been also implicated in central nervous system affective and cognitive pathologies such as bipolar disorder, schizophrenia, and attention deficit and hyperactivity disorder (ADHD). The first studies of genetically engineered mice targeting components of the DA system focused on motor behavior and on the action of addictive drugs. However, in the course of the last 20 years (the first KO relevant to the DA system to be generated were those of the D1 receptors in 1994), we have seen an increasing shift in the use of these mutants: from tools to unravel the pharmacology of addiction integrated to in vivo models to study DA-related affective and cognitive disorders.

Key words Breeding strategies, Genetic drift, Knockout mice, Transgenic mice, Dopamine receptor, Dopamine transporter, Behavior, Cognitive tests, Locomotor activity

1 Introduction

In this chapter, we will focus on the analysis of dopaminergic (DA-ergic) mutants (D1, D2, D3, D4, D5, DAT, COMT) along this later line (Table 1). We will present first the mutants to be discussed and methodological considerations in the production and breeding of genetically modified mice, followed by the phenotype analysis of these mice across three behavioral dimensions: (1) locomotor activity, (2) prepulse inhibition of the startle response (PPI), and (3) cognitive function.

2 Production of DA-ergic Mutant Mice

Knockout mice for the different DA-related genes (metabolizing enzymes, receptors, and transporter) have been generated using homologous recombination techniques [1] that will not be detailed here. Most of these genes were constitutively removed, those for

Mario Tiberi (ed.), *Dopamine Receptor Technologies*, Neuromethods, vol. 96,
DOI 10.1007/978-1-4939-2196-6_17,

Table 1
Mutant mice discussed in this chapter

Mutant mice	Mutation	Main references
• D1 receptor		
D1-KO (strain 1)	Knockout of the D1 gene	[2, 33, 60, 61, 79, 80]
D1-KO (strain 2)	Knockout of the D1 gene	[3, 63, 81]
• D2 receptor		
D2-KO (strain 1)	Knockout of the D2 gene	[4]
D2-KO (strain 2)	Knockout of the D2 gene	[5, 31, 53, 62, 79, 80]
D2L-KO	Knockout of the D2L isoform	[6, 7, 77, 82]
D2-Tg	D2 receptor overexpression	[65]
• D3 receptor		
D3-KO (strain 1)	Knockout of the D3 gene	[8, 33, 79, 81, 83]
D3-KO (strain 2)	Knockout of the D3 gene	[9]
D3-KO (strain 3)	Knockout of the D3 gene	[53]
• D4 receptor		
D4-KO	Knockout of the D4 gene	[10, 38, 53, 84]
• D5 receptor		
D5-KO	Knockout of the D5 gene	[11]
• Combined receptor KO		
D1/D3-KO	Breeding D1-KO and D3-KO	[33, 61]
D2/D3-KO	Breeding D2-KO and D3-KO	[37]
• Dopamine transporter		
DAT-KO (strain 1)	Knockout of DAT	[12, 40, 42, 46, 64, 85–88]
DAT-KO (strain 2)	Knockout of DAT	[43, 55–57]
DAT-KD	Knockdown of DAT	[89]
DAT-CI	DAT cocaine insensitive	[45, 90]
DAT-TG	DAT overexpression	[91, 92]
• COMT		
COMT-KO	Knockout of COMT	[93, 94]
COMT mutant	COMT-Val overexpression	[70]

the DA receptors D1 [2, 3], D2 [4, 5], D2L [6, 7], D3 [8, 9], D4 [10], and D5 [11], the DA transporter (DAT) [12], the vesicular monoamine transporter 2 [13, 14], the limiting-step synthesizing enzyme tyrosine hydroxylase [15], and the DA-degrading enzyme catechol-O-methyltransferase (COMT) [16]. More recently, conditional recombination, using the Cre-Lox system, has been also utilized to generate the D1 and D5 receptors [17], the D2 receptor [18, 19], and the VMAT2 gene knockout [20].

Strategies for general phenotyping of transgenic and knockout mice are not different from any characterization of behavioral and physiological function, but often an interesting phenotype has to be uncovered using a "screening" strategy to reveal any unsuspected consequence. This has been largely documented in the past [21–23] and it will not be debated here. We will mostly focus in the next sections on tests specifically designed to evaluate cognitive functions; such specific tests are usually not comprised in these former behavioral batteries.

3 Breeding Strategies

Phenotyping analysis of transgenic or knockout mice requires the production of a large number of animals, most often through breeding in your own animal facility. The reader, who wants to start a production of genetically modified mice, will be no better informed than going through the Jackson Laboratory Handbook on Genetically Standardized Mice (http://jaxmice.jax.org/literature/handbook.html), into which all procedures and advices are provided.

We will below go through what we believe is the most important in order to generate reliable and reproducible data. There are two essential matters that should be respected: (1) the control of the genetic background and (2) the use of an appropriate breeding strategy.

3.1 Control of the Genetic Background

For empirical reasons of germline transmission and for historical reasons of embryonic stem (ES) cell production, homologous recombination techniques have longtime use ES cells from 129SV origin [24–26] that have been largely disseminated in the scientific community. This genetic strain was commonly used to study embryonic development, but rapidly appears as having quite a poor repertoire of integrated responses commonly used to assess brain functions [27]. On the other hand, neuroscientist widely used mice of the C57BL/6 genetic background, and this later has been used for backcrossing of knockout strains, following some basic recommendations [28]. It is essential to keep in mind that the genetic background of your knockout mouse will have a direct role in the phenotype that you'll be observing. This is obvious for considering mood disorder-related phenotypes, for example, [27, 29, 30], but it will also play a role even in locomotor or addictive behaviors linked to DA transmission [31, 32]. Therefore, starting with your first recombined founder, on a 129SV background, it will be necessary to start backcrossing on C57BL/6 mice. The first generation (F1) will provide 50 % genes from each parent. The F1 will again be crossed with a C57BL/6, giving rise to an F2 with 25 % 129SV genes and 75 % C57BL/6 genes. Each successive breeding will decrease by 50 % the 129SV gene load (Table 2). Therefore, the F10 will theoretically comprise less than 0.1 % of genes originating from the A129SV strain. Of course, because of the recombination event rate, the immediate vicinity of the recombined gene will still

Table 2
Percent of 129SV and C57BL/6 mouse strain genes after breedings

Generation	% of 129SV genes	% of C57BL/6 genes
F1	50	50
F2	25	75
F3	12.5	87.5
F4	6.25	93.75
F5	3.125	96.875
…………		
F10	0.09765625	99.90234375

be in a 129SV cluster, but it should comprise only a small number of genes. A definitive map could however be obtained, if necessary, using a precise mapping with specific single-nucleotide polymorphisms (SNPs) for each strain. But most often backcross breeding for ten generations (that would take about 2 years) is more than enough for any phenotyping analysis.

Once this is established, the most important problem in the long-term management of a mouse colony is the occurrence of a genetic drift. This drift results from spontaneous mutations that will eventually become fixed in the strain when sisters and brothers are crossed. It is therefore likely that after several generations of inbreeding, there will be several permanent genetic changes. What should be done to avoid this as much as possible is to: (1) freeze strains very early, using embryos or sperm congelation, and (2) keep outbreeding strains with fresh mice.

Freezing the strain will allow the experimenter not only to disseminate it more easily to collaborators abroad but also to be able to restart the colony in case of a general contamination (something always likely to happen) or if the experimenter suspects a genetic drift because some of the characteristic phenotypes of the strain under study are no longer observed. Most animal facilities now have the technical knowledge for reviving frozen material. If having the choice, the experimenter may use sperm congelation that is less expensive than embryo freezing.

3.2 How to Use the Best Breeding Strategy

To keep breeding systematically with commercially available C57BL/6 is the better manner to avoid genetic drift, and it will also permit to continue the genetic enrichment on this strain. Therefore, the experimenter should always keep one or two cages with breeding pairs composed of a fresh C57BL/6 mouse and one heterozygote for a given mutation. The "F1" heterozygotes offspring that will be produced will now be used to generate the experimental colony.

To perform behavioral experiments always requires a large number of mice, and producing them is both time and money consuming. Therefore, it is essential to make the right choices for producing these animals. To produce enough knockout mice for a given gene, we always establish heterozygote breeding pairs that are born from our maintenance cages described just above. From such mating, we expect 25 % of WT mice, 25 % of homozygote knockout mice, and 50 % of heterozygote mice. On average, using C57BL/6 mice, we obtain 8 pups per litter, which makes 2 KO and 2 WT animals. You will have to set the number of breeding pairs according to the number of experimental mice you will needed. For example, if you would need 20 KO mice of about the same age, you will need to have 10 breeding cages. Usually we let the males and females together, meaning that we will have new litters every 2 months.

We always mate F1 heterozygote mice, and we refresh the couples when the genitors are 1 year old, even if they are still having good size litters. This should just be a routine to install, in order to avoid any problems. Importantly, we never mate F2 mice, again to avoid any possibility of genetic differences.

When following strictly these procedures, the only true control group that should be used to compare the KO mice is the group of WT littermate mice. There is always a remote possibility that the exact locus of the recombined allele (that should still be from 129SV origin) may interfere; however, if this is suspected, there are now plenty of alternative strategies (viral rescue, backcross on another strain, shRNA), beyond the scope of this chapter, that may be utilized for control experiments.

4 Phenotyping of DA-ergic Mutants

4.1 Locomotor Activity

Locomotor activity can be measured by automatic photocell cages or scored by observation. Being relatively easy to quantify, locomotor activity is widely assessed in mutant mice. However, for a correct interpretation of the results, it is important to examine separately distinct components of locomotor activity.

Three components of locomotor activity can be distinguished: (1) horizontal locomotor activity, (2) vertical locomotor activity (rearings), which in drug naïve animals is considered a measure of exploratory activity, and (3) stereotypies which reflect perturbed perseverative patterns of behavior.

Another important distinction is of a temporal rather than spatial nature. Locomotor activity is a measure of testing environment (transparent cages or open field without bedding) that is novel to the mouse. When locomotion is plotted as a function of time, wild-type mice show high amounts of locomotion in the beginning of the testing period with a steep decline over time, indicating within-session habituation. Similarly, when locomotor activity is tested repeatedly, the amount of locomotion during the first exposure is higher than that in subsequent exposures, indicating between-session habituation. Thus, it is necessary to distinguish (1) locomotor activity recorded in the first minutes of the first exposure to the, up to that point unknown to the animal, testing environment, which indicates response to novelty; (2) total locomotor activity in a habituated context, recorded after 15–20 min of preexposure to the testing environment, which reflects total locomotion; and (3) within- or between-session habituation of locomotor activity, reflected as lower counts as a function of time.

4.1.1 Typical/Anticipated Locomotor Activity Results for DA Receptor Mutants

Locomotor activity has been extensively assessed in DA receptor mutants. With the exception of D5-KO mice, which show normal locomotor activity, the literature reports significant changes in locomotor activity at baseline in D1, D2, D3, and D4 mutants. These changes are receptor specific and as analyzed below indicate (1) a bidirectional implication of dopamine receptors in locomotor activity and (2) a differential implication of DA receptors in novelty-induced hyperlocomotion and in total amounts of locomotor activity.

The effects of D1 disruption on locomotor activity have been reported independently by two groups. These two groups generated two different D1-KO strains in 1994 [2, 3]. In these initial studies, one group assessed locomotion for 2 h and reported increased locomotor activity in D1-KO compared with WT mice [3]. The other group [2] reported somehow different results showing no differences in horizontal locomotor activity and decreases in vertical locomotor activity. However, these results were obtained with a different experimental design since locomotion was measured solely upon a 15 min exposure in an open field [2], a paradigm that reflects response to novelty and exploratory behavior rather than general locomotor activity. Subsequent studies with the mice generated by Drago et al. in a 2 h exposure paradigm showed a profile similar to the one reported by Xu et al. that is spontaneous hyperactivity [33]. This hyperactivity persisted after repeated exposure to the testing cage [33]. Combined together, these results suggest a differential role of D1 receptors in novelty-induced hyperlocomotion (diminished in D1-KO) and in overall levels of locomotor activity (increased in D1-KO). In addition reduced rearings in the exploratory open-field context reported by Drago et al. suggest that D1 inactivation impairs motivational aspects of behavior. Furthermore, persistent hyperlocomotion and retarded habituation in D1-KO [33] suggest alterations in attentive and cognitive domains. Indeed, as shown with mutants that impair cholinergic homeostasis, lack of habituation can be considered as a form of perseverative behavior indicative of deficits in cognitive function [34, 35].

Disruption of the D2 receptor results to hypoactivity. Although an initial study has proposed D2-KO mice to be bradykinetic and Parkinsonian-like, this result has not been replicated by other groups. However, hypoactivity has been well documented in D2-KO, which showed lower horizontal activity and lower rearing counts but normal patterns of habituation [31]. Interestingly, these mice demonstrated decreased initiation of spontaneous movement [31]. Examination of another D2-KO strain also showed that when behavioral activation is assessed in novelty situations, D2-KO demonstrate lower horizontal locomotor activity and lower rearing counts than wild-type littermates, suggesting altered motivational behavior [36]. Overall, hypoactivity in D2-KO is reminiscent of that induced by administration of D2-like antagonists (namely, antipsychotic drugs) in WT mice, suggesting a predominant role of D2 receptors in mediating the hypokinetic effects of antipsychotics. This notion was further confirmed by the fact that the hypolocomotor effects of haloperidol were absent in D2-KO mice [31]. Similarly to the D2-KO, a reduced level of locomotion and of rearings was seen with the D2L-KO mice [6].

Interestingly, disruption of the D3 receptor leads to hyperactivity and not hypoactivity, despite the fact that the D3 receptor belongs to the D2-class subtypes. Like the D1-KO discussed above, D3-KO were also shown to be hyperactive. However, hyperactivity in D3-KO markedly differs from that of D1-KO to almost the opposite phenotype. D3-KO mice showed increased locomotor activity upon a 15 min exposure in an open field [8]. A similar hyperactivity during the first 5 min of exposure to an actimeter [9] was seen with another D3 strain. However, upon prolonged exposure to the testing environment, this hyperlocomotor effect of the D3 ablation is lost and total locomotor activity does not differ between D3-KO and WT mice [9, 33]. Moreover, in marked contrast with D1-KO, upon prolonged and/or repeated exposure to the testing cage, D3-KO display rapid habituation responses both within (prolonged exposure [9]) and between sessions (repeated exposure [33]). Thus, the transient hyperactivity seen in D3-KO suggests a role of the D3 receptor in the locomotor response to novelty but not in overall levels of locomotion.

It should be noted that the analysis of double mutants, such as D2/D3-KO and D1/D3-KO, reveals alterations of the overall locomotor phenotypes characteristic for either D1 or D2 mutants, that is, hyperactivity and impaired habituation in D1/D3-KO [33] and hypoactivity in D2/D3-KO [37], further confirming the predominant role of D1 and D2 receptors in shaping general locomotion patterns.

As is the case for the D3 receptor, studies with D4-KO also show a preferential implication of the D4 receptor in the response to novelty. Thus, D4-KO mice show a diminished response to novelty in exploratory tests [38] but no change in overall levels of locomotor activity [38, 39].

4.1.2 Typical/Anticipated Locomotor Activity Results for DAT Mutants

Spontaneous hyperlocomotion is the most obvious characteristic of DAT-KO mice and has been directly related to the uncontrollable and persistent elevation of DA striatal levels [12]. This dramatically increased spontaneous locomotor activity shows little decrease over time indicating disrupted locomotor habituation and is evident in both phases of the light-dark cycle [12]. Hyperlocomotion in DAT-KO is characterized by a marked perseverative motor pattern and is accompanied by severe stereotypies [12, 40].

Locomotor hyperactivity in DAT-KO can be reversed by an acute administration of either typical or atypical antipsychotics [41], suggesting that this model of persistent hyperdopaminergia has a strong face validity to psychomotor activation seen in psychotic bipolar and schizophrenic patients.

Strikingly, it has been also shown that hyperactivity in DAT-KO can also be reduced by psychostimulants such as D-amphetamine and methylphenidate [42]. D-amphetamine and methylphenidate are used to reduce hyperactivity in ADHD patients; several genetic studies show an association between a polymorphism in the noncoding regions of the DAT gene and ADHD, suggesting that DAT-mediated processes could significantly contribute to the pathogenesis of this disorder. Another link between DAT function and ADHD has been proposed upon the fact that DAT-KO mice show increased impulsivity [43]. Therefore, hyperactivity in DAT-KO has been proposed as a simple model in which the effects of ADHD pharmacological agents can be assessed [41]. Recently, we showed that non-stimulant compounds with proven clinical therapeutic efficacy for the treatment of ADHD, such as the NET inhibitor reboxetine, also reduce hyperlocomotion in DAT-KO [44]. Hyperactivity relevant to ADHD has also been studied in another strain of DAT mutants, the DAT-CI (cocaine insensitive) mice. DAT-CI mutants display significant spontaneous hyperactivity that can be reversed by psychostimulants and NET inhibitors [45].

Overall, pharmacological studies with antipsychotics and ADHD compounds in DAT-KO led to propose that reduction of locomotor hyperactivity in these mice could constitute a high-throughput behavioral readout of high predictive validity for agents aiming at these pathologies. Along this line, we have shown that CB1 receptor blockade, by acute or chronic administration of the CB1 antagonist AM251, failed to reduce spontaneous locomotor hyperactivity in DAT-KO in accordance with lack of antipsychotic activity of CB1 antagonists in humans [46–48]. On the other hand, we have shown that experimental compounds such as the two structurally distinct D3 receptor antagonists SB-277011A and U99194 also dampen hyperactivity in DAT-KO [44].

4.2 Prepulse Inhibition (PPI)

PPI can be defined as the ability of a low-intensity stimulus ("prepulse") to diminish the startling, responding to subsequent stimuli that are identical in nature but of higher intensity as compared to the prepulse ("pulses"). The use of PPI as an experimental paradigm for understanding schizophrenia comes from clinical observations with schizophrenia patients unable to optimally filter or "gate" irrelevant, intrusive sensory stimuli [49]. The proposed construct of "gating deficits" in schizophrenia has been extended to deficient inhibition of both sensory and cognitive information [50]. Sensorimotor gating deficits in schizophrenics (indexed by reduced PPI) have been well characterized and considered as an endophenotype in human genetic studies. PPI also occurs in mice and can be easily measured in automated startle cages with well-validated experimental protocols and conditions [51].

Because all psychotogenic drugs (such as DA agonists and NMDA antagonists and serotoninergic compounds) decrease PPI in rodents [51], this paradigm has been widely use in pharmacological studies to screen for putative antipsychotic medications in animal models. Based upon the observation that PPI has a strong genetic component in mice, it was suggested that it may be a useful behavioral phenotype to explore in genetic mouse models. However, it should be noted that (1) PPI disruption in humans as an isolated measure is not a diagnostic test for schizophrenia and (2) a direct link of PPI disruption in animal models with positive symptom of schizophrenia is a misconception. These limitations aside, PPI measurement in genetically engineered mice can be a relevant tool to better understand the role of DA circuits in sensorimotor gating and has been assessed in most if not all DA mutants.

4.2.1 Typical/Anticipated PPI Results for DA-ergic Mutants

None of the dopamine receptor KO mice (D1, D2, D3, D4, D5) showed an altered PPI phenotype at baseline [52]. Nevertheless, studies with dopamine receptor mutant mice that were challenged pharmacologically with DA drugs revealed a complex pharmacological profile of dopamine agonists in the disruption of PPI.

Comprehensive studies [52] compared the effects of three distinct compounds: D-amphetamine, cocaine, and the D1/D2 agonist apomorphine across different DA receptor KO mice. Surprisingly, amphetamine disruption of the PPI was still present in these mice, while apomorphine and cocaine effects were completely absent.

The opposite profile was seen with D2-KO, in which amphetamine effects were completely abolished, cocaine effects partially attenuated, and apomorphine effects not altered. Thus, it appears that direct (apomorphine) or indirect (cocaine) DA agonists disrupt PPI primarily via D1 receptors; in contrast D2 receptors mediate amphetamine-induced PPI deficits and are only partially involved in the effects of cocaine.

In addition to the predominant roles of D1 and D2 receptors, a partial involvement of the D3 receptors was revealed in the effects of cocaine only, given that a high dose of cocaine produced an exacerbated effect in D3-KO. On the other hand, the effects of amphetamine were intact in D3-KO, D4-KO [53], and D5-KO [11], confirming the major implication of D2 receptors in the PPI altering function of the drug.

In contrast to the DA receptor KO that show no alteration of PPI at baseline, DAT-KO mice present a marked reduction of PPI at baseline [40], indicating severe deficits in sensorimotor gating. In accordance with the hyperdopaminergic hypothesis of inadequate filtering, the PPI deficit can be attributed to persistent hyperdopaminergic state of these mice [12]. Interestingly, the PPI deficit was reversed by the D2 antagonist raclopride [40] but not with the D1 antagonist SCH23390, indicating a predominant role of the D2 receptors in mediating this effect. Relevant to this observation, and in accordance with the D2-antagonist profile of all known antipsychotics, it was subsequently shown that the PPI deficit in DAT-KO mice was corrected by atypical antipsychotic drugs [54]. A marked PPI deficit at baseline was also reported in another DAT-KO strain, produced independently by another group [55]. Surprisingly, in this strain PPI deficits were corrected by cocaine and methylphenidate [56]. This effect that is different from that of D2 antagonist [57] is reminiscent of the paradoxical calming effects of psychostimulants on DAT spontaneous hyperlocomotion [42] that was discussed above. Because of differences in the two DAT-KO mouse strains, it remains to be examined whether psychostimulants are also effective against PPI deficits in the first DAT-KO strain [12]. Nevertheless, at this point caution is needed in pharmacological studies with the DAT-KO mice, since there might be false-positive hits when using them as a screen for future antipsychotics based on locomotor activity and PPI measures.

4.3 Cognitive Testing

Growing evidence indicates that a wide variety of genetic mutations and polymorphisms of different components of the DA system impact cognition. Furthermore affective and psychotic DA-related mental disorders are characterized by profound cognitive deficits and may thus be implicated in various aspects of these mental disorders. Important differences between human and mouse brain structure and function notwithstanding, behavioral tests and paradigms in mice have contributed critical information about brain mechanisms involved in cognitive processes, and a significant effort has been underway in the last decade to streamline batteries of tests with translational value that can be validated both in humans and animal models, through dedicated committees (e.g., MATRICS).

It has been widely acknowledged that many behavioral tasks in mice successfully translate to specific neuropsychological tests in humans and allow to studying multiple aspects of cognitive functions. These have been reviewed extensively [22, 58].

It is important to note that distinct tests have been developed in order to specifically explore different domains of cognition, including attention, learning, and memory. It has been repeatedly underscored that when studying mutant mice it is critical to examine multiple aspects of cognition by using an extensive battery of tests; however this is seldom the current practice.

In this chapter, we summarize discoveries of genetic modifications in mice that target different components of the dopaminergic system (D1, D2, D3, D4, D5, DAT, COMT) and that impact cognition.

However, a complete comparative survey of cognitive functions in DA mutants is at this point impossible. Indeed, DA mutants have not been systematically screened for all different domains of cognitive function. Some cognitive functions have not been adequately addressed in DA mutants and, conversely, some mutants have been studied only marginally. For instance, D4-KO mice have been principally evaluated in the Go-NoGo- and 5-choice tests, which assess attention [59]; in contrast attention has not been studied well in the other DA-ergic mutants. In addition, even for cognitive domains that have been more widely explored, few groups have assessed the same behaviors under the same conditions across different genes. Subsequently, missing information and methodological differences and inconsistencies often hinder meaningful interpretations.

We will therefore focus on the domains that have been mostly studied with DA-ergic mutants, which are (a) spatial learning and memory, (b) working memory, and (c) executive function.

4.3.1 Spatial Learning and Memory

Spatial learning and memory tests in rodents are designed to assess the ability of the animal to learn the location of a reward based on spatial information in the environment. A number of tests such as the Barnes maze and the radial arm maze utilize palatable food as a reward; in the case of the Morris water maze that we will detail below, the reward is a platform that allows the animal to escape from an aversive and stressful situation that is being immerged into the water.

Thus, in the Morris water maze (MWM), mice learn to escape from the water by swimming to a small stainless steel platform. The apparatus consists of a large circular stainless steel pool (usually 150 cm diameter) filled with water maintained at room temperature. The water is made opaque using a colored substance, e.g., aqueous emulsion of the same color as the platform. A video tracking system is usually used to monitor activity.

There are two complementary versions of the test: the hidden and the visible platform versions that respectively assess spatial versus nonspatial learning and memory (Fig. 1). In the spatial version of the test, the platform is hidden just below (1–2 cm) of the water

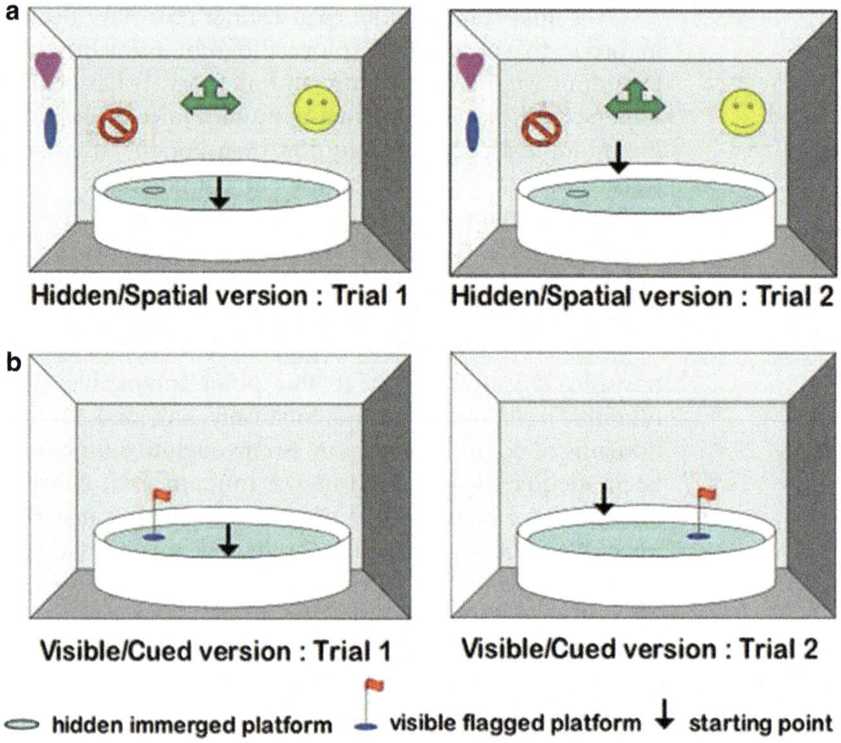

Fig. 1 *Spatial and cued versions of the Morris water maze.* Panel (**a**) hidden/spatial version. Presence of salient colorful cues on the walls. The platform is hidden below the water level. The placement of the cues and of the platform does not change between trials. The starting point does. Panel (**b**) visible/cued version. There are no cues on the walls. The platform is clearly visible and flagged above the water. Both the position of the platform and the starting point randomly change between trials

surface, which is rendered opaque so the mouse cannot directly see it. In the cued version of the test, the platform is visible and distinctly marked just above the water.

Hidden platform version: During the training phase of the spatial learning version of the Morris water maze, mice learn the position of a small hidden platform, using prominent distal extramaze cues arranged in the room around the pool. The position of the platform is fixed through the training phase but the position from which the mouse is placed in the platform is changed between trials. The extramaze visual cues, which are also fixed, allow the animal to progressively form a spatial cognitive map using the cues to mark the location of the platform (Fig. 1a).

Each trial starts with the mice facing the interior wall of the pool and ends when they climb onto the platform or after a maximum searching time (cutoff 60 or 90 s). The time it takes for the mouse to find the platform and the distance traveled before finding the platform are scored, and the animal is left on the platform for

20–30 s before being removed to its cage. Animals that do not find the platform inside the time limit are given the maximum time score but are guided and placed on the platform for 20–30 s. Mice are usually given 1 or 2 four-trial blocks per day and the experiment continues for 5–7 days, depending on the learning curve of the mice. As the mouse acquires the task, the time elapsed (latency) and the distance traveled before finding the platform should diminish progressively.

Probe trial: The probe trial is the ultimate spatial memory test of the water maze (WM) paradigm. It is administered to animals that have mastered the WM spatial task and have learned the correct location of the platform. The trial is administered usually 1 day after the last acquisition trial to evaluate memory per se. In this trial the platform is removed and the mice are left to swim throughout the 60 or 90s cutoff time. Mice that correctly remember the location of the platform tend to swim at the area surrounding the platform. The distance traveled in this area (or quadrant of the pool in which the platform was located) in comparison to the distance traveled in the other areas (or quadrants) is computed as an indicator of mnemonic performance.

Cued-platform version: This version of the WM does not assess spatial cognition but rather general skills and/or the capacity to form a coherent associative escape strategy. During the cued version of the water maze, mice are trained to find a platform made visible by a distinct "cue" such as a flag, a small ball, or any distinctive sign of bright color fixed directly above the platform. Spatial cues in the room are absent. The training procedure is identical to that of the spatial version except that the platform's position and the animal's starting position vary for each trial (Fig. 1b).

Typical/Anticipated Water Maze Results for DA-ergic Mutants

Among the different DA receptor KO, the D1-KO [60], D3-KO [61], as well as the double D1+D3-KO [61], the D5-KO [11], and the D2-KO [62] have been studied in the MWM; there are no data with the D4-KO. Notably, no deficit was seen with the D3-KO or the D5-KO. In contrast, D1-KO and D2-KO mice displayed marked deficits in both spatial and cued versions of the MWM.

For wild-type mice, latency to escape (i.e., time to find the platform) decreased significantly with trials, indicating learning to use spatial indicators to locate the platform. D1-KO mice had significantly longer escape latencies as compared to WT and in the probe test they showed no spatial preference for the quadrant in which the platform was previously located. In the visual cued version, D1-KO performed better than in the spatial version but were still impaired in comparison to the WT mice [60]. The deleterious effects of D1 invalidation on the WM performance were also carried along in the double D1+D3-KO mice, which were also impaired in both versions of the WM [61].

These results suggest an important role for the D1 receptor in spatial learning and memory. Studies with another strain of D1-KO further strengthen this conclusion by showing that D1-KO are also markedly impaired in the Barnes maze another test of spatial learning and memory [63].

Recently, we have shown that the D2-KO are also profoundly impaired in the spatial version of MWM, suggesting that D1 receptor integrity is not sufficient to ensure optimal performance in this task. Furthermore, this spatial learning deficit was reproduced by local injections of the D2 antagonist sulpiride in the temporal hippocampus, showing a direct role of the D2 receptor expressed in this hippocampal subregion. We also shown that the genetic deletion of the presynaptic D2 receptor was sufficient to reproduce this learning deficit [62].

Interestingly, the hyperdopaminergic DAT-KO mice were also impaired in the WM. The DAT-KO mice show profound deficits in the cued version of the MWM [47, 64]. This phenotype may be directly linked to their hyperdopaminergic state because drugs that increase synaptic dopamine (such as D-amphetamine and the DAT inhibitor GBR 12395) also disturb performance in the WM in WT mice [47].

In the spatial version of MWM, DAT-KO mice also show a diminished capacity to learn the task as compared to WT mice. Notably, the spatial impairment of DAT-KO has been shown in another spatial test, the radial arm maze [42]. To this point it is unknown if an overstimulation of D1 receptors contributes in the WM impairment seen in DAT-KO. An involvement of D2 receptors is more likely since WM deficits in the DAT-KO can be partially corrected by the typical antipsychotic D2 antagonists haloperidol and sulpiride.

On the other hand, mice that overexpress D2 receptors in the striatum (D2-Tg mice) show a normal performance in the WM [65]. Thus it is suggested that WM deficits in DAT-KO are mediated by extra-striatal D2 receptors, probably hippocampal D2 receptors, which were shown to be involved in the modulation of neurochemical circuits underlying cognitive deficits [66]. Collectively, these data indicate that for spatial learning, too much DA levels (as for the DAT-KO mice) or too little DA signalization (D1-KO and D2-KO mice) is responsible for learning deficits.

4.3.2 Working Memory

In contrast to long-term (or reference) memory which stores information over prolonged periods of time (several hours to several years and to a lifetime), working memory (operating at a short-term scale) recruits information for rehearsal, elaboration, recoding, and comparison in order to solve a current problem [67].

Working memory can be assessed in the same paradigms that are used to evaluate reference memory; the procedure is thus modified to include a short-term interval. Thus, working memory can be assessed in the WM [68]. In that case the paradigm typically consists of a two- or four-trial block/day in which the hidden trial is located in one of the four quadrants and randomly relocated to another quadrant on each of the subsequent days. The mouse obtains information on the location of the platform during trial 1 that is recorded in working memory and will be of help to find the platform in subsequent trials. The working memory is measured as the reduction in distance traveled between subsequent trials rather than between subsequent days.

However, deficits in working memory are usually assessed on alternation tasks. In these tasks mice must alternate between two (or more) responses from one trial to the next, with a short delay separating the trials. Most commonly T-maze, Y-maze, or radial maze is used in two different types of experimental paradigms: spontaneous alternation and delayed non-match to sample. In the spontaneous alternation design, the animal is free to select and explore any arm of the maze and no delay is imposed between choices. Reentry into an already visited arm during a trial is recorded as an error. In the delayed non-match to sample design, the animal has to retrieve a palatable reward by following an alternation rule. In the simplest case of a choice between two arms, the animal learns that when arm A is rewarded in the first visit, arm B will be subsequently rewarded. During the training phase, a short delay of 4–5 s separates the two visits. During the test phase, the animal has to remember which arm was rewarded and retain the information for a larger 15–300 s delay (corresponding to working memory) so as to visit correctly the opposite arm during the second (post-delay) visit.

Typical/Anticipated Working Memory Results for DA-ergic Mutants

The D1-KO has been tested in a Y-maze spontaneous alternation paradigm [60] and their performance was similar to that of the WT. The D2- and D3-KO were tested in a T-maze spatial delayed alternation paradigm. Both the D2- and D3-KO showed impaired working memory at the 15 and 20 s delay as compared to the WT mice [69]. While there are no results with D4- and D5-KO, it is interesting to note that in a delayed T-maze test, the COMT-TG mice, which model a hypodopaminergic state, showed deficits in working memory. Mice required significantly more days to reach the learning criteria during the training phase [70] and they performed worse at subsequent delays. Surprisingly, the D2-TG mice showed a similar phenotype [65], namely, difficulty in acquiring the task during the 4 s delay training phase, which was interpreted by the authors as a deficit in working memory.

In contrast to these D2 overexpressing mice, the COMT-KO mice which model a persistent hyperdopaminergic state, evidenced between others as increased anxiety and disorganized behavior, acquired the test more rapidly, a sign of an actual improvement in spatial working memory [70].

It is therefore evident that these two models are very dissimilar. Further studies are needed to clarify the role of hyperdopaminergia in working memory.

4.3.3 Executive Function

Executive functions are high-order cognitive processes that subserve in the planning and control of goal-directed behaviors. Executive functions encompass the ability to form problem-solving strategies and the ability to transfer skills beyond initial learning to new situations with different demands, by inhibiting and shifting responding when necessary.

Impaired executive function is a trait symptom of DA-related psychopathologies, highly correlated with the severity of their prognosis.

In mice, executive function can be evaluated with a now classic test, the Attentional Set-Shifting Test (ASST) which is a rodent adaptation of the Wisconsin Card Sorting Test (WCST), a test used to evaluate executive function in humans.

In the ASST the mouse learns to retrieve a reward by focusing on a sole perceptual feature of a complex stimulus. In the most common variation of the test, the animal is placed with a choice of two locations, one of which contains the hidden reward. These can be two bowls that differ in texture and odor of the contained medium [71] or two arms of a maze that differ in their texture and color [72].

The animal has first to learn that only one dimension is relevant when searching for the reward (e.g., odor only when both odor and medium can vary from trial to trial). In that case the reward is always hidden in a bowl that is sprinkled with the arbitrarily chosen correct odor (correct exemplar).

The animal is subsequently requested to solve a series of problems by performing different discriminations: (1) reversal, where the correct and incorrect exemplars inside the relevant discrimination (odor) are reversed; (2) intra-dimensional shift, where a new discrimination has to be learned still based on the previously relevant perceptual dimension (odor); and (3) extra-dimensional shift, where in order to learn a new discrimination attention has to be shifted toward the previously irrelevant dimension (medium becomes relevant if odor was the previous relevant dimension).

We will describe below the ASST protocol as it is performed in our laboratory [47]. The test is conducted according to the protocol of Birrell and Brown for rats [71] that we have adapted for mice.

Habituation and Training

Training starts a week (7 days) before the test. Food reward pellets are given to the mice in their home cage to familiarize them with the taste and odor. On the first 2 days, mice are used to 20 min of handling. After that, mice are placed in the test box with the cups for two sessions of 20 min each to allow exploration of the box.

Food rewards are put in both cups. On the third day of training, cups are filled with sawdust and food pellets are put on top of them and gradually buried deeper and deeper during consecutive sessions until the mice learn to dig properly for food. Food deprivation starts on day 5 of habituation. Mice are maintained at 80–85 % of their initial weight.

Apparatus

The apparatus is rectangular, made of transparent plexiglass (dimensions 30 × 20 × 20 cm). At the two ends of the box are placed two distinctly separated plastic cups (3 × 4 × 5 cm), one of which is baited with a small piece of cereal (30 mg; choco pops, Nestle).

Figure 2 shows the order of discriminations, which is the same for all mice. In the simple discrimination (SD), the bowls differed along one of the two dimensions (odor or digging medium) only. For the compound discrimination (CD), a second dimension is introduced, but the relevant dimension and correct and incorrect exemplars remain unchanged. For the reversal (CDR), all exemplars and the relevant dimension remain unchanged, but the previously correct stimulus is now incorrect. For both intra-dimensional (ID) shift and extra-dimensional (ED) shift, new exemplars of both dimensions are used (a total change design). In the ID shift, the relevant dimension is the same as before, whereas in the ED shift, the mouse has to shift attention to the previously irrelevant dimension. A mouse should complete successfully six consecutive trials in order to achieve each discrimination task.

Each of the above trials evaluates different cognitive capacities that are thought to rely on different brain structures. Thus, the reversal learning involves the orbitofrontal cortex (OFC) and dorsomedial striatum, while set-shifting preferentially mobilizes medial cortex structures (anterior cingulate, prelimbic, and infralimbic cortex) [71, 73]. Therefore, the evaluation of reversal learning and set-shifting within the same task can provide an informative framework to study the neurotransmitter circuits involved.

While both reversal and ED/ID shifts evaluate cognitive flexibility, the ID/ED shift offers a translational measure relevant to human performance on the WCST [74], which requires the subject to alter the response strategy and use previously irrelevant information to solve the new set of problems. In control subjects, the ID/ED shift requires more trials to criterion than an ID shift (Fig. 2). If the number of trials to solve the EDS problem is not significantly greater than the previous IDS problem, then the data are interpreted as the lack of formation of the attentional set.

The only DA mutants for which ASST data are reported in the literature are the D2-KO and D3-KO and the COMT-TG mice bearing the Val/Val polymorphism.

The most striking result is an inhibitory role of the D3 receptor in cognitive flexibility evidenced by a better performance of the D3-KO mice in reversal discrimination [75]. In contrast, the

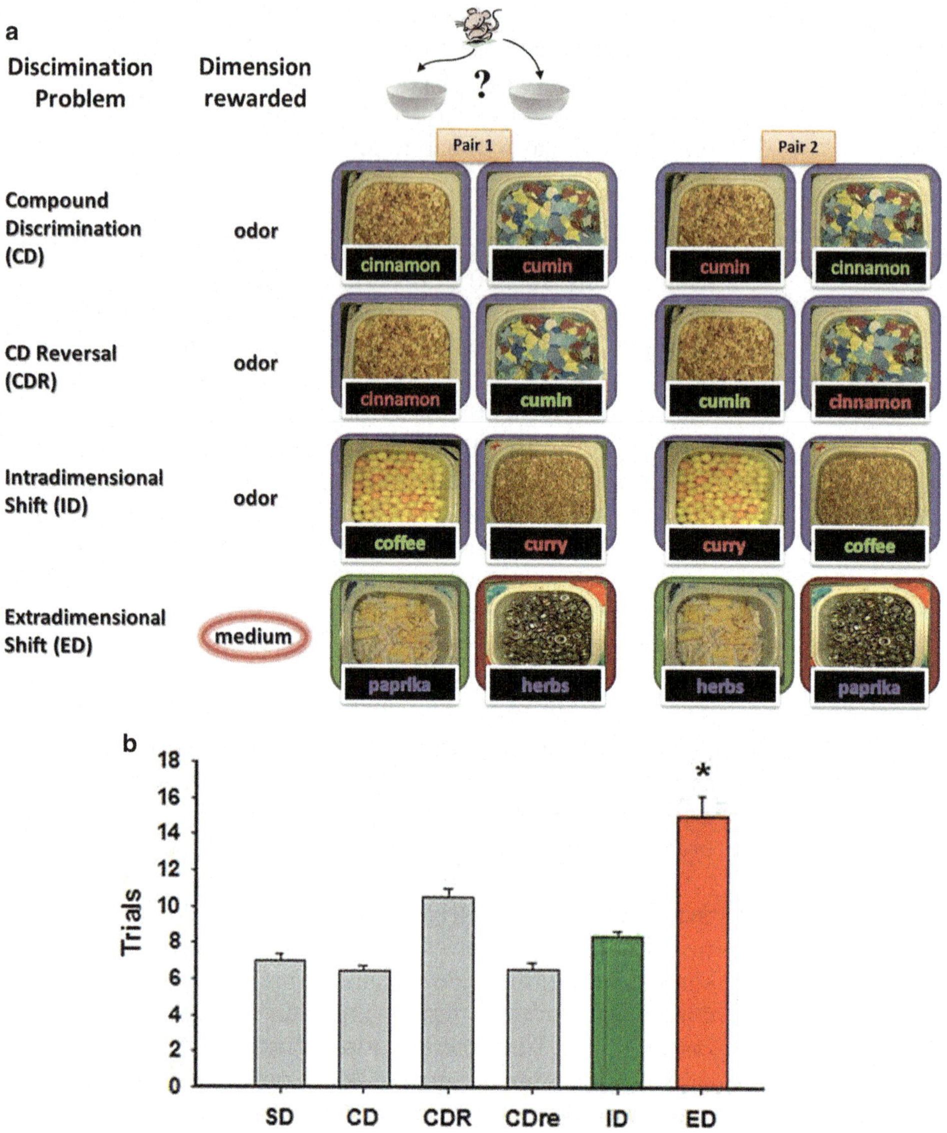

Fig. 2 *ASST different sessions with odor as the first relevant dimension*. Panel (**a**) examples of combinations into stimulus pairs are shown for a shifting from odor to digging medium. The correct exemplar is shown in *green*, paired with either exemplar from the irrelevant dimension (*purple*). On every trial (CD, CDR, ID, ED), the pair of stimuli differs for both the relevant and irrelevant dimensions. Panel (**b**) the bars represent the mean of the number of trials for each session. To validate the session, the mouse must find six times consecutively the hidden reward. The ID session presents no difficulty to the mice. In the ED session, in comparison with ID session, the mice need significantly more trials to find the new discrimination rule associated with the hidden reward

D2-KO mice were slower to learn the CD, suggesting deficits in the initial acquisition of the task-governing rules but did not show deficits in an ID reversal task [75]. However, in another study, the same group showed a deficit in an ED reversal task for D2-KO [76].

It should be noted that Glickstein et al. [75] reported no difference for D3-KO and D2-KO in the ED shift; however in this study there was no difference in the number of trials to complete the ED as compared to the ID for WT control mice making the interpretation of the results uneasy. Even so, it clearly emerges that invalidated D2 function impairs cognitive processes. Indeed, mice invalidated for the D2L isoform are also impaired in a conditioned avoidance learning and memory paradigm [77]. Thus, while detrimental for associative learning, and working memory, invalidated D2 function seems to have less effect on the ED shift.

In contrast to the D2-KO, the COMT-TG mice which model the Val/Val polymorphism, resulting in a higher COMT activity and thus a higher rate of DA catabolism, showed a selective impairment in the ED shift with no impairment in reversal learning [70]. Pharmacological studies suggest that this phenotype could result from a compromised D1-like receptor activation [78]. However, this result has yet to be confirmed with relevant genetic models.

5 Conclusion

Table 3 recapitulates the findings pertaining to cognitive function in DA mutants that were discussed here. A clear differential role emerges for the D1 and D2 receptors: both are required for optimal performance but their invalidation appears to affect preferentially different cognitive functions. In contrast to D1-KO, D2-KO, and, to a lesser extent, D3-KO, cognitive function in D4-KO and D5-KO remains to be fully appreciated.

Surprisingly, this is also the case for hyperdopaminergic mice. Studies published so far suggest that constitutive hyperdopaminergia is deleterious for spatial and nonspatial learning and memory, while working memory remains unaffected. However, this is a highly speculative concept since different mutants were tested in different domains, spatial and "cued" learning, and memory was tested solely in DAT-KO mice, while working memory was tested in COMT-KO only. Further studies with both mutants are warranted as are studies exploring executive function in hyperdopaminergic mice. It is evident that hyperlocomotion, stereotypies, and increased anxiety to novelty hinder cognitive testing in hyperdopaminergic mice. However, studies with these mutants would be invaluable in merging the hyperdopaminergia hypothesis of affective and psychotic disorders such as bipolar disorder and schizophrenia with the profound cognitive dysfunction seen in these disorders that still today handicaps prognosis and constitutes a huge unmet medical need.

Table 3
Summary of cognitive function defects in DA-ergic mutant mice

	DA levels or DA receptor expression/ function								
	Decreased signaling						Increased signaling		
	D1-KO	D2-KO	D3-KO	D4-KO	D5-KO	COMT-TG	D2-TG	COMT-KO	DAT-KO
Spatial learning and memory									
WM	↓↓	↓	=		=		=		↓
Barnes	↓								
Radial maze									↓
Associative learning									
WM visible cued version	↓		=						↓↓
Avoidance learning		↓	+						
Working memory									
Spontaneous alternation	=								
Delayed alternation (T-maze)		↓	↓			↓	↓	+	
Executive function									
CD: compound discrimination		↓	=		=				
CDR: reversal learning		=	+		=				
EDR: reversal of ED		↓							
ED: extra-dimensional shift		=?	=?			↓			

Defects identified in cognitive function of DA-ergic mutant mice are indicated with the following symbols: ↓ impaired; + ameliorated; = no difference; =? no difference and unclear. Empty cells indicate "no data available"

References

1. Hall B et al (2009) Overview: generation of gene knockout mice. Current protocols in cell biology/editorial board, Juan S Bonifacino [et al.] Chapter 19:Unit 19 12 19 12 11–17
2. Drago J et al (1994) Altered striatal function in a mutant mouse lacking D1A dopamine receptors. Proc Natl Acad Sci U S A 91(26): 12564–12568
3. Xu M et al (1994) Dopamine D1 receptor mutant mice are deficient in striatal expression of dynorphin and in dopamine-mediated behavioral responses. Cell 79(4):729–742
4. Baik JH et al (1995) Parkinsonian-like locomotor impairment in mice lacking dopamine D2 receptors. Nature 377(6548):424–428
5. Kelly MA et al (1997) Pituitary lactotroph hyperplasia and chronic hyperprolactinemia in dopamine D2 receptor-deficient mice. Neuron 19(1):103–113
6. Wang Y et al (2000) Dopamine D2 long receptor-deficient mice display alterations in striatum-dependent functions. J Neurosci 20(22):8305–8314
7. Usiello A et al (2000) Distinct functions of the two isoforms of dopamine D2 receptors. Nature 408(6809):199–203
8. Accili D et al (1996) A targeted mutation of the D3 dopamine receptor gene is associated with hyperactivity in mice. Proc Natl Acad Sci U S A 93(5):1945–1949
9. Xu M et al (1997) Dopamine D3 receptor mutant mice exhibit increased behavioral sensitivity to concurrent stimulation of D1 and D2 receptors. Neuron 19(4):837–848

10. Rubinstein M et al (1997) Mice lacking dopamine D4 receptors are supersensitive to ethanol, cocaine, and methamphetamine. Cell 90(6):991–1001
11. Holmes A et al (2001) Behavioral characterization of dopamine D5 receptor null mutant mice. Behav Neurosci 115(5):1129–1144
12. Giros B et al (1996) Hyperlocomotion and indifference to cocaine and amphetamine in mice lacking the dopamine transporter. Nature 379(6566):606–612
13. Fon EA et al (1997) Vesicular transport regulates monoamine storage and release but is not essential for amphetamine action. Neuron 19(6):1271–1283
14. Wang YM et al (1997) Knockout of the vesicular monoamine transporter 2 gene results in neonatal death and supersensitivity to cocaine and amphetamine. Neuron 19(6):1285–1296
15. Zhou QY et al (1995) Targeted disruption of the tyrosine hydroxylase gene reveals that catecholamines are required for mouse fetal development. Nature 374(6523):640–643
16. Gogos JA et al (1998) Catechol-O-methyltransferase-deficient mice exhibit sexually dimorphic changes in catecholamine levels and behavior. Proc Natl Acad Sci U S A 95 (17):9991–9996
17. Sarinana J et al (2014) Differential roles of the dopamine 1-class receptors, D1R and D5R, in hippocampal dependent memory. Proc Natl Acad Sci U S A 111(22):8245–8250
18. Anzalone A et al (2012) Dual control of dopamine synthesis and release by presynaptic and postsynaptic dopamine D2 receptors. J Neurosci 32(26):9023–9034
19. Bello EP et al (2011) Cocaine supersensitivity and enhanced motivation for reward in mice lacking dopamine D2 autoreceptors. Nat Neurosci 14(8):1033–1038
20. Narboux-Neme N et al (2011) Severe serotonin depletion after conditional deletion of the vesicular monoamine transporter 2 gene in serotonin neurons: neural and behavioral consequences. Neuropsychopharmacology 36(12):2538–2550
21. Crawley JN (1999) Behavioral phenotyping of transgenic and knockout mice: experimental design and evaluation of general health, sensory functions, motor abilities, and specific behavioral tests. Brain Res 835(1):18–26
22. Crawley JN (2008) Behavioral phenotyping strategies for mutant mice. Neuron 57(6): 809–818
23. McIlwain KL et al (2001) The use of behavioral test batteries: effects of training history. Physiol Behav 73(5):705–717
24. Doetschman TC et al (1985) The in vitro development of blastocyst-derived embryonic stem cell lines: formation of visceral yolk sac, blood islands and myocardium. J Embryol Exp Morphol 87:27–45
25. Li E et al (1992) Targeted mutation of the DNA methyltransferase gene results in embryonic lethality. Cell 69(6):915–926
26. Nagy A et al (1993) Derivation of completely cell culture-derived mice from early-passage embryonic stem cells. Proc Natl Acad Sci U S A 90(18):8424–8428
27. Crawley JN et al (1997) Behavioral phenotypes of inbred mouse strains: implications and recommendations for molecular studies. Psychopharmacology (Berl) 132(2):107–124
28. Silva AJ et al (1997) Mutant mice and neuroscience: recommendations concerning genetic background. Banbury conference on genetic background in mice. Neuron 19(4):755–759
29. Farley S et al (2012) Increased expression of the Vesicular Glutamate Transporter-1 (VGLUT1) in the prefrontal cortex correlates with differential vulnerability to chronic stress in various mouse strains: effects of fluoxetine and MK-801. Neuropharmacology 62(1):503–517
30. Holmes A et al (2003) Abnormal anxiety-related behavior in serotonin transporter null mutant mice: the influence of genetic background. Genes Brain Behav 2(6):365–380
31. Kelly MA et al (1998) Locomotor activity in D2 dopamine receptor-deficient mice is determined by gene dosage, genetic background, and developmental adaptations. J Neurosci 18(9):3470–3479
32. Morice E et al (2004) Phenotypic expression of the targeted null-mutation in the dopamine transporter gene varies as a function of the genetic background. Eur J Neurosci 20(1): 120–126
33. Karasinska JM et al (2005) Deletion of dopamine D1 and D3 receptors differentially affects spontaneous behaviour and cocaine-induced locomotor activity, reward and CREB phosphorylation. Eur J Neurosci 22(7):1741–1750
34. Bales KR et al (2006) Cholinergic dysfunction in a mouse model of Alzheimer disease is reversed by an anti-A beta antibody. J Clin Invest 116(3):825–832
35. Tzavara ET et al (2003) Dysregulated hippocampal acetylcholine neurotransmission and impaired cognition in M2, M4 and M2/M4 muscarinic receptor knockout mice. Mol Psychiatry 8(7):673–679
36. Viggiano D et al (2003) Dopamine phenotype and behaviour in animal models: in relation to

attention deficit hyperactivity disorder. Neurosci Biobehav Rev 27(7):623–637
37. Jung MY et al (1999) Potentiation of the D2 mutant motor phenotype in mice lacking dopamine D2 and D3 receptors. Neuroscience 91(3):911–924
38. Dulawa SC et al (1999) Dopamine D4 receptor-knock-out mice exhibit reduced exploration of novel stimuli. J Neurosci 19(21):9550–9556
39. Kruzich PJ et al (2004) Dopamine D4 receptor-deficient mice, congenic on the C57BL/6J background, are hypersensitive to amphetamine. Synapse 53(2):131–139
40. Ralph RJ et al (2001) Prepulse inhibition deficits and perseverative motor patterns in dopamine transporter knock-out mice: differential effects of D1 and D2 receptor antagonists. J Neurosci 21(1):305–313
41. Gainetdinov RR, Caron MG (2003) Monoamine transporters: from genes to behavior. Annu Rev Pharmacol Toxicol 43:261–284
42. Gainetdinov RR et al (1999) Role of serotonin in the paradoxical calming effect of psychostimulants on hyperactivity. Science 283(5400):397–401
43. Yamashita M et al (2013) Impaired cliff avoidance reaction in dopamine transporter knockout mice. Psychopharmacology (Berl) 227(4): 741–749
44. Barth V et al (2013) In vivo occupancy of dopamine D3 receptors by antagonists produces neurochemical and behavioral effects of potential relevance to attention-deficit-hyperactivity disorder. J Pharmacol Exp Ther 344(2):501–510
45. Napolitano F et al (2010) Role of aberrant striatal dopamine D1 receptor/cAMP/protein kinase A/DARPP32 signaling in the paradoxical calming effect of amphetamine. J Neurosci 30(33):11043–11056
46. Tzavara ET et al (2006) Endocannabinoids activate transient receptor potential vanilloid 1 receptors to reduce hyperdopaminergia-related hyperactivity: therapeutic implications. Biol Psychiatry 59(6):508–515
47. El Khoury MA et al (2012) Interactions between the cannabinoid and dopaminergic systems: evidence from animal studies. Prog Neuropsychopharmacol Biol Psychiatry 38(1):36–50
48. Meltzer HY et al (2004) Placebo-controlled evaluation of four novel compounds for the treatment of schizophrenia and schizoaffective disorder. Am J Psychiatry 161(6):975–984
49. Geyer MA et al (2001) Pharmacological studies of prepulse inhibition models of sensorimotor gating deficits in schizophrenia: a decade in review. Psychopharmacology (Berl) 156(2–3): 117–154
50. Swerdlow NR et al (2008) A novel rat strain with enhanced sensitivity to the effects of dopamine agonists on startle gating. Pharmacol Biochem Behav 88(3):280–290
51. Svenningsson P et al (2003) Diverse psychotomimetics act through a common signaling pathway. Science 302(5649):1412–1415
52. van den Buuse M (2010) Modeling the positive symptoms of schizophrenia in genetically modified mice: pharmacology and methodology aspects. Schizophr Bull 36(2):246–270
53. Ralph RJ et al (1999) The dopamine D2, but not D3 or D4, receptor subtype is essential for the disruption of prepulse inhibition produced by amphetamine in mice. J Neurosci 19(11):4627–4633
54. Barr AM et al (2004) The selective serotonin-2A receptor antagonist M100907 reverses behavioral deficits in dopamine transporter knockout mice. Neuropsychopharmacology 29(2):221–228
55. Sora I et al (1998) Cocaine reward models: conditioned place preference can be established in dopamine- and in serotonin-transporter knockout mice. Proc Natl Acad Sci U S A 95(13):7699–7704
56. Yamashita M et al (2006) Norepinephrine transporter blockade can normalize the prepulse inhibition deficits found in dopamine transporter knockout mice. Neuropsychopharmacology 31(10):2132–2139
57. Arime Y et al (2012) Cortico-subcortical neuromodulation involved in the amelioration of prepulse inhibition deficits in dopamine transporter knockout mice. Neuropsychopharmacology 37(11):2522–2530
58. Young JW et al (2009) Using the MATRICS to guide development of a preclinical cognitive test battery for research in schizophrenia. Pharmacol Ther 122(2):150–202
59. Young JW et al (2011) The effect of reduced dopamine D4 receptor expression in the 5-choice continuous performance task: Separating response inhibition from premature responding. Behav Brain Res 222(1):183–192
60. El-Ghundi M et al (1999) Spatial learning deficit in dopamine D(1) receptor knockout mice. Eur J Pharmacol 383(2):95–106
61. Karasinska JM et al (2000) Modification of dopamine D(1) receptor knockout phenotype in

mice lacking both dopamine D(1) and D(3) receptors. Eur J Pharmacol 399(2–3):171–181

62. Rocchetti J et al (2014) Presynaptic D2 dopamine receptors control long-term depression expression and memory processes in the temporal hippocampus. Biol Psychiatry pii: S0006-3223(14)00166–8. doi: 10.1016/j.biopsych.2014.03.013.
63. Ortiz O et al (2010) Associative learning and CA3-CA1 synaptic plasticity are impaired in D1R null, Drd1a–/– mice and in hippocampal siRNA silenced Drd1a mice. J Neurosci 30(37):12288–12300
64. Morice E et al (2007) Parallel loss of hippocampal LTD and cognitive flexibility in a genetic model of hyperdopaminergia. Neuropsychopharmacology 32(10):2108–2116
65. Kellendonk C et al (2006) Transient and selective overexpression of dopamine D2 receptors in the striatum causes persistent abnormalities in prefrontal cortex functioning. Neuron 49(4):603–615
66. Tzavara ET et al (2003) Biphasic effects of cannabinoids on acetylcholine release in the hippocampus: site and mechanism of action. J Neurosci 23(28):9374–9384
67. Rodriguiz RM, Wetsel WC (2006) Assessments of cognitive deficits in mutant mice. In: Levin ED, Buccafusco JJ (eds) Animal models of cognitive impairment. CRC, Boca Raton, FL, Chapter 12
68. Terry AV Jr (2009) Spatial navigation (water maze) tasks. In: Buccafusco JJ (ed) Methods of behavior analysis in neuroscience, 2nd edn. CRC, Boca Raton, FL, Chapter 13
69. Glickstein SB et al (2002) Mice lacking dopamine D2 and D3 receptors have spatial working memory deficits. J Neurosci 22(13): 5619–5629
70. Papaleo F et al (2008) Genetic dissection of the role of catechol-O-methyltransferase in cognition and stress reactivity in mice. J Neurosci 28(35):8709–8723
71. Birrell JM, Brown VJ (2000) Medial frontal cortex mediates perceptual attentional set shifting in the rat. J Neurosci 20(11): 4320–4324
72. Floresco SB et al (2006) Multiple dopamine receptor subtypes in the medial prefrontal cortex of the rat regulate set-shifting. Neuropsychopharmacology 31(2):297–309
73. Ragozzino ME (2007) The contribution of the medial prefrontal cortex, orbitofrontal cortex, and dorsomedial striatum to behavioral flexibility. Ann N Y Acad Sci 1121:355–375
74. Grant DA, Berg EA (1948) A behavioral analysis of degree of reinforcement and ease of shifting to new responses in a Weigl-type card-sorting problem. J Exp Psychol 38(4): 404–411
75. Glickstein SB et al (2005) Mice lacking dopamine D2 and D3 receptors exhibit differential activation of prefrontal cortical neurons during tasks requiring attention. Cereb Cortex 15(7):1016–1024
76. De Steno DA, Schmauss C (2009) A role for dopamine D2 receptors in reversal learning. Neuroscience 162(1):118–127
77. Smith JW et al (2002) Dopamine D2L receptor knockout mice display deficits in positive and negative reinforcing properties of morphine and in avoidance learning. Neuroscience 113(4):755–765
78. Floresco SB (2013) Prefrontal dopamine and behavioral flexibility: shifting from an "inverted-U" toward a family of functions. Front Neurosci 7:62
79. Doherty JM et al (2008) Contributions of dopamine D1, D2, and D3 receptor subtypes to the disruptive effects of cocaine on prepulse inhibition in mice. Neuropsychopharmacology 33(11):2648–2656
80. Ralph-Williams RJ et al (2002) Differential effects of direct and indirect dopamine agonists on prepulse inhibition: a study in D1 and D2 receptor knock-out mice. J Neurosci 22(21):9604–9611
81. Xing B et al (2010) Dopamine D1 but not D3 receptor is critical for spatial learning and related signaling in the hippocampus. Neuroscience 169(4):1511–1519
82. Hranilovic D et al (2008) Emotional response in dopamine D2L receptor-deficient mice. Behav Brain Res 195(2):246–250
83. Xing B et al (2012) The dopamine D1 but not D3 receptor plays a fundamental role in spatial working memory and BDNF expression in prefrontal cortex of mice. Behav Brain Res 235(1):36–41
84. Helms CM et al (2008) D4 receptor deficiency in mice has limited effects on impulsivity and novelty seeking. Pharmacol Biochem Behav 90(3):387–393
85. Jones SR et al (1999) Loss of autoreceptor functions in mice lacking the dopamine transporter. Nat Neurosci 2(7):649–655
86. Jones SR et al (1998) Profound neuronal plasticity in response to inactivation of the dopamine transporter. Proc Natl Acad Sci U S A 95(7): 4029–4034
87. Spielewoy C et al (2000) Increased rewarding properties of morphine in dopamine-transporter knockout mice. Eur J Neurosci 12(5):1827–1837

88. Spielewoy C et al (2000) Behavioural disturbances associated with hyperdopaminergia in dopamine-transporter knockout mice. Behav Pharmacol 11(3–4):279–290
89. Zhuang X et al (2001) Hyperactivity and impaired response habituation in hyperdopaminergic mice. Proc Natl Acad Sci U S A 98(4):1982–1987
90. Chen R et al (2006) Abolished cocaine reward in mice with a cocaine-insensitive dopamine transporter. Proc Natl Acad Sci U S A 103(24): 9333–9338
91. Ghisi V et al (2009) Reduced D2-mediated signaling activity and trans-synaptic upregulation of D1 and D2 dopamine receptors in mice overexpressing the dopamine transporter. Cell Signal 21(1):87–94
92. Salahpour A et al (2008) Increased amphetamine-induced hyperactivity and reward in mice overexpressing the dopamine transporter. Proc Natl Acad Sci U S A 105(11):4405–4410
93. Huotari M et al (2002) Brain catecholamine metabolism in catechol-O-methyltransferase (COMT)-deficient mice. Eur J Neurosci 15(2):246–256
94. Huotari M et al (2002) Effect of dopamine uptake inhibition on brain catecholamine levels and locomotion in catechol-O-methyltransferase-disrupted mice. J Pharmacol Exp Ther 303(3):1309–1316

Chapter 18

Optogenetic Regulation of Dopamine Receptor-Expressing Neurons

T. Chase Francis and Mary Kay Lobo

Abstract

Optogenetics has provided neuroscientists with the tools to control activity of specific neurons within a circuit. Optogenetic manipulation of dopamine receptor-containing neurons in the striatum holds great potential in understanding and treating a number of neuropsychiatric and neurological disorders. Coupling optogenetics with cell subtype-specific transgenic mouse lines permits dissection of dopamine receptor 1 (D1)- and dopamine receptor 2 (D2)-enriched circuits including the mesolimbic reward circuit and the basal ganglia circuit. This has led to multiple new insights into the function of dopamine receptor-expressing neurons in motivational and motor behaviors. This article discusses techniques to express microbial opsins in dopamine receptor-expressing neurons and to optogenetically activate or silence these neurons within the striatum in awake, behaving animals.

Key words Optogenetics, Dopamine receptors, Channelrhodopsin, Halorhodopsin, Cre-inducible AAVs, BAC transgenic animals, Striatum, Nucleus accumbens, Medium spiny neurons

1 Introduction

1.1 Targeting Opsins to D1 Receptor- and D2 Receptor-Expressing Neurons

Optogenetics has revolutionized the field of neuroscience by permitting in vivo circuit-level control of neuronal firing in awake, behaving animals. The advantage of current optogenetic tools is that they allow for a single-component approach involving a light-activated protein within a single gene [1, 2]. Genes for microbial opsins, including the light-activated cation channel, channelrhodopsin 2 (ChR2), the enhanced 3rd-generation halorhodopsin chloride pump (eNpHR3.0), or the archaerhodopsin hydrogen pump (ArchT), can be packaged into viruses [1, 2]. Opsin viruses can be injected into selective brain regions to infect selective neuronal cell subtypes, and neuronal activity can be controlled in a spatiotemporal specific manner using light [1, 2]. These circuit-selective approaches for in vivo control of neuronal

Mario Tiberi (ed.), *Dopamine Receptor Technologies*, Neuromethods, vol. 96,
DOI 10.1007/978-1-4939-2196-6_18,

activity during behavioral paradigms have resulted in many new insights into behavioral function [1–5].

Optogenetics has been particularly advantageous for understanding the function of the two striatal projection medium spiny neurons (MSNs), those enriched in dopamine receptor 1 (D1) versus dopamine receptor 2 (D2) [5, 6]. These two neurons are heterogeneously intermixed within dorsal striatum and ventral striatum (the latter referred to as nucleus accumbens (NAc)) [7]. The ability to identify and manipulate genetics and function in these two neurons was impossible until recent development of bacterial artificial chromosome (BAC) transgenic animals [8–10]. BAC transgenic mice expressing Cre recombinase, using D1, D2, or adenosine receptor 2A (A2A, a D2-MSN-enriched gene) BACs, have been used in combination with Cre-inducible adeno-associated viruses (AAVs) to selectively express opsin molecules in these D1- and D2-expressing MSNs [11–15]. These double inverted open (DIO) reading frame AAVs contain an opsin gene tagged with a fluorescent protein (FP) in reverse, and this opsin-FP is flanked by two sets of incompatible lox sites (Fig. 1). In the presence of Cre, the opsin-FP is flipped into the correct orientation allowing selective expression of the opsin-FP in D1- or D2-expressing neurons, which allows for precise in vivo spatiotemporal control of these selective neuron populations using light.

1.2 Insights into Function of D1 Receptor- and D2 Receptor-Expressing Neurons

A number of studies have uncovered distinct roles of striatal MSNs in behavioral function. The dorsal striatal D1-MSNs and D2-MSNs were previously hypothesized to play balanced but antagonist roles in motor function through their direct and indirect pathways of the basal ganglia (BG) [16, 17]. Recently this hypothesized model

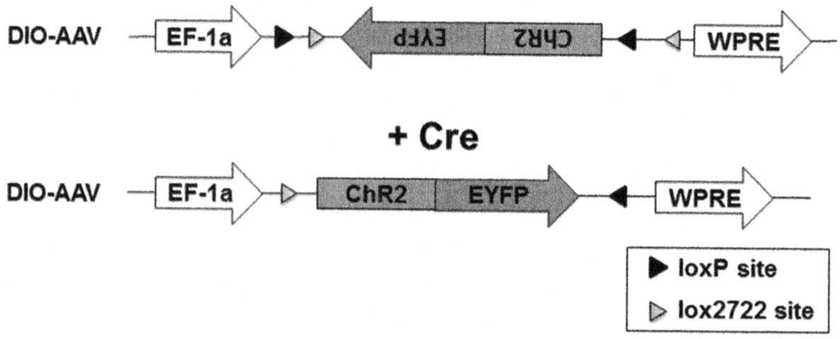

Fig. 1 Schematic representation of an adeno-associated virus with a double inverted open (DIO) reading frame. In this vector, ChR2-EYFP is in reverse and flanked by two incompatible lox sites, loxP (*black arrowhead*) and lox 2722 (*gray arrowhead*). In the presence of Cre recombinase, an event occurs between two compatible lox sites causing ChR2-EYFP to flip in the correct orientation. This is followed by a second event resulting in excision between loxP and lox2722 leaving two incompatible lox sites to prevent further inversions

in motor function was proven accurate by optogenetic activation (using ChR2) of these neuron populations [11]. However, the function of the D1-MSNs and D2-MSNs in NAc, a critical brain region for reward and motivation, was not as clear. Recent studies using optogenetic control (also using ChR2) of NAc MSN subtypes revealed that activating D1-MSNs promotes reward for a psychostimulant, cocaine, while activating D2-MSNs blocks this reward [13]. Other optogenetic studies or other methodologies to alter neuronal activity in the D1- and D2-expressing MSNs are consistent with opposing roles for these MSN subtypes in reinforcement, action value, and drug abuse [12, 14, 15, 18–21].

In this chapter, we establish a simple method for optogenetic activation or inhibition of D1 or D2 receptor-expressing cell subtypes. Our methods provide a procedure for producing implantable fibers, injecting viral constructs into mice expressing Cre in D1- or D2-expressing cells, and activating striatal MSN subtypes during behavior. The approach outlined focuses on published optogenetic paradigms in NAc D1- and D2-expressing neurons [13, 18]. However, modifications can be made effortlessly in these methods, including stimulation paradigms, targeting of different striatal regions or other brain areas expressing D1 or D2 receptors, or optic fiber targeting (cell bodies vs. terminals), permitting a wide range of experimental flexibility to suit the individualized research goals.

2 Materials

2.1 Stereotaxic Equipment

1. Isoflurane Anesthesia System by E-Z Systems (EZ-7000 System, EZ-150C) and ReFresh carbon filters (EZ-258, World Precision Instruments).
2. Stereotaxic frame (Model 902 Dual Set Up, Small Animal), ear cups (921 Zygoma Ear Cup), and mouse nose bar (Model 926-B) produced by Kopf Instruments.
3. Syringe holders (Model 1772) and bracket clamps (Model 1770-C) by Kopf Instruments.
4. Sterile latex surgical gloves (MDS194134, Medline).
5. Puralube vet ointment (14590500, Dechra).
6. Implantable fiber holder (Doric, SCH_1.25).
7. Hamilton syringes (84851) with 33 gauge needles (7762-06).
8. Various surgical instruments: small scissors, scalpel, forceps, etc. (FST).
9. Ideal Microdrill and drill bits set (58609, Stoelting).
10. Andis Ultraedge hair clippers (64430, Andis).
11. Cotton tip applicators 6″ (6553, Curity).

12. Betadine surgical scrub (NDC 67618-150-17, Purdue Products).
13. Bupivacaine analgesic.
14. Vetbond skin adhesive (3 M).
15. Grip cement combo (675570, Dentsply).
16. Reflex 7 skin closure system + clips (RF7 Kit, FST) for multiple surgeries (see step 12 of Sect. 3.2.1).
17. Bone anchor screws (MD-1310, BASInc).

2.2 Optogenetic Equipment

1. Blue Laser + Coupler (BL-473-00100-CWM-SD-05-LED-0, OEM). *See* **Note 1** about other light delivery systems.
2. Green Laser + Coupler (GR-532-00100-CWM-SD-05-LED-0, OEM).
3. Waveform generator (33220A, Agilent). Alternatively, non-expensive alternatives are available such as circuit boards made by Arduino.
4. Light meter (PM100D) and sensor for meter (S130C, Thorlabs).
5. Protective eyewear (NR-ARG-EN207-33, OEM Laser).
6. Polishing plate 9.5″ × 13.5″ (6.0 CTG913), polishing pad 9″ × 13″, 50 durometer (7.0 NRS913), and lapping film for polishing at sizes 0.30 μm (2.0 LFG03P), 1 μm (3.0 LFG1P), 3 μm (4.0 LFG3P), and 5 μm (5.0 LFG5P) produced by Thorlabs.
7. FC and SC polishing disk (8.0 D50-FC, Thorlabs).
8. Fiber stripping tool for 125–135 μm diameter fiber (9.0 T06S13, Thorlabs).
9. Optic fiber for ferrules: 105 μm core, 125 μm cladding, acrylate jacketed (AFS105/125Y, Thorlabs).
10. PFP LC 1.25 mm OD multimode ceramic zirconia ferrules (MM-FER2007C-1260, Precision Fiber Products (PFP)).
11. PFP Ceramic split sleeve, 1.5 mm ID (SM-CS125S, PFP).
12. Epo-Tek 353ND Epoxy-8oz bottle (ET-353ND-8OZ, PFP).
13. Miller CS-30-W carbide-tip fiber-optic scribe (M1-46124, PFP).
14. Chemtronics V-Groove and ferrule cleaning swabs (38542F, PFP).
15. Heat shrink tubing: length (12″), inner diameter (3/32″), thin size wall (26003-12, FTZ industries).
16. Loctite super glue (Loctite).
17. Helping Hand vice grips with magnifier (7521, Michigan Industrial Tools).

3 Methods

The methods below are not an exhaustive list of differing viral surgery methods, fiber implantation strategies, or stimulation parameters, but rather provide an example of how to perform a simple optogenetic experiment in NAc D1- and D2-expressing neurons. Furthermore, for detailed methods on use of optical fibers in conjunction with electrophysiological arrays for in vivo recording combined with optogenetics in D1-MSNs and D2-MSNs, refer to this study by Kravitz et al. [22]. The timeline provided here was created to provide an overview of an optogenetic experiment in NAc D1- and D2-expressing medium spiny neurons (Fig. 2).

3.1 Implantable Fiber Production

Several methods exist for activation or inhibition of opsin-expressing cell subtypes. Originally, fibers were guided through cannula implants into the brain. However, a new protocol enables investigator to construct chronic implantable fibers and fiber-optic patch cords [23]. The protocol below specifically provides a methodology for constructing implantable fibers. For methodology on optical patch cord construction, refer to Ref. [23]. Fibers can range in diameter and have varying numerical aperture (NA) [23]. We recommend using fibers with a 105 μm diameter core for mouse studies with an NA of 0.22. The patch cord used should match the implantable ferrule fiber and NA to prevent light power

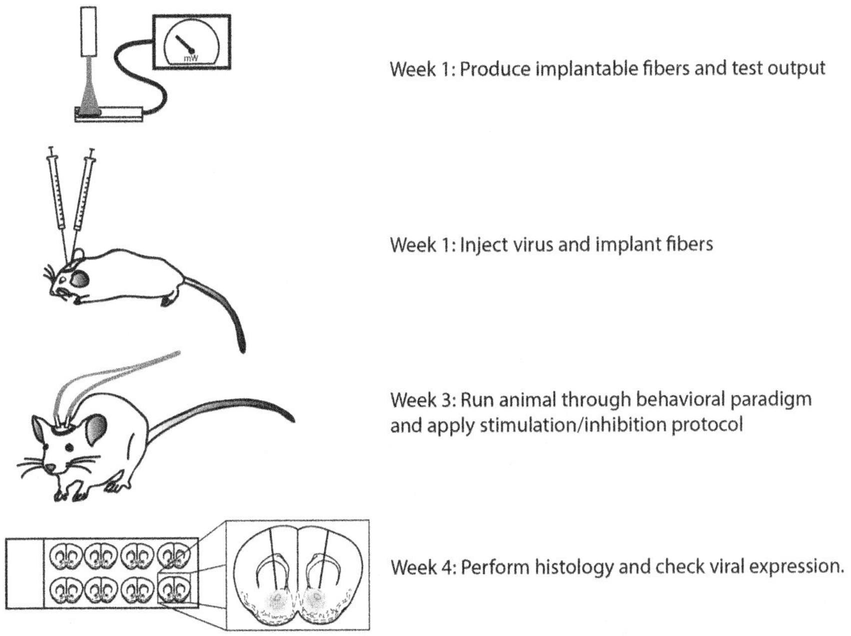

Fig. 2 Timeline of surgeries and optogenetic manipulation

attenuation. It is possible for the implantable fiber to be made with a larger diameter and larger NA than the patch cord used with little to no light loss. Fibers with larger or smaller NAs can be used for different tissue volumes. These fibers can be cut at various lengths to target different striatal depths. As an alternative, implantable fibers or patch cords can be purchased from various vendors (e.g., Thor Labs, Doric Lenses). However, in-house construction of implantable fibers and patch cords is more cost-effective. An abbreviated protocol for production of fibers to target the NAc is provided below:

1. Strip 3 cm of the fiber to expose the 105 μm inner core being careful to remove excess outer core material from the surface of the exposed inner core of the fiber. Refer to Ref. [23] for a diagram of the implantable fiber.
2. Cut the un-stripped portion of the fiber with a diamond knife at least 2 cm from the edge of the stripped region (the exposed inner core).
3. Mix the epoxy solution at a ratio of 1 part hardener to 10 parts resin and load the solution in a 1 mL syringe with a blunted 25 gauge needle.
4. Secure the ceramic ferrule in a vise or vise grip (helper hands) with a weighted base. Dispense the epoxy solution into the flat, non-conical side of the ceramic ferrule until a small amount of epoxy is observed on the opposite, conical side. Ensure that the blunted needle is flat atop the ferrule to avoid epoxy from accumulating outside of the ferrule. Then feed the stripped fiber into the ferrule until the entire stripped portion is within the ferrule allowing excess stripped fiber to exit the conical portion of the ferrule.
5. Heat the epoxy for 30 s with a heat gun at low heat until epoxy is black and hardened. WARNING: The ferrule and the vise are very hot. Wait at least 1 min before removing the ferrule from the holder.
6. Using the diamond knife, score a portion of the stripped fiber at 1 mm below the bottom conical side of the ferrule. Bend the stripped portion of the fiber to the side in order to break off excess fiber. To prevent light loss, it is important to score the fiber at the base of the ferrule before removing the excess fiber. If this is not done, the fiber could break inside of the ferrule.
7. Cut the non-stripped portion of the fiber to the appropriate length by scoring the fiber at 4 mm (for targeting the NAc) with a diamond knife making sure not to cut through the fiber completely. Next, pull on the portion of the non-stripped fiber just beyond the area scored with hemostats until the fiber cleanly breaks off.

8. Polish the conical side of the ceramic ferrule with a set of 4 lapping film/polishing papers. The fiber should be secured in a set of hemostats and kept completely flat against the polishing paper. Make around 20–30 small circular patterns on each grade paper until smooth. Polish the fiber on the coarsest lapping paper to the finest paper in this order: 5 μm (black), 3 μm (pink), 1 μm (green), and 0.3 μm (white).
9. Attach a patch cord to the laser. WARNING: It is important to wear safety glasses before turning on the laser to prevent laser light from entering eyes. To align the patch cord, place the end of the patch cord 1 mm away from the sensor, holding the end in place with the helper hands. Twist the alignment knobs to alter the position of the coupler, ensuring the coupler remains tightly in place (not loose). Alter the position of the coupler until the maximum light reading is observed. When the maximal output is obtained, record the power value measured.
10. Check the output of the fiber by attaching a patch cord affixed to a laser to the ferrule. First, attach a ceramic sleeve over the ferrule at the end of the patch cord, covering about half of the ferrule. Next, using hemostats, feed the conical portion of the ferrule into the sleeve until flush with the end of the patch cord ferrule. The overall light output should emit from the fiber as a cone and a smooth circle should be observed when shining the tip of the fiber on a flat surface. Next, check the output of the fiber by holding the tip of the ferrule perpendicular to the light meter at a distance of 1 mm. The transmittance from the tip of the implantable fiber should not be less than 20 % of the light output from the end of the patch cord to ensure adequate light penetration. A record of fiber transmittance can be kept in order to calculate the amount of light output from each fiber. These records can be used to adjust laser intensity in order to maintain a consistent output of light across animals when using different patch cords.

3.2 Surgical Methods

The methods described below are intended for bilateral viral injection followed by cannula implantation. A number of different opsins can be injected for stimulation or inhibition of cell subtypes. Such variants include the traditional ChR2, eNpHR3.0, ArchT, or the fast-spiking channelrhodopsins, long-lasting depolarization opsins, and G-protein-coupled receptor opsins (i.e., OptoXRs). For a more complete review of available opsins, see the review by Yizhar et al. [2] and the methods study by Mattis et al. [24]. For optogenetic methods using a cannula guide in place of implantable fibers, refer to the methods study by Zhang et al. [25]. The methods outlined below are intended for the original ChR2 or eNpHR3.0 opsins. Validation of virus expression and function should be performed before performing in vivo optogenetic behavioral tasks (*see* **Note 2**).

3.2.1 Viral Injection

1. Anesthetize D1-Cre or D2-Cre mice based on an approved Institutional Animal Care and Use Committee (IACUC) protocol using the E-Z Isoflurane Anesthesia System. Ensure that the animal is under isoflurane anesthesia by observing deepened breaths, a depressed rate of breathing (one deep breath every 1–2 s), and a lack of response to a tail/foot pinch. Anesthesia should be induced at 3 % isoflurane and maintained at 1.5 % isoflurane. *See* **Note 3** for more information on alternative anesthetics. When performing surgery from this point, it is important to maintain a sterile surgical environment throughout the surgery (e.g., using sterile latex gloves, needles, etc.) as to prevent unwanted infections in mice undergoing surgery.
2. Shave the animal's scalp using 0.25 mm blade metal hair clippers until the scalp is bald and exposes a 1 cm margin from the incision region. Then place the animal's nose and teeth into the nose piece surrounded by the isoflurane nose cone aspirator.
3. After applying Puralube ointment to the eyes of the mouse to prevent desiccation, place Zygoma ear cups onto the sides of the mouse's head (near the ears) and then make sure the head is level on all planes and secured within the cups attached to the stereotaxic frame. See step 9 for more on leveling the head.
4. Wipe the scalp with a sterile cotton tipped applicator dipped in 10 % Betadine solution. Next, scrub the scalp with an applicator soaked in 70 % ethanol to sterilize the scalp. Repeat these steps a second time to ensure the scalp is sterilized.
5. Inject ~0.1 mL of 0.25 % bupivacaine topical analgesic under the scalp to numb the incision area. Wait at least 2 min before proceeding.
6. Make a 2 cm incision down the midline of the scalp using a scalpel or small scissors and push the skin away from the midline with a sterile cotton tipped applicator to expose the skull.
7. Apply a hydrogen peroxide solution (3–10 %) to the skull to visualize the bregma and to dry the skull. It is important that the skull is dry for later fiber implantation steps and curing of the dental cement.
8. Load syringes attached to the stereotaxic frame with bracket clamps with 1 μL water and make a 1 μL air bubble. Load 0.5–1.0 μL (this can depend on titer of virus and the spread desired) of the DIO-AAV. Note: Do not inject more than 1.0 μL into the brain to prevent lesioning.
9. Align the needle tips to the bregma and record the coordinates. To determine if the top of the skull is level, first lift the needle tips off the bregma and move one of the needles 1.5–2 mm in the lateral (Lat) direction, drop the needle down to skull level, and record the dorsal/ventral (D/V) coordinates.

The difference in the new D/V coordinates and the D/V coordinates at the bregma should be ~0.2 mm or less. Follow the same steps for the opposite lateral direction and the anterior/posterior (A/P) direction. If the difference is greater than 0.2 mm, adjust the head along all planes until the difference is reduced. When the head is adjusted, return the needles to the original coordinates at the bregma and record new coordinates. For NAc localization, use the following coordinates: A/P, +1.6; Lat, +1.5; and D/V, −4.4, at an angle of 10°. The angle is meant to allow for simultaneous bilateral injection of virus. Note that viral injections do not need to be limited to the NAc or striatal regions (*see* **Notes 4** and **5**). Raise the needles from the skull and move the needles to the adjusted coordinates.

10. Drill burr holes with the Microdrill just below the needle tip. Then slowly lower the needles into the burr holes taking around 30–60 s to lower to calculated D/V coordinate. Slowly lower one needle at a time into the brain until the needle is at the final A/P coordinate. Lowering slowly ensures a reduction in possible tissue lesioning.
11. Begin injecting virus slowly at a rate of 1 μL per min (or slower) until the entire volume of the virus is dispensed. Allow for the needles to remain in the skull for 10 min before removal. Note: If it is necessary to inject greater than 0.5 μL of virus solution, raise the needles 0.5 mm up the A/P axis after injecting 0.5 μL and dispense the rest of the virus solution at 1 μL per min.
12. Following the 10 min post-infusion period, slowly raise the needles out of the skull. Proceed to step 1 of Sect. 3.2.2 if implanting fibers immediately. Some behaviors may not be conducive to fiber implantation immediately due to the risk of head cap loss. If this is the case, suture the scalp or use staples using the Reflex 7 skin closure system in order to close up the skull. It is not recommended to use VetBond if the fiber will be implanted at a later date. VetBond can stick to the skull cap making it difficult to find a flat and clean surface for the fiber implantation. However, if only virus is being injected for opsin validation, VetBond is an appropriate suture method.

3.2.2 Fiber Implantation

1. Proceeding from the viral injection steps, remove syringes from the stereotaxic setup and load ferrules into implantable cannula holder. Add the cannula holders to the setup maintaining an angle of 10°.
2. Drill a third burr hole in the skull lateral from the midline and posterior to the bregma for a skull screw. Screw the skull screw into the new burr hole making a maximum of 2.5 turns. Apply Loctite glue around the screw. Ensure that the skull is dry so that the glue can easily adhere to the skull.

3. Lightly score the skull with the Microdrill or a scalpel. This scoring provides a better grip for the dental cement.
4. Attach the fiber to the fiber holder and fix the holder to the stereotaxic setup. Lower one of the implantable fibers over the previously drilled burr hole, so the tip of the fiber is at skull level and record the coordinates. Subtract –4.0 (or the length the fiber was cut) to obtain the new D/V coordinates for the fiber. Add Loctite glue to the bottom of the ceramic portion of the implantable fiber (avoid getting glue on the fiber) and slowly lower the fiber into the brain taking 30–60 s to lower the fiber to the calculated D/V coordinates. If the ceramic portion of the fiber remains above the skull, add more glue to help secure the fiber to the skull.
5. After waiting 2–3 min for the glue to dry, loosen the cannula holder and raise it from the fiber. Be careful not to move the fiber while removing the holder. Repeat step 3 for the other side if implanting fibers bilaterally.
6. Prepare the grip dental cement based on given instructions. Most dental cement will be at the correct consistency if gel-like. Aspirate dental cement into a 1 mL disposable syringe and add a large gauge needle (22 gauge or larger) to the syringe. Apply cement evenly around the skull making sure to surround implantable fibers on all sides and fully cover the skull screw. Note: It is important to leave at least 0.5 mm of the ceramic portion of the implantable fiber exposed above the dental cement so the patch cord can be fully attached to the fiber.
7. House animals separately or with a divider in cage to prevent conspecific removal of the newly made head cap. Mouse cages should be placed on heating pads to allow for body temperature regulation following surgery.

3.3 Light Stimulation/Inhibition

A variety of stimulation or inhibition paradigms may be used based on cell firing properties and desired outcomes. This protocol bases its stimulation parameters off functionally relevant frequencies at a predetermined power to modulate cocaine-mediated reward and locomotor outcomes in D1- and D2-MSNs in the NAc [13, 18]. Manipulation of cell subtypes with light should occur 2 weeks after an AAV injection. This time point is when the AAV is expressed in cell bodies. For terminal stimulation a 3–4-week period is necessary to allow AAV expression in terminals [25]. If using a herpes simplex virus (HSV), which we do not discuss in the protocol, stimulation can occur at day 2–4, post viral surgery, when the HSV is optimally expressed [13].

1. Attach sleeves to the end of the patch cord and add a cover (e.g., heat shrink wrap, tape, etc.) that will block the light from the animal's eyes. It is important to prevent the light from

reaching the animals eyes to reduce potential photoelectric artifacts (*see* **Note 6**).

2. Scruff the animal firmly and place an index finger on the back of the head cap just above where the skull screw was attached. This grip will prevent the animal from dramatic movements that lead to head cap loss.
3. Gently feed the sleeve on the patch cord onto the implantable fiber with a subtle twisting motion (no more than a 10° turn) until the patch cord cannot be fed further. The ceramic ferrule from the patch cord should be flush with the ceramic ferrule of the implantable fiber.
4. Stimulation and inhibition parameters were based on [13] and [18]. Parameters should be adjusted to produce the desired outcome on cell firing (*see* **Note 7**). Furthermore, the power of the laser should be adjusted accordingly (*see* **Note 8**).
   (a) *For Stimulation:* Based on maximal firing rates of medium spiny neurons, 10 Hz 473 nm blue light stimulation for 3 min on and 5 min off for a total of 30 min at 2–4 mW power was performed [13]. This protocol was used in a conditioned place preference paradigm during the 30 min conditioning period. This paper also used the same frequency at 3 min on and 2 min off over 15 min to examine downstream molecular adaptations after optogenetic activation of MSN subtypes. More recently a stimulation protocol, also 10 Hz with 473 nm blue light stimulation, for 30 s on and 30 s off for 60 min at 2–4 mW power was used during a cocaine-induced locomotor activity paradigm [18]. The stimulation protocol may be repeated daily as necessary. We have used these stimulations paradigms repeatedly for 2–5 days. Care should be taken to use similar light output across days (*see* **Note 9**).
   (b) *For Inhibition:* Reference [18] used the halorhodopsin chloride pump variant eNpHR3.0 to inhibit D1-expressing neurons during cocaine-induced locomotor activity combined with downstream molecular profiling. To start, use continuous inhibition with 532 nm green light illumination for 60 min at 2–4 mW power from the tip of the fiber [18]. The inhibition protocol may be repeated as necessary. We have used this inhibition paradigm repeatedly for 5 days. *See* **Notes 10** and **11** about the caveat of eNpHR3.0 rebound firing and optional optogenetic and non-optogenetic tools for silencing neurons.
5. To remove the patch cord, scruff the mouse by once again placing a finger on the back of the head cap. Using a gentle twisting motion, pull up in a direction completely parallel with the fiber until the patch cord ferrule is free. Note: The sleeve may come

off the end of the patch cord. This result is common and the sleeve should be left on the fiber implanted in the skull of the animal. It is important to examine for bedding or other material lodged in the sleeve before each stimulation session. If this is the case, obstructions can be removed with small tweezers. Only remove the sleeve as a last resort as this action may potentially dislodge the implantable fiber or head cap.

6. After experiments are complete, mice should be perfused and viral expression should be checked. Note the location and spread of the virus expression and examine for the tract marks of the implantable fibers.

4 Notes

1. There are a variety of light sources that may be used for optogenetics and a number of companies that supply these lasers. For a review on types of light sources used in optogenetics, *see* [2].
2. All opsins should be validated prior to in vivo optogenetic manipulation. Immunohistochemical analysis will ensure the viral transduction is specifically localized to the cells of interest and the spread of the virus is located specifically in the region of interest. The volume of injected virus can be adjusted to control the extent of virus spread. Electrophysiological validation will provide the researcher with confidence that the light-driven channel and not an artifact produced the resultant behavioral outcome.
3. Alternative anesthetics (e.g., ketamine/xylazine) may be used in place of isoflurane anesthesia.
4. Optogenetic regulation of dopamine receptor neurons is not limited to striatal regions. Dopamine receptors are found in prefrontal cortex, midbrain, and other regions. A recent study used optogenetics to inhibit the prefrontal cortex during fixed-interval timing performance [26].
5. Opsins can also be transduced in afferent brain regions, and implantable fibers can be placed in projection regions to stimulate or inhibit dopamine receptor neuron expressing inputs.
6. Light scattered away from the patch cord during optogenetic manipulation may affect neural activity by reaching light-sensing organs (e.g., the retina) and therefore may affect behavior [22].
7. Stimulation protocols should be optimized to produce the behavior of interest in the specific cell subtype of interest. It is recommended that the researcher begins with a stimulation protocol that mimics endogenous firing properties of the cell subtype of interest.

8. The strength of the laser should be adjusted to an adequate power. Note that some opsins require a higher light power [24]. Irradiance values (mW/mm^2) can be predicted by using an online calculator [27] derived from Aravanis et al. [28]. The irradiance of the light should not exceed 50 mW/mm^2 at 0.5 mm from the tip of the fiber to prevent tissue damage.
9. If performing repeated stimulation, patch cords should be cleaned between animals with a V-Groove and ferrule cleaning swab and repolished before each stimulation or inhibition session.
10. It should be noted that inhibition through halorhodopsin pumps has the potential to cause a spike rebound following inhibition. Other opsins such as the hydrogen pump ArchT diminishes this post-inhibitory rebound.
11. Alternative inhibition systems exist that are often less temporally precise including inhibitory designer receptors exogenously activated by designer drugs (DREADDs) developed by Bryan Roth [29].

References

1. Bernstein JG, Boyden ES (2011) Optogenetic tools for analyzing the neural circuits of behavior. Trends Cogn Sci 15(12):592–600
2. Yizhar O, Fenno FE, Davidson TJ et al (2011) Optogenetics in neural systems. Neuron 71(1):9–34
3. Yizhar O (2012) Optogenetic insights into social behavior function. Biol Psych 71: 1075–1080
4. Tye KM, Diesseroth K (2012) Optogenetic investigation of neural circuits underlying brain disease in animal models. Nat Rev Neurosci 13(4):251–266
5. Lenz JD, Lobo MK (2013) Optogenetic insights into striatal function and behavior. Behav Brain Res 255:44–54
6. Kravitz AV, Kreitzer AC (2012) Striatal mechanisms underlying movement, reinforcement, and punishment. Physiol (Bethesda) 27(3): 167–177
7. Gerfen CR (1992) The neostriatal mosaic: multiple levels of compartmental organization. Trends Neurosci 15(4):133–139
8. Yang XW, Model P, Heintz N (1997) Homologous recombination based modification in Escherichia coli and germline transmission in transgenic mice of a bacterial artificial chromosome. Nat Biotechnol 15(9):859–865
9. Gong S, Zheng C, Doughty ML et al (2003) A gene expression atlas of the central nervous system based on bacterial artificial chromosomes. Nature 425:917–925
10. Gong S, Doughty M, Harbaugh CR et al (2007) Targeting cre recombinase to specific neuron populations with bacterial artificial chromosome constructs. J Neurosci 27(37): 9817–9823
11. Kravitz AV, Freeze BS, Parker PRL et al (2010) Regulation of parkinsonian motor behaviours by optogenetic control of basal ganglia circuitry. Nature 466:622–626
12. Kravitz AV, Tye LD, Kreitzer AC (2012) Distinct roles for direct and indirect pathway striatal neurons in reinforcement. Nat Neurosci 15(6):816–818
13. Lobo MK, Covington HE III, Chaudhury D et al (2010) Cell type-specific loss of BDNF signaling mimics optogenetic control of cocaine reward. Science 330(6002):385–390
14. Tai L-H, Lee AM, Benavidez N et al (2012) Transient stimulation of distinct subpopulations of striatal neurons mimics changes in action value. Nat Neurosci 15:1281–1289
15. Bock R, Shin JH, Kaplan AR (2013) Strengthening the accumbal indirect pathway promotes resilience to compulsive cocaine use. Nat Neurosci 16:632–638
16. Albin RL, Young AB, Penny JB (1989) The functional anatomy of basal ganglia disorders. Trends Neurosci 12(10):366–375

17. Graybiel A (2000) The basal ganglia. Curr Biol 10(14):R509–R511
18. Chandra R, Lenz JD, Gancarz AM et al (2013) Optogenetic inhibition of D1R containing nucleus accumbens neurons alters cocaine-mediated regulation of Tiam1. Front Mol Neurosci 6(13):1–8. doi:10.3389/fnmol.2013.00013
19. Ferguson SM, Eskenazi D, Ishikawa M et al (2011) Transient neuronal inhibition reveals opposing roles of indirect and direct pathways in sensitization. Nat Neurosci 14(1): 22–24
20. Ferguson SM, Phillips PEM, Roth BL et al (2013) Direct-pathway striatal neurons regulate the retention of decision-making strategies. J Neurosci 33(28):11668–11676
21. Hikida T, Kimura K, Wada N et al (2010) Distinct roles of synaptic transmission in direct and indirect striatal pathways to reward and aversive behavior. Neuron 66(6): 896–907
22. Kravitz A, Owen SF, Kreitzer AC (2013) Optogenetic identification of striatal projection neuron subtypes during in vivo recordings. Brain Res 1511:21–32
23. Sparta D, Stamatakis AM, Phillips JL et al (2012) Construction of implantable optical fibers for long-term optogenetic manipulation of neural circuits. Nat Protoc 7:12–23
24. Mattis J, Tye KM, Ferenczi EA et al (2012) Principles for applying optogenetic tools derived from direct comparative analysis of microbial opsins. Nat Methods 9:159–172
25. Zhang F, Gradinaru V, Adamantidis AR et al (2010) Optogenetic interrogation of neural circuits: technology for probing mammalian brain structures. Nat Protoc 5:439–456
26. Narayanan NS, Land BB, Solder JE et al (2012) Prefrontal D1 dopamine signaling is required for temporal control. Proc Natl Acad Sci U S A 109(50):20726–20731
27. Deisseroth K (2013) Predicted irradiance values: model based on direct measurements in mammalian brain tissue. http://www.stanford.edu/group/dlab/cgi-bin/graph/chart.php
28. Aravanis AM, Wang LP, Zhang F et al (2007) An optical neural interface: in vivo control of rodent motor cortex with integrated fiberoptic and optogenetic technology. J Neural Eng 4:S143–S156
29. Arumbruster BN, Li X, Pausch MH et al (2007) Evolving the lock to fit the key to create a family of G protein-coupled receptors potently activated by an inert ligand. Proc Natl Acad Sci U S A 104(12):5163–5168

Chapter 19

Characterization of D3 Dopamine Receptor Agonist-Dependent Tolerance Property

Samantha R. Cote and Eldo V. Kuzhikandathil

Abstract

Among dopamine receptors, the signaling function of D3 dopamine receptor subtype has been particularly difficult to study in vivo. In vivo signaling studies are hampered by the limited expression profile of the D3 receptor and lack of selective ligands. We and others have extensively characterized D3 receptor signaling in heterologous expression systems. These studies have provided insight into properties of the D3 receptor that might explain the difficulty of studying receptor function in vivo. The D3, but not the closely related D2 receptor, exhibits a tolerance property wherein repeated agonist stimulation results in a progressive decrease in signaling response. The D3 receptor also exhibits a slow response termination (SRT) property wherein the termination of D3 receptor-induced signaling response terminates very slowly following the removal of agonist. The role of these D3 dopamine receptor properties in physiology and pathophysiology is not known; however, with an understanding of the mechanisms underlying these properties, new approaches are being developed to clarify the in vivo role of D3 receptor tolerance and SRT properties. We recently showed that the D3 receptor tolerance and SRT properties are agonist dependent. In this chapter we describe tools and methods to study D3 receptor signaling function and properties in vivo. We describe the use of the *drd3*-EGFP reporter mouse model and novel non-tolerance causing D3 receptor agonist to study D3 receptor signaling function in vivo and the role of its signaling properties on locomotor behavior.

Key words Dopamine receptor, Tolerance, Desensitization, Transgenic mice, Mitogen-activated protein kinase, Agonists, Locomotor activity

1 Introduction

D3 dopamine receptor subtype belongs to the family of D2-like dopamine receptors which include D2, D3, and D4 dopamine receptor subtypes. Studies in heterologous expression systems have shown that, by and large, the D2-like receptors couple to same signaling effectors which include adenylate cyclase, ion channels, and mitogen-activated protein kinase (MAPK) [1–3]. In addition, these receptors bind dopamine, albeit with slightly

Mario Tiberi (ed.), *Dopamine Receptor Technologies*, Neuromethods, vol. 96,
DOI 10.1007/978-1-4939-2196-6_19, © Springer Science+Business Media New York 2015

different affinities, and have overlapping expression patterns [1–3]. This raises a question as to why the three D2-like dopamine receptor subtype genes have been retained during evolution. Our studies suggest that the answer might lie in not what but rather how the individual D2-like dopamine receptor subtypes couple to signaling pathways [4–6]. We have reported that despite coupling to similar effectors, the D2 and D3 dopamine receptor subtypes have different signaling properties [4]. Specifically, the D3 dopamine receptor subtype but not the D2 receptor exhibits a tolerance property wherein repeated agonist stimulation results in a progressive loss of signaling response. In addition, the D3 receptor exhibits a SRT property wherein the termination of the signaling response after the removal of agonist is significantly slower than at D2 receptors. We have shown that the tolerance and SRT properties of the D3 receptor are mediated by distinct structural domains and receptor conformations [5, 6]. More recently we have reported that the two D3 receptor properties are agonist dependent [7, 8]. While dopamine and several classical D3 receptor agonists induce tolerance and SRT, a new class of atypical D3 receptor agonists which include *cis*-8-OH-PBZI, FAUC 73, and ES609 do not [7]. The new agonists are pharmacological tools that can be used to study the role of D3 receptor tolerance and SRT properties in vivo. In this chapter we describe the use of *cis*-8-OH-PBZI to study tolerance and SRT in vivo.

The overlapping expression pattern of dopamine receptor subtypes in the brain, and the limited expression of D3 dopamine receptor, makes it difficult to study the receptor signaling function in vivo; therefore, it is first necessary to identify an in vivo model system to study the signaling function and properties of the D3 receptor. In order to assess D3 receptor signaling function, we have recently described the use of the *drd3*-EGFP reporter mice [9]. This novel transgenic mouse model was developed by the Gene Expression Nervous System Atlas (GENSAT) project. These transgenic mice express the enhanced green fluorescent protein (EGFP) in cells natively expressing wild-type D3 receptor (Fig. 1). It is important to note that the *drd3*-EGFP mice do not express a D3 receptor-EGFP fusion protein. The *drd3*-EGFP mice have a transgene in which the EGFP reporter gene is under the control of the D3 receptor gene promoter. This allows for expression of EGFP fluorescent protein in cells that natively express the D3 receptor mRNA and a means of determining activation of signaling proteins (MAPK) and transcription factors (c-Fos) following D3 receptor stimulation [10]. In this chapter we describe the use of this mouse model and immunohistochemical methods to assess D3 receptor signaling function and properties.

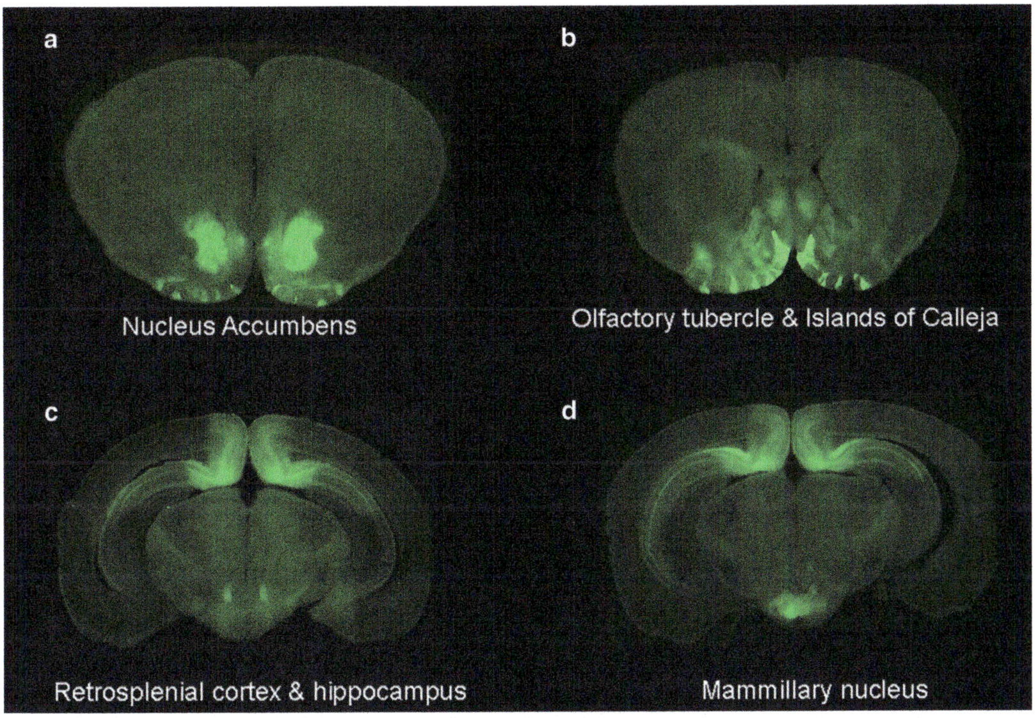

Fig. 1 Representative images of coronal sections of the brain obtained from an adult *drd3*-EGFP mouse. The fluorescence intensity indicates the density of fluorescent cells and EGFP-filled fibers in the nucleus accumbens (**a**); olfactory tubercle and islands of Calleja (**b**); retrosplenial granular cortex, dorsal subiculum, and hippocampus (**c**, **d**); and mammillary bodies (**d**). The freshly cut live 150 μm sagittal section was placed on glass coverslips and visualized using an inverted fluorescence microscope and appropriate EGFP filters

2 Materials

2.1 Animals

The *drd3*-EGFP mice were originally obtained from the GENSAT program at Rockefeller University, and a local breeding colony has been established at Rutgers-New Jersey Medical School. All experimental procedures and drug treatments were performed in accordance with and following approval by the Institutional Animal Care and Use Committee.

2.2 Drugs and Solutions

1. (+)-(4aR,10bR)-3,4,4a,10b-Tetrahydro-4-propyl-2H,5H-[1] benzopyrano[4,3-b]-1,4-oxazin-9-ol hydrochloride (PD128907) (catalog# 1243, Tocris, Minneapolis, MN, USA).
2. *cis*-8-Hydroxy-3-(n-propyl)-1,2,3a,4,5,9b-hexahydro-1H-benz[e]indole hydrobromide (PBZI) (catalog# P0618, Sigma-Aldrich, St. Louis, MO, USA).

2.3 Immunohistochemistry

1. Monoclonal anti-phospho-p44/p42 MAPK antibody produced in rabbit (catalog# 4370, Cell Signaling Technology, Danvers, MA, USA).

2. Goat anti-rabbit Alexa Fluor® 594 secondary antibody (catalog# A11037, Invitrogen).
3. 10× phosphate-buffered saline (catalog# 70011-044, Invitrogen).
4. Sodium fluoride (catalog# S7920, Sigma).
5. 20 % paraformaldehyde (catalog# 15713, Electron Microscopy Sciences, Hatfield, PA, USA).
6. Sucrose (catalog# S5016, Sigma).
7. Tissue-Tek® OCT™ compound (catalog# 62550-01, Electron Microscopy Sciences).
8. Superior™ Blocking Buffer-Dry Blend in TBS (catalog# 786-657, G-Biosciences, St. Louis, MO, USA).
9. Triton X-100 (catalog# T8787, Sigma).
10. Tween®-20 (catalog# P9416, Sigma).
11. 12-well Costar tissue culture plates (catalog# 07-200-81, Fisher Scientific, Pittsburgh, PA, USA).
12. 15 mm Netwell™ Insert with 74 μm mesh size polyester membrane (catalog #3477, Corning Costar, Tewksbury, MA, USA).
13. Shandon Cryotome™ FE Cryostat (catalog# A78900002, Thermo Fisher Scientific Inc., Waltham, MA, USA).
14. MX35 premier + microtome blade (catalog# 3051835, Thermo Fisher Scientific).
15. Superfrost Plus; white microscope slides (catalog# 12-550-15, Fisher Scientific, Pittsburgh, PA, USA).
16. Cover glasses, circles, 15 mm, thickness 0.09–0.12 mm (catalog# 633011, Carolina Biological, Burlington, NC, USA).
17. ProLong® Gold Antifade Reagent with DAPI (catalog# P36935, Invitrogen, Grand Island, NY, USA).

2.4 Locomotor Activity

PAS-Open Field system and data analysis software (San Diego Instruments, San Diego, CA, USA).

3 Methods

3.1 D3 Dopamine Receptor Tolerance Property in drd3-EGFP Mice Using an Immunohistochemical Approach

The experimental design is shown in Fig. 2.

1. Adult, 2-month-old male *drd3*-EGFP mice are group housed (4 or less mice/cage) with littermates in a temperature- and humidity-controlled environment. Mice are maintained on a 12 h light (on at 7 a.m.)/dark (off at 7 p.m.) cycle. Mice are given standard rodent chow and water ad libitum.
2. Split 2–3-month-old male *drd3*-EGFP mice into five groups: saline/saline; saline/0.05 mg/kg PD128907; 0.5 mg/kg PD128907/0.05 mg/kg PD128907; saline/10 mg/kg PBZI; 20 mg/kg PBZI/10 mg/kg PBZI. (*See* **Note 1**.)

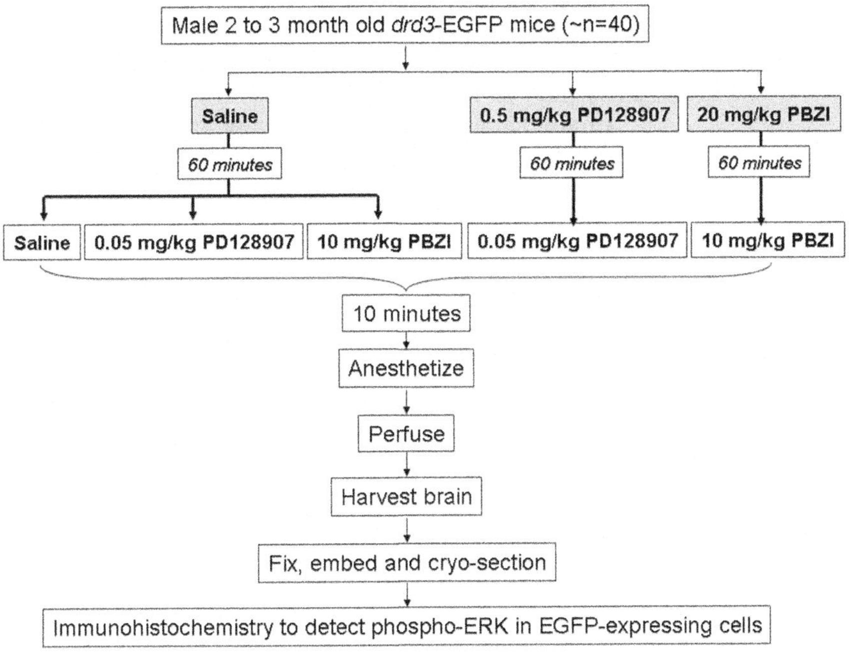

Fig. 2 Experimental scheme used for determining the tolerance property in the D3 receptor-MAPK signaling pathway using the *drd3*-EGFP mice. The mice were administered two sequential injections that were 60 min apart—injection #1 (*gray rectangles with bold lettering*) and injection #2 (*open rectangles with bold lettering*)

3. Weigh the animals and use the weight to calculate the amount of drug(s) needed for each dose. The drugs are diluted in saline to a concentration such that the injection volume for the above doses is ~0.1 mL. (*See* **Note 2**.)
4. Prepare enough 1×PBS and 4 % paraformaldehyde (PFA) for subsequent trans-cardiac perfusion and brain harvest. 30 mL 1×PBS and 30 mL of 4 % PFA are needed per animal. An additional 50 mL of 4 % PFA is needed for overnight fixation of brain tissue. The PFA should be made in 1×PBS. Once these solutions are prepared, they should be kept on ice. (*See* **Note 3**.)
5. The general protocol for this experiment is to treat *drd3-EGFP* mice one at a time with the first injection of agonist or vehicle and 60 min later to treat the animals again with the same agonist or vehicle. 10 min after the second injection, the tissue harvest should begin. This is due to the fact that we see maximal D3 receptor-mediated MAPK activation 10 min after agonist administration. (*See* **Note 4**.)
6. Inject the animal subcutaneously with saline, 0.5 mg/kg PD128907, or 20 mg/kg PBZI and place back inside home cage for 60 min. Saline, PD128907, and PBZI are injected subcutaneously. Inject the animals under the skin at the back of the neck with a 28 g 1/2 in. 100 cc insulin syringe. (*See* **Note 5**.)

7. After 1 h, inject the animal with saline, 0.05 mg/kg PD128907, or 10 mg/kg PBZI and place back in the home cage for about 10 min.
8. After 10 min, anesthetize the animal. We recommend using sodium pentobarbital at a concentration of 50 mg/kg intraperitoneally using insulin syringes. The animal should be under deep anesthesia within a few minutes.
9. Once animal is anesthetized, work quickly to expose the heart. Use forceps to lift up the skin at the abdomen and cut through the skin and muscle with scissors. Using the same scissors, cut up toward the chest and then cut out toward the sides with scissors pointing away from the organs (away from the midline). Cut the ribs and hold the xiphoid process with forceps and cut any remaining connective tissue to expose the heart.
10. In order for the solutions to perfuse through the whole body, insert the 20 g 1 in. needle of the catheter from the gravity syringe(s) into the left ventricle.
11. As soon as the needle is inserted into the left ventricle, nick the right atrium with scissors.
12. IMMEDIATELY turn on the 1×PBS line and allow perfusion. After perfusing 30 mL of PBS, switch to the 4 % PFA line and perfuse another 30 mL. (*See* **Note 6**.)
13. After the perfusion is complete, all the blood should be cleared and the tissue should be fixed. The brain is dissected out of the skull.
14. Store the brain overnight in the 4 % PFA at 4 °C. We recommend using a 50 mL polypropylene conical tube filled to ~50 mL with 4 % PFA.
15. The next day, prepare a 30 % sucrose solution (per 50 mL, 15 g of sucrose in 50 mL of 1×PBS). You will need 50 mL per fixed brain.
16. Store at 4 °C in the dark for 3–4 days or until the brain sinks to the bottom of the container. When the brain sinks, it means it has become denser than the solution because it has absorbed some of the sucrose.
17. Once the brain has sunk, it is embedded using Tissue-Tek OCT compound and frozen at −20 °C for same day cryo-sectioning.
18. Prepare the solutions needed for immunohistochemistry: 1×PBS; permeabilization solution (0.3 % TritonX100, 0.05 % TWEEN20, and 10 mM NaF diluted in 1×PBS); blocking buffer—we recommend using "Superior Blocking Buffer" and this is prepared according to manufacturers' directions.
19. Cryo-section the embedded brain coronally. Cut 20 μm thick sections and collect midbrain sections that contain the striatum

in 1×PBS. We recommend collecting the sections in 12-well dishes with Netwell mesh cups filled with ice-cold 1×PBS. You can collect multiple sections/well. The Netwell mesh cups allow for easy transfer of tissue to new dishes for washing and staining.

20. After sectioning, wash the sections three times with 3 mL of cold 1×PBS for 5 min using gentle agitation.
21. Transfer the cups and tissue to new dishes and add 3 mL of permeabilization solution/well. Incubate for 30 min with gentle agitation.
22. After the 30 min incubation, transfer the cups and tissue to new dishes and add 3 mL of the blocking solution/well. Incubate for 30 min with gentle agitation.
23. After the 30 min incubation, transfer the cups and tissue to new dishes and add 1 mL of anti-phospho-p44/p42 MAPK antibody diluted 1:500 in blocking buffer. Incubate the sections overnight (~15 h) with gentle agitation at 4 °C.
24. The next day, transfer the cups containing the tissue to new dishes containing 3 mL of ice-cold permeabilization solution. Wash the tissue for 5 min with gentle agitation at room temperature. Next, wash for 10 min with ice-cold 1×PBS with gentle agitation. We recommend washing with 1×PBS with 0.05 % Tween20 first to decrease nonspecific antibody binding and background.
25. Transfer the cups to a new dish and add 1 mL of goat anti-rabbit Alexa Fluor® 594 at a 1:1,000 concentration diluted in blocking buffer. Incubate for 1 h at room temp in the dark.
26. Transfer the cups containing the tissue to new dishes containing 3 mL of ice-cold permeabilization solution. Wash the tissue for 5 min with gentle agitation at room temperature. Next, wash for 10 min with ice-cold 1×PBS with gentle agitation.
27. Before mounting a section for fluorescent imaging, wash with sterile dH_2O. This is recommended to reduce any interaction between salt from the PBS and the mounting media.
28. Mount slices onto glass slides and add a drop of the Prolong® antifade mounting media with DAPI before placing the coverslip over the section.
29. D3 receptor-expressing cells will endogenously express EGFP. Those cells that also have phospho-ERK will exhibit red fluorescence due to the antibody staining. Take images of the striatum using filters to observe EGFP, Alexa Fluor® 594, and DAPI nuclear stain. We recommend taking multiple images from serial sections for each animal.
30. After imaging, determine the number of D3 receptor-expressing cells by counting the number of green cells/image.

Of those cells, count the number that also exhibit red fluorescence (i.e., those that express phospho-ERK). Determine the percentage of cells with green and red fluorescence. The percent co-localization can be compared between saline, the tolerance, and non-tolerance causing agonists. Double injections of tolerance causing agonist PD128907 will cause a decrease in the percent co-localization between EGFP and phospho-ERK compared to a single injection. No difference in percent co-localization is expected between a single and double injection of non-tolerance causing agonist PBZI.

31. The percentage co-localization between EGFP (D3 receptor) and phospho-ERK represents D3 receptor-mediated increase in phosphorylation of ERK. This data can be graphed and analyzed using the recommended Sigma Plot v11 software.

3.2 Examining the D3 Dopamine Receptor Tolerance Property in drd3-EGFP Mice Using a Behavioral Approach

The experimental design is shown in Fig. 3.

1. Adult, 2-month-old male *drd3*-EGFP mice are group housed (4 or less mice/cage) with littermates in a temperature- and humidity-controlled environment. Mice are maintained on a 12 h light (on at 7 a.m.)/dark (off at 7 p.m.) cycle. Mice are given standard rodent chow and water ad libitum.
2. Split 2–3-month-old male *drd3-EGFP* mice into five groups: saline/saline; saline/0.05 mg/kg PD128907; 0.5 mg/kg PD128907/0.05 mg/kg PD128907; saline/10 mg/kg PBZI; 20 mg/kg PBZI/10 mg/kg PBZI.
3. Weigh the animals and use the weight to calculate the amount of drug(s) needed for each dose. The drugs are diluted in saline to a concentration such that the injection volume for the above doses is ~0.1 mL.

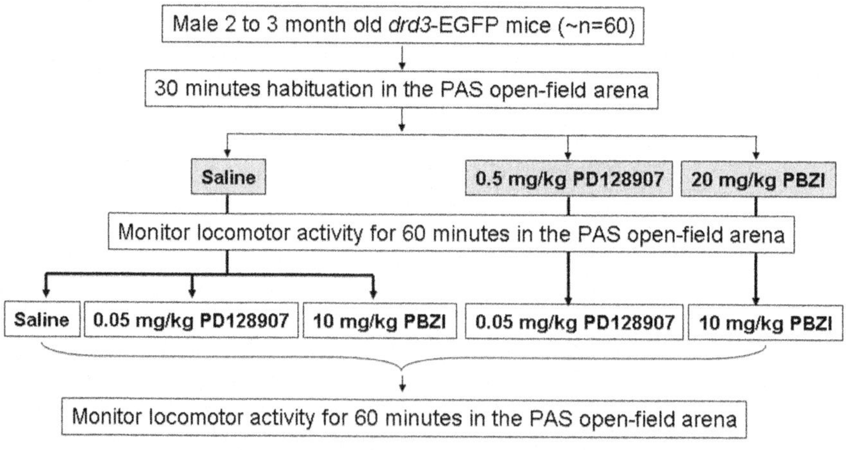

Fig. 3 Experimental scheme used for determining agonist-induced behavioral tolerance in locomotor activity using the *drd3*-EGFP mice. The mice were administered two sequential injections that were 60 min apart—injection #1 (*gray rectangles with bold lettering*) and injection #2 (*open rectangles with bold lettering*)

4. Turn on the open-field photobeam activity system (PAS) and the attached computer. Open the PAS activity recording software for measurement of D3-mediated locomotor activity. Create a new database, under the "File" tab. This database will be used for all the open-field recordings. Once the database is created, create a new session under the "Session" tab. Once a new session is created, select "Edit this session" under the "Session" tab and make the following edits: In the description session under the "Session Data" tab, write a detailed description of what tests are being run, drugs used, animal strain, gender, etc. for your records. Next, select the "Chambers" tab. For each open-field arena that will be used, fill in the Subject ID and the Subject variables. These descriptors are important and will help you identify what data is from what animal in the Master Session file. Next, select the "Start/Stop Control" tab and specify the interval length and the number of intervals per phase: Use an interval length of 10 min, or 600 s. There will be three intervals for the 30 min habituation phase, six intervals for the first drug injection phase, and six to nine intervals for the second drug injection recording phase. Count the total number of intervals needed and enter this number. Ensure the following boxes are checked if the system being used has multiple arenas: Under the Manually enable phase section, check all boxes. Under the Disable phase section, check all boxes. Under phases per session, select Multiple phases/end session manually. Under the "then start phase" and "then end phase" sections, ensure "Immediately" is selected. After all the appropriate edits have been made, select save and close. (*See* **Notes 7–9**.)
5. The software is now prepared to collect data. Place the animal in the center of the arena (repeat this if using multiple arenas) and then select "Run this session" under the session tab, and then hit "Start all." The open-field test is now starting and the data is being recorded. Leave the recording room and allow the animals' activity to be recorded for the 30 min habituation phase. (*See* **Note 8**.)
6. Prepare the drugs while the animals are habituating. PD128907 and PBZI need to be injected subcutaneously. To inject the animals, use the skin at the back of the neck and a 28 g 1/2 in. 100 cc insulin syringe. You will be injecting either saline, 0.5 mg/kg PD128907, or 20 mg/kg PBZI.
7. After the 30 min habituation phase is complete, hit Stop All. This will stop the recording, so you can inject the animals and record the next phase when you are ready. Inject the animal(s), place them back into the center of the arena, and hit Start All. The second phase (first drug recording phase) is now recording. Allow 60 min of recording for this phase.

8. Prepare the drugs for the third recording phase while the animals are in the second phase. You will be injecting either saline, 0.05 mg/kg PD128907, or 10 mg/kg PBZI.
9. After this phase is complete, hit Stop All again and administer the second injection. When injections are finished and animals are back in the arena, hit Start All and allow another 60–90 min of recording. Once the recording is finished, select End Session.
10. After you end the session, select "View report for this session" under the "Session" tab. Select the export tab and ensure all the data boxes are checked and hit Export. The data is now exported into the data file and can be analyzed. You will use the PAS reporter software to export the locomotor activity or distance traveled data to Microsoft Excel, where it can then be graphed and statistically analyzed using other software. We recommend Sigma Plot v11 for graphing and statistical analysis.
11. Open the PAS reporter software and "Enable the content" by clicking the options button. Now the Main Switchboard should be visible. First select "Table Linker." Here you can select the database file where all of your data is located. Open all of the files. As you open all of the files, you will see them appear on the left-hand side under Tables.
12. Next, select the "Distance, Speed and Resting Calculator." For each session that you ran, you will need to specify a resting length criterion. This value will be used to determine how many times the animal was inactive, or resting. We suggest a resting length of 4 s. Enter the numerical value and hit calculate. Do this for every session you wish to analyze. You will see new files appear to your left.
13. Now, the locomotor activity has been calculated and can be exported to Microsoft Excel. For open-field locomotor activity, you need to export the IS file to your left. Double-click on this file and then select the "External Data" Tab and select "Excel" under Export. Ensure the "Export data with formatting and layout" is checked. It is also recommended to check the "Open the destination file after the export operation is complete."
14. Once the Excel file containing all of the data is open, it will have the following information that will be used for graphing and analysis: The Chamber Number column represents the open-field arena number. The Phase Number represents the recording phase. Phase 1 is the habituation period, Phase 2 is the first drug injection recording phase, and Phase 3 is the second drug injection recording phase. The interval number represents the time interval within each recording phase. Recall that each interval is 10 min. The Subject ID is the descriptor you gave the animal before the recording began. Lastly, the

locomotor activity, or distance traveled in cm/time interval, is the Total XY Distance tab. This is the information that will be graphed and analyzed.

15. You can now graph and analyze the data using appropriate software. We recommend Sigma Plot v11. It is also recommended to plot the data in a line graph with time on the *X* axis and distance on the *Y*. Error bars representing SEM should be used. This software also allows you to run statistical analyses when appropriate. You will want to compare the agonist-induced locomotor activity between agonists to ensure that both D3 receptor agonists are inducing similar D3-mediated locomotor effect. The agonist-induced locomotor activity difference between the first and second doses should be compared to assess the effect of the D3 receptor tolerance property on *drd3*-EGFP locomotor activity.

4 Notes

1. Male mice from different litters should not be group housed with each other to prevent aggressive behaviors; therefore, it is recommended that animals be individually housed.
2. Mice should be weighed on the day of experiments prior to injections for accurate dosing.
3. Paraformaldehyde is toxic and hazardous and should be handled in a chemical fume hood. Animal perfusion should also be performed in a chemical fume hood to minimize exposure. The perfusion should be done in a pan with high sides and the waste perfusion solution should be collected as chemical hazardous waste and disposed according to local regulations.
4. As the mice are typically euthanized and the brains collected one at a time, it is important to ensure that the drug treatment is performed in a separate room from the room where the animals will be euthanized and perfused.
5. Drugs are diluted in either 0.9 % saline or in 1× PBS, though the former is preferred as some drugs might precipitate in PBS.
6. Commercially available pump systems can be purchased, or a gravity perfusion system can be used to perform the transcardiac perfusion. We use a home-made gravity system that uses separate 60 cc syringe reservoirs with individual switch valve and a common manual flow-control valve. The reservoir and valves are connected with PE tubing with a connector at the end which fits a 20 g ½ in. disposable needle.
7. The locomotor arena needs to be wipe cleaned thoroughly with deionized water in between animals.

8. The locomotor arenas should be placed in a room that is insulated so as to minimize sensory disturbances (auditory and olfactory). Spatial cues within the room should not be altered.
9. The behavior studies described in this chapter were performed during the day between 8 a.m. and 12 p.m. It is important that all behavior assessments in experiments that are done over multiple days be performed at the same time of the day.

Acknowledgments

The work reported in the chapter supported by the F. M. Kirby Foundation, UMDNJ Foundation, and NIH grant (MH082376) to E.V.K. S.R.C. was supported by a NIH T32 predoctoral training grant (NS 51157-5).

References

1. Ahlgren-Beckendorf JA, Levant B (2004) Signaling mechanisms of the D3 dopamine receptor. J Recept Signal Transduct Res 24(3): 117–130
2. Neve KA, Seamans JK, Trantham-Davidson H (2004) Dopamine receptor signaling. J Recept Signal Transduct Res 24(3):165–205
3. Beaulieu JM, Gainetdinov RR (2011) The physiology, signaling, and pharmacology of dopamine receptors. Pharmacol Rev 63(1): 182–217
4. Kuzhikandathil EV, Westrich L, Bakhos S, Pasuit J (2004) Identification and characterization of novel properties of the human D3 dopamine receptor. Mol Cell Neurosci 26(1): 144–155
5. Westrich L, Kuzhikandathil EV (2007) The tolerance property of human D3 dopamine receptor is determined by specific amino acid residues in the second cytoplasmic loop. Biochim Biophys Acta 1773(12):1747–1758
6. Westrich L, Gil-Mast S, Kortagere S, Kuzhikandathil EV (2010) Development of tolerance in D3 dopamine receptor signaling is accompanied by distinct changes in receptor conformation. Biochem Pharmacol 79(6): 897–907
7. Kuzhikandathil EV, Kortagere S (2012) Identification and characterization of a novel class of atypical dopamine receptor agonists. Pharm Res 29(8):2264–2275
8. Gil-Mast S, Kortagere S, Kota K, Kuzhikandathil EV (2013) An amino acid residue in the second extracellular loop determines the agonist-dependent tolerance property of the human D3 dopamine receptor. ACS Chem Neurosci 4(6):940–951
9. Li Y, Kuzhikandathil EV (2012) Molecular characterization of individual D3 dopamine receptor-expressing cells isolated from multiple brain regions of a novel mouse model. Brain Struct Funct 217(4):809–833
10. Cote SR, Kuzhikandathil EV (2014) In vitro and in vivo characterization of the agonist-dependent D3 dopamine receptor tolerance property. Neuropharmacology 79:359–367

Chapter 20

Dopamine D1 and D2 Receptors in Chronic Mild Stress: Analysis of Dynamic Receptor Changes in an Animal Model of Depression Using In Situ Hybridization and Autoradiography

Dariusz Zurawek, Agata Faron-Górecka, Maciej Kuśmider, Joanna Solich, Magdalena Kolasa, Kinga Szafran-Pilch, Katarzyna Kmiotek, Piotr Gruca, Mariusz Papp, and Marta Dziedzicka-Wasylewska

Abstract

Depression is a multifaceted illness that involves altered monoamine neurotransmission. Many monoamine receptor subtypes (e.g., dopamine D1 and D2) demonstrate altered expression levels in depressed patients and animal models of depression. Currently, there are an increasing number of molecular and biochemical studies on the mechanism of stress resilience. In this chapter, we describe a chronic mild stress (CMS) procedure along with in situ hybridization and autoradiography protocols to study changes in brain dopamine receptor expression of rats subjected to CMS. Chronic mild stress procedure (CMS) is one of few behavioral animal models of depression, and this model has good construct, face, and predictive validity. Moreover, approximately 30 % of rats exposed to stress regimen are stress resilient. There are numerous biochemical techniques that allow to measure changes in receptor density and the mRNA expression level. Receptor-specific radioligand binding measures concentration and visualizes the spatial distribution of the receptor proteins. In situ hybridization is a specific probe-based semiquantitative histochemical technique that can be used to visualize the spatial distribution of RNA sequences in tissue slices.

Key words Chronic mild stress, Stress resilience, Autoradiography, In situ hybridization, Dopamine D2 receptor, Dopamine D1 receptor, mRNA, Rat, Anhedonia

1 Introduction

Stress and adversity are the main causes of many psychopathologies, such as depression. The molecular and neurobiological mechanisms that underlie the susceptibility to stress have been investigated for decades. However, the exact phenotype and molecular markers of depression have not yet been elucidated. Many complex studies have revealed that depression is a multifaceted illness that involves

Mario Tiberi (ed.), *Dopamine Receptor Technologies*, Neuromethods, vol. 96,
DOI 10.1007/978-1-4939-2196-6_20, © Springer Science+Business Media New York 2015

neuronal atrophy [1–3] and regulates the expression of specific genes [4] and neuropeptides [1, 4, 5]. Moreover, stress alters monoamine neurotransmission [2, 3]. Individuals with depression exhibit disrupted levels of dopamine, serotonin, noradrenalin, and their specific receptors in many brain circuits [3, 6, 7]. These observations have led to the discovery of many antidepressant drugs regulating monoamine neurotransmission through specific monoamine transporter blockade. Although antidepressants (e.g., tricyclic antidepressants, selective serotonin reuptake inhibitors, serotonin–noradrenaline reuptake inhibitors) alter synaptic monoamine levels immediately [8], their therapeutic and behavioral effects appear after chronic administration [9]. Monoamines in the brain act through their transmembrane protein receptors, which belong to the GPCR family. The GPCRs transduce the signal from the extracellular environment into the cell. Many monoamine receptor subtypes (e.g., dopamine D1 and D2, serotonin 5-HT1a, and beta-1 adrenergic) demonstrate altered expression levels in depressed patients and animal models of depression. It is important to keep in mind that these correlations between behavior and receptor expression changes are not straightforward. Therefore, further studies on monoamine receptor expression and their common, complex, and time-dependent relationships are needed to fully characterize the normal and pathological states.

Many studies have investigated the biochemical and neurobiological changes that underlie vulnerability to stress, but relatively few studies have focused on targets that promote resilience to stress. Psychiatric and psychological research has been performed on stress-resilient people for more than 40 years [10]. Molecular and neurobiological studies on the mechanism of resilience to stress are rising. Behavioral animal models of depression (and psychological studies in humans) have revealed that resilience to stress is an active, adaptive process that involves many brain circuits and biochemical changes [10, 11]. Haglund et al. [11] defined resilience to stress as “the ability to successfully adapt to stressors, maintaining psychological well-being in the face of adversity.”

The chronic mild stress paradigm (CMS) is described as the animal behavioral model of depression, with good concept validities [9, 12]. CMS produces anhedonia, a core symptom of major depression, which is related to impairments in the reward system [12]. In the CMS model, stressed rats exhibit anhedonia by reducing their consumption of palatable sucrose solutions as compared to nonstressed rats. The induced stressed conditions, such as prolonged mild stress exposure, mimic natural factors in the etiology of depression [12]. The chronic administration of antidepressant drugs (e.g., imipramine and desipramine) reinstates normal sucrose consumption in stressed animals. This evidence can be interpreted as a reversion of depressive-like symptoms by chronic antidepressant therapy.

There are few approaches to measure rat hedonic behavior: e.g., sucrose/saccharin preference, palatable food intake, and brain reward stimulation [12]. Measuring method selection depends on individual experience and convenience of the procedure. Sucrose/saccharin consumption and preference are widely used methods. In our laboratory we use sucrose consumption method and it will be described in this chapter. There are some controversies concerning CMS procedure, especially some scientific groups reported "anomalous" behavioral effects following this procedure [12] which manifested in increased sucrose/saccharin consumption or sucrose/saccharin preference. In our behavioral experiment, we observed as well that approximately 30 % of rats exposed to variable, unpredictable mild stressors do not develop decreased sucrose consumption. Very often in "classical" approach to CMS paradigm, this group of animals is simply rejected from the further study as behaving "anomalously." However, this group of animals may be, in fact, resilient to stress in the CMS model [7, 13]—since the stress procedure does not cause any behavioral deficits (anhedonia) observed as unaltered sucrose consumption in this population [7, 13]. The reader should consider that although the CMS model can be complicated, time-consuming, and difficult to establish, the results obtained are definitely worth of all hard work and few attempts.

In situ hybridization (ISH) was developed in late 1960s and has become useful molecular technique that allows the measurement of mRNA expression level. It is widely used in molecular basic research and clinical trials [14, 15] and is based on the pairing of short, labeled antisense oligonucleotide probes with respective nucleic acid sequences in tissue or cell [14]. Probes can be labeled with radioactive (often ^{35}S or ^{125}I-gamma-ATP) tail, fluorescent marker, or chromogen [16]. In this chapter we focus only on radiolabeled probe-based in situ hybridization which—despite the fact that it requires longer processing time and stringent disposal methods—it is much more sensitive than non-radiolabeled approach [14]. It can detect low copy number of mRNA in tissue or even in single cell. ISH, in contrast to other molecular techniques, allows the visualization of spatial distribution of target of interest in morphologically preserved tissue or cell. Although ISH technique is a semiquantitative method, it is able to detect even slight differences in mRNA expression level in small anatomical region or population of specific cells. These alterations may not be detected using other molecular techniques which work on tissue or cell homogenate because of dilution by cells or neighboring tissue regions not carrying specific changes [16]. ISH can be used on frozen, formalin-fixed, and paraffin-embedded tissue slices. To perform valuable, sensitive, and truthful mRNA ISH assay, it is crucial to use a mixture of few probes that are complementary to different and specific regions of target mRNA and are designed with great care. Oligonucleotide probes should be between 18 and 50 bases and with GC content between 40 and 60 %. Increased GC content causes increased nonspecific binding. Probes should be complementary only to mRNA of interest. Available online software, like OligoWiz 2.3.0 and OligoAnalyzer 3.1, can help designing appropriate probes. If a complementarity to other off-targets is higher than 70 %, the probe is excluded from ISH assay. Probe complementarity can be checked using Basic Local Alignment Search Tool (BLAST) available online. We recommend reading the detailed and comprehensive chapter about appropriate probe design and application previously published by Marinela Aquino de Muro [17].

Quantitative autoradiography of bound radiolabeled ligand is a sensitive biochemical tool. It has been widely used over two decades, especially in pharmacological studies to determine the spatial distribution of receptors in human and animal brains. This approach retains the morphology of the tissue and allows the quantitative measurements of receptor expression levels even in small brain structures. Moreover, it is a simple method that can be adapted to the study many different types of receptors. Autoradiography is based on specific binding of radioisotope-labeled (often ^{3}H or ^{125}I) compound to the receptor of interest. Photographic detection system and analysis of resulting images is the most common approach to determine receptor concentration and spatial distribution. It should be noted that modified autoradiographic technique may produce images even at the electron-microscopic level. The choice of a highly selective radiolabeled ligand for the receptor type under study is the most important aspect and should be considered very carefully during planning of the experiment. Most of commercially available radioligands are selective to different types of receptors but very few of them are highly selective to only one receptor subtype. Radioligand specificity may depend on the radioligand concentration used in experiment. In autoradiographic experiments the concentration of specific radioligand used is based on its dissociation constant parameter (Kd). Kd describes the affinity of a ligand for a receptor. In brief, it refers to the ligand concentration occupying 50 % of the receptors. The interaction between radioligand and different types of receptors can be investigated using ligand–receptor saturation and competition binding assays, which will not be covered here as it has been previously described in details in the literature. In this chapter we focus on macroscopic and spatial visualization as well as quantitative measurements of the dopamine D1 and D2 receptor expression levels in cryodissected brain slices from rats subjected to CMS.

2 Materials

2.1 Chronic Mild Stress Procedure (CMS) and Tissue Preparation

1. Male Wistar rats (Charles River, Germany). This rat strain is recommended by us because we have been successful in performing CMS tests on it. However, other rat and mouse strains can be used as well.
2. Housing cages equipped with bottles.
3. Housing room with 12 h light-dark cycle, constant temperature (22 ± 2 °C), and humidity (50 ± 5 %) conditions.
4. Rat fodder.
5. Sawdust bedding.
6. Stroboscopic illuminator.
7. Dry ice.
8. n-Heptane.
9. Jung CM 3000 cryostat microtome (Leica, Germany).
10. Gelatin-coated microscope slides (LabScientific Inc., Livingston, NJ, USA).
11. Tissue-freezing medium (Leica, Germany).

2.2 Radioligand Binding to Dopamine D1 and D2 Receptors and Analysis of Autoradiograms

1. Rat brain tissue slices (thickness: 12 μm) mounted on gelatin-coated microscope slides and cut on cryostat by a standard procedure.
2. H_2O purified by Milli-Q ultrafiltration system (Merck-Millipore, Germany).
3. Tritium-labeled domperidone ([^{3}H]domperidone) in ethanol (American Radiolabeled Chemicals, St. Louis, USA. Store at −20 °C, radiation hazardous).
4. Tritium-labeled SCH23390 ([^{3}H]SCH23390) in ethanol (Perkin Elmer, USA. Store at −20 °C, radiation hazardous). Experimenter using radiation materials should wear appropriate personal protective equipment. Hazardous radiation materials should be treated according to rules of the Institutional Radiation Compliance Office.
5. (+)Butaclamol (Sigma Aldrich, St. Louis, USA. Store at 4–8 °C).
6. Cis-(*Z*)-flupentixol (Sigma Aldrich, USA. Store 4–8 °C).
7. Ascorbic acid (Merck, USA).
8. Bovine albumin (Sigma Aldrich, USA. Store at 4–8 °C).
9. Ethanol 99.8 % pure p.a.
10. 50 mM Tris–HCL buffer (pH 7.4).
11. NaCl (POCh, Poland).
12. EDTA (POCh, Poland).

13. $CaCl_2$ (POCh, Poland).
14. $MgCl_2$ (POCh, Poland).
15. KCl (POCh, Poland).
16. Beckman 6500 LS scintillation counter.
17. Polypropylene scintillation vials 4 ml.
18. High-flashpoint LSC-cocktail Ultima Gold (Perkin Elmer, USA).
19. Fuji Imaging Plate BAS-TR2025 for bio-imaging analyzer (FujiFilm, Japan).
20. FujiFilm BAS-5000 Phosphoimager (FujiFilm, Japan).
21. FujiFilm BAS Cassette 2025 (FujiFilm, Japan).
22. Autoradiographic [^{3}H] microscales RPA506 (GE Healthcare, UK).

2.3 *In Situ* Hybridization of Radiolabeled Probes Complementary to D1 and D2 Receptor mRNA Expressed in Rat Brain Tissue

2.3.1 Radiolabeling of the Probes

1. Sterile and RNase-free Eppendorf tubes 1, 5 ml.
2. RNase-free water.
3. Custom-made oligonucleotides complementary to 4–51, 766–813, and 848–901 bases of the rat dopamine D2 receptor mRNA (Life Technologies, USA). Resuspend lyophilized oligos in sterile RNase-free water and store at −20 °C.
4. Custom-made available oligonucleotides complementary to 13–60, 520–567, and 664–711 bases of the rat dopamine D1 receptor mRNA (Life Technologies, USA). Resuspend lyophilized oligos in sterile RNase-free water and store at −20 °C.
5. [^{35}S]Deoxyadenosine 5′-(alpha-thio)triphosphate ([^{35}S] α-dATP, Hartman Analytic, Germany. Store at −20 °C; use up to 3 months, radiation hazardous).
6. Commercially available 5× reaction buffer for Terminal Deoxynucleotidyl Transferase (Thermo Scientific, USA. Store at −20 °C).
7. Terminal transferase enzyme (Fermentas, Lithuania. Store at −20 °C).
8. TE (Tris–EDTA) buffer pH 7.4 (Sigma Aldrich, USA).
9. tRNA 25 mg/ml (Sigma Aldrich, USA. Store at −20 °C).
10. Phenol.
11. Chloroform and isoamyl alcohol mixture (50:1).
12. Isoamyl alcohol.
13. Ethanol 99.8 %—pure p.a.
14. Ethanol 70 %—pure p.a.
15. 4 M NaCl in sterile RNase-free water.
16. 1 M Dithiothreitol (DTT)—prepare fresh before use.
17. Beta liquid scintillation counter (e.g., Beckman 6500 LS scintillation counter).

18. High -flashpoint liquid scintillation counting cocktail (LSC) Ultima Gold (Perkin Elmer, USA).
19. 4-ml polypropylene scintillation vials.

2.3.2 Prehybridization

1. PBS (Phosphate buffered saline pH 7.4).
2. 4 % formaldehyde solution in PBS (prepare fresh before use). Formaldehyde is toxic. Do not inhale. Store and work with in fume hood.
3. TEA (Triethanolamine, Sigma Aldrich, USA).
4. NaCl.
5. Acetic anhydride (Sigma Aldrich, USA).
6. Ethanol 99.8 %—pure p.a.—basic.
7. Chloroform. Do not inhale. Store and work with in fume hood.
8. 0.1 % diethylpyrocarbonate (DEPC)-treated water (Sigma Aldrich, USA). DEPC inactivates enzymes degrading RNA. DEPC-treated water must be autoclaved to inactivate DEPC.

2.3.3 Hybridization

1. SSC buffer 20× solution (saline-sodium citrate buffer pH 7.4–7.6) (Sigma Aldrich, USA).
2. Formamide (Sigma Aldrich, USA).
3. 5× Denhardt's solution (BioChemica, USA).
4. Yeast tRNA (25 mg/ml, Sigma Aldrich, USA. Store at −20 °C).
5. ssDNA 10 mg/ml (Sigma Aldrich, USA. Store at −20 °C).
6. Dextran sulfate sodium salt MW ~500,000 (Sigma Aldrich, USA).
7. 0.1 % DEPC-treated water (Sigma Aldrich, USA).
8. 1 M DTT (prepare fresh before use).
9. A mixture of radiolabeled probes (*see* Sect. 2.3.1).

2.3.4 Washing and Film Developing

1. SSC buffer 20× solution (saline-sodium citrate buffer pH 7.4–7.6) (Sigma Aldrich, USA).
2. Formamide (Sigma Aldrich, USA).
3. 0.1 % DEPC-treated water (Sigma Aldrich, USA).
4. Ethanol 99.8 %—pure p.a.
5. Fuji Imaging Plate BAS-SR for bio-imaging analyzer (FujiFilm, Japan).
6. FujiFilm BAS-5000 Phosphoimager (FujiFilm, Japan).
7. FujiFilm BAS Cassette 2025 (FujiFilm, Japan).

3 Methods

3.1 Chronic Mild Stress Procedure (CMS) and Tissue Preparation

1. Purchase the animals (e.g., Wistar rats, initial weight approximately 70–100 g) and bring them into the laboratory approximately 2 months (7–8 weeks) before the start of stressing procedure. During these 2 months perform adaptation to laboratory conditions (2 first weeks) and sucrose consumption training procedure (about 5–6 weeks). Purchase more rats (approximately 20 %) than you expect to use in the CMS experiment because not all of the animals will meet the test criteria after the adaptation period and baseline test (*see* **Note 1**). All behavioral experiments received Local Bioethics Committee permission.
2. Allow the animals to adapt to the new laboratory conditions. Initially, keep the animals grouped 10 per cage (T4, measuring approximately 60 cm/40 cm) until they gain at least 140 g of body weight (usually 7–10 days). Then, transfer the rats into smaller cages (T3, measuring approximately 45 cm/30 cm), and house them singly throughout the duration of the experiment. House animals with water and food provided ad libitum, and maintain the animals on a 12-h light/dark cycle with constant temperature (22 ± 2 °C) and humidity (50 ± 5 %).
3. After rats gain approximately 200 g of body weight, begin training them to consume the 1 % palatable sucrose solution. Training consists of several 1-h tests, initially performed twice weekly (for 2 weeks) and then once weekly. Present the sucrose solution in the home cage (do not remove cages from the breeding room) following overnight (14-h) food and water deprivation. During the training and tests, rats should only have access to water bottles containing the sucrose solution. The solution should be freshly prepared prior to each test by dissolving 10 g sucrose per 1 l of tap water (22 ± 1 °C). For sucrose presentation, use different bottles from those used to provide water in the home cages. We recommend using 300-ml plastic bottles with stainless steel ball-tip nozzles (*see* **Note 2**).
4. Measure the sucrose intake by weighing the bottles containing sucrose solution at the beginning and end of every test (*see* **Note 3**).
5. Continue the adaptation tests until the consumption of sucrose solution becomes stable (usually 7–8 tests). Based on the sucrose intake of the final adaptation test (named the baseline test), divide the animals randomly into two matched groups: control and to-be-stressed (*see* **Note 1**). These two groups must be separated from each other and housed in different rooms. Control rats cannot have contact with stressed rats even once at any time during the experiment. The animals

should have free access to food and water except during the 14-h overnight food and water deprivation before each sucrose test. It is good to have at least 8–10 animals per group. This number may depend on individual needs or further experiments.

6. Subject the stressed group of animals to the weekly chronic mild stress regimen. Each week, the stress regime consists of the following:
   - One period of water deprivation.
   - Two periods of food deprivation two periods of 45° cage tilt.
   - Two periods of intermittent illumination (lights on and off every 2 h).
   - One period of soiled cage (250 ml of water in the sawdust bedding). After soiling period, change the sawdust bedding to fresh one. Rats must be kept always in the same cage. Do not mix cages.
   - One period of paired housing (*see* **Note 4**).
   - Two periods of low-intensity stroboscopic illumination (150 flashes/min).
   - Two periods of no stress.

Apply overnight cage tilting as the last stressor preceding decapitation, regardless of the established weekly stress schedule.

7. All of the stressors listed above have duration of 10–14 h. Arrange the stress schedule according to your experimental design. Remember that food and water deprivation precedes each sucrose consumption test. *See* Fig. 1 for the example of a stress schedule that can be used in CMS model.
8. For consecutive weeks, perform the weekly chronic mild stress procedure and monitor the sucrose intake (in all groups of rats) once per week. The number of weeks depends on individual experiment design. We generally perform 5–7-week CMS. Keep the same pattern of performing sucrose consumption test as it was during adaptation period (i.e., the same day of week and the same time of a day). A representative example of the effect of CMS on sucrose consumption is shown in Fig. 2.
9. Decapitate the animals and collect the tissues 24 h after the last scheduled sucrose test of the CMS procedure. Tissue collection time may be longer than 24 h as it has been revealed that behavioral deficits caused by CMS procedure can last for few weeks. Tissue collection time depends on individual experimental design. The whole brains and blood samples will be used for further biochemical and molecular studies.

	Tuesday	Wednesday	Thursday	Friday	Saturday	Sunday	Monday
Day 8AM- 8PM	SUCROSE TEST after test Food deprivation (10am-8pm)	Stroboscopic ilumination (8am-8pm)	Light on and off every 2 h (8am-8pm)	Stroboscopic ilumination (8am-6pm)	Food deprivation (10am-8pm)	Light on and off every 2 h (8am-8pm)	No stress (8am-6pm)
Night 8PM- 8AM	Cage tilting (8pm-8am)	No stress (8pm-8am)	Water deprivation (8pm-8am)	Grouping (8pm-10am)	Cage tilting (8pm-8am)	Sawdust wetting (8pm-8am)	Food and water deprivation (6pm-8am) before sucrose test

Fig. 1 Representative weekly stress schedule that can be adapted to chronic mild stress model. Each stressor should last 10–14 h

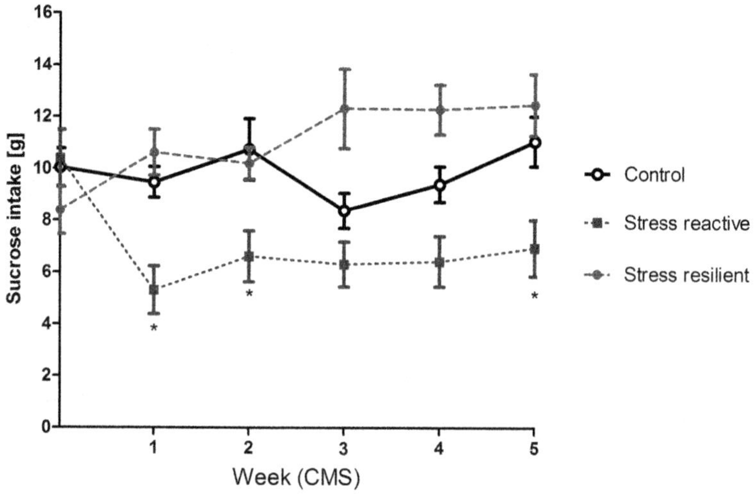

Fig. 2 Effects of a 5-week CMS procedure on sucrose consumption. The *y*-axis represents weekly sucrose consumption values in three different groups of rats (*n* = 10 for each group). Point 0 represents the final baseline test. The stress-resilient group of animals is distinguished retrospectively following the CMS experiment on the basis of sucrose consumption data. The *x*-axis represents sucrose intake [g]. The data are analyzed using a repeated measures ANOVA, * $p < 0.05$ vs. control

10. Immerse the whole brains in n-heptane cooled with dry ice for approximately 20 s, and store at –80 °C until use.
11. Cut the frozen rat brains (including brain structures of interest) into coronal sections (12 μm thickness), and mount them

on gelatin-coated microscope slides. The prepared tissue slices are suitable for radioligand binding and in situ hybridization assays.

In approximately 70 % of stressed animals, the CMS procedure causes a gradual decrease in sucrose consumption compared to the baseline test and the non-stressed controls. This effect gains statistical significance after 1–2 weeks of stress. The remaining 30 % of stressed animals exhibit unchanged sucrose intake relative to their baseline test and to the non-stressed controls. These animals are classified as stress resilient. The effect of stress (or lack of this effect) is typically stable throughout the study. The CMS model may be adapted to investigate the therapeutic effects of various chronic antidepressants [8, 9]. Moreover, CMS is a well-validated model. CMS causes additional behavioral impairments, such as blunted reward during brain stimulation, decreased sexual and aggressive behaviors, and altered sleep architecture [11].

3.2 Radioligand Binding to Dopamine D1 and D2 Receptors and Analysis of Autoradiograms

An appropriate and selective radiolabeled ligand should always be chosen if available. Autoradiography provides a single snapshot of what happened in the tissue during the experimental conditions. To investigate dynamic receptor changes using this technique, it is important to collect tissue samples at several experimental time points. It is important to note that for one autoradiographic assay, you need two parallel tissue sections from the same animal. One section is incubated with the radioligand mixture to measure total binding. The parallel section is incubated in the radioligand mixture enriched with a non-radiolabeled compound specific to the assayed receptor to measure nonspecific binding.

3.2.1 [^{3}H]Domperidone Binding to Dopamine D2 Receptors in Rat Brain Tissue

1. Prepare (or use commercially available) 50 mM Tris–HCl buffer (pH 7.4).
2. Prepare an ion-enriched buffer containing: 50 mM Tris–HCl, 120 mM NaCl, 1 mM EDTA, 1.5 mM, $CaCl_2$, 4 mM $MgCl_2$, and 5 mM KCl.
3. Add [^{3}H]domperidone to the ion-enriched buffer to a final concentration of 0.4 nM. This concentration refers to the dissociation constant (Kd) for the [^{3}H]domperidone–dopamine D2 receptor. The Kd can be measured using a ligand–receptor saturation binding assay or can be found in the literature [18]. To calculate the required radioligand concentration, use the following equation:

$$C = \frac{\text{DPM}}{\text{specific activity}(\text{Ci}/\text{mmol}) \times 2{,}220 \times \text{mL}}$$

where C is the radioligand concentration in nM, DPM is disintegrations per minute (calculated using a Beckman 6500 LS scintillation counter), the *radioligand specific activity* (in Ci/mmol) is provided by the manufacturer, 2,220 is a derivative that comes from dependence that 1 Ci (Ci) is $2{,}22 \times 10^{12}$ DPM, and ml is sample volume (in ml) used for radioactivity measurements.

4. Add 100 μl of [^{3}H]domperidone mixture to the polypropylene scintillation vial together with 4 ml of high-flashpoint LSC-cocktail Ultima Gold (Perkin Elmer, USA).
5. Measure radioactivity using a Beckman 6500 LS Scintillation Counter, and calculate [^{3}H]domperidone concentration using the equation provided above (*see* **Note 5**).
6. Prepare a 1 mM (+) butaclamol stock solution in ethanol (*see* **Note 6**).
7. Preincubate all tissue slices in 50 mM Tris–HCl buffer (pH 7.4) at room temperature for 15 min. This step rehydrates the cut tissue and removes potential endogenous dopamine.
8. Incubate the preincubated tissue slices in ion-enriched buffer containing 0.4 nM [^{3}H]domperidone for 2 h at 25 °C to obtain the total radioligand binding.
9. Incubate in parallel preincubated tissue slices in ion-enriched buffer containing 0.4 nM [^{3}H]domperidone and 10 μM (+) butaclamol for 2 h at 25 °C to obtain nonspecific radioligand binding.
10. Terminate the incubation by washing the sections twice in 50 mM Tris–HCl buffer (pH 7.4) for 5 min at 4 °C.
11. Wash the slices for 5 min at 4 °C in purified water.
12. Dry the tissue sections overnight under a gentle stream of air.
13. Load the radiolabeled slices into a FujiFilm BAS Cassette.
14. Add an autoradiographic [^{3}H] microscale RPA506 onto an additional empty microscope slide.
15. Place the Fuji Imaging Plate against the slices for 7–10 days. Keep the loaded cassette in a refrigerator (*see* **Note 7**).
16. Develop the imaging plate using a FujiFilm BAS-5000 Phosphoimager (FujiFilm, Japan).
17. Quantify and calibrate the results using the [^{3}H] microscale, and analyze the autoradiograms using the ImageGauge or MultiGauge v 3.0 software (FujiFilm, Japan).
18. Calculate the specific radioligand binding to dopamine D2 receptors by subtracting nonspecific binding values in adjacent brain slices from the total binding values. Representative autoradiograms showing [^{3}H]domperidone binding sites are shown in Fig. 3. Figure 4 shows a representative diagram of the analyzed and normalized [^{3}H]domperidone binding results.

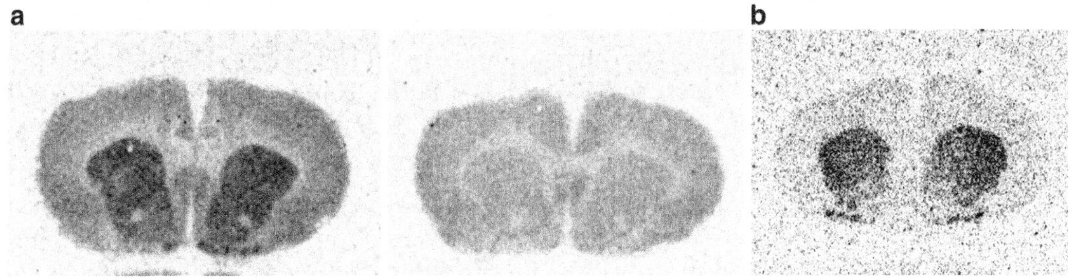

Fig. 3 (**a**) Representative autoradiograms of specific (*left picture*) and nonspecific (*right picture*) [^{3}H]domperidone binding to the dopamine D2 receptor. (**b**) Representative autoradiogram of in situ hybridized dopamine D2 mRNA

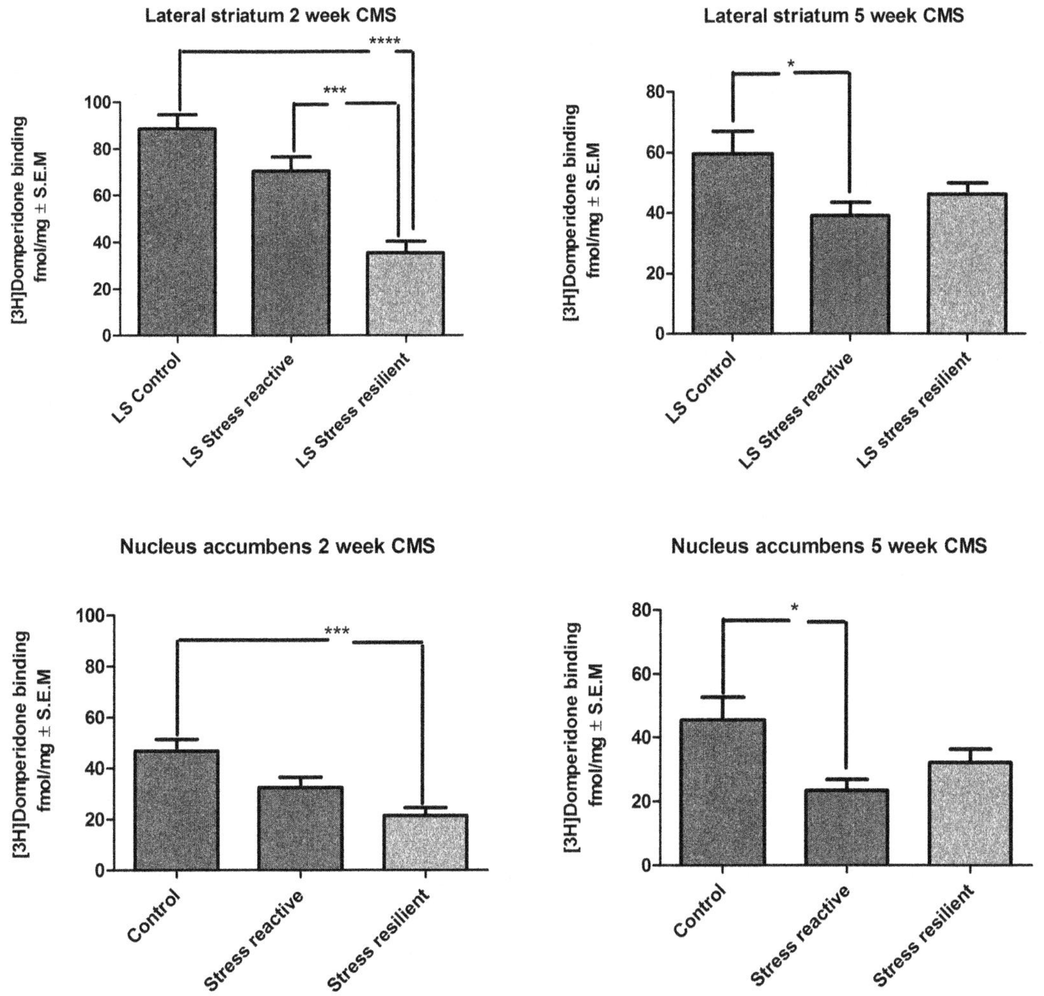

Fig. 4 Representative diagram showing [^{3}H]domperidone specific binding to dopamine D2 receptors in the lateral striatum following a 2-week CMS procedure (**a**) and a 5-week CMS procedure (**b**) in three selected groups of rats ($n=10$ for each group in each time point). The lower images demonstrate changes in dopamine D2 receptor levels in the nucleus accumbens septi ($n=10$ for all three groups of rats in each time point) following a 2-week CMS procedure (**c**) and a 5-week CMS procedure (**d**). This evidence indicates that the CMS procedure may change [^{3}H]domperidone binding to dopamine D2 receptor density in stress-resilient group of rats contributing to plastic response to stress. The data are from Zurawek et al. [7]

3.2.2 [^{3}H]SCH23390 Binding to Dopamine D1 Receptors in Rat Brain Tissue

1. Prepare (or use commercially available) 50 mM Tris–HCl buffer (pH 7.4).
2. Prepare an ion-enriched buffer containing 50 mM Tris–HCl, 120 mM NaCl, 5 mM KCl, 2 mM $CaCl_2$, 1 mM $MgCl_2$, 0.1 % ascorbic acid, and 0.1 % bovine albumin.
3. Add [^{3}H]SCH23390 to the ion-enriched buffer (1 nM final concentration). This concentration refers to the dissociation constant value (Kd) for the [^{3}H]SCH23390–dopamine D1 receptor. The Kd can be measured using a ligand–receptor saturation binding assay or can be found in the literature [19]. To calculate the required radioligand concentration, use the equation described in Sect. 3.2.1 (**step 3**).
4. Add 100 μl of [^{3}H] SCH23390 in ion-enriched buffer to polypropylene scintillation vial together with 4 ml of high flashpoint LSC-cocktail Ultima Gold (Perkin Elmer, USA).
5. Measure the radioactivity using a Beckman 6500 LS scintillation counter, and calculate the [^{3}H] SCH23390 concentration using the equation mentioned above (*see* Sect. 3.2.1).
6. Prepare a 1 mM *cis*-(*Z*)-flupentixol stock solution in water (*see* **Note 8**).
7. Preincubate all tissue slices in ion-enriched buffer (*see* **step 2**) at room temperature for 10 min. This step rehydrates the cut tissue and removes potential endogenous dopamine.
8. Incubate the preincubated tissue slices in ion-enriched buffer containing 1 nM [^{3}H]SCH23390 for 1 h at 25 °C to obtain total radioligand binding.
9. Incubate parallel preincubated tissue slices in ion-enriched buffer containing 1 nM [^{3}H]SCH23390 and 5 μM *cis*-(*Z*)-flupentixol for 1 h at 25 °C to obtain nonspecific radioligand binding.
10. Terminate the incubation by washing the brain sections twice in 50 mM Tris–HCl buffer (pH 7.4) for 5 min at 4 °C.
11. Wash the brain slices for 5 min at 4 °C in purified water.
12. Dry the tissue sections overnight under a gentle stream of air.
13. Load the radiolabeled slices into a FujiFilm BAS Cassette.
14. Add an autoradiographic [^{3}H] microscale RPA506 onto an additional, empty microscope slide.
15. Place the Fuji Imaging Plate against the slices for 7–10 days. Keep the loaded cassette in a refrigerator (*see* **Note** 7).
16. Develop the imaging plate using a FujiFilm BAS-5000 Phosphoimager (FujiFilm, Japan).
17. Quantify and calibrate the results using the [^{3}H] microscale, and analyze the autoradiograms using ImageGauge or MultiGauge v 3.0 software (FujiFilm, Japan).

Fig. 5 (**a**) Representative autoradiograms of specific (*left picture*) and nonspecific (*right picture*) [^{3}H]SCH23390 binding to the dopamine D1 receptor. (**b**) Representative autoradiogram of in situ hybridized dopamine D1 mRNA

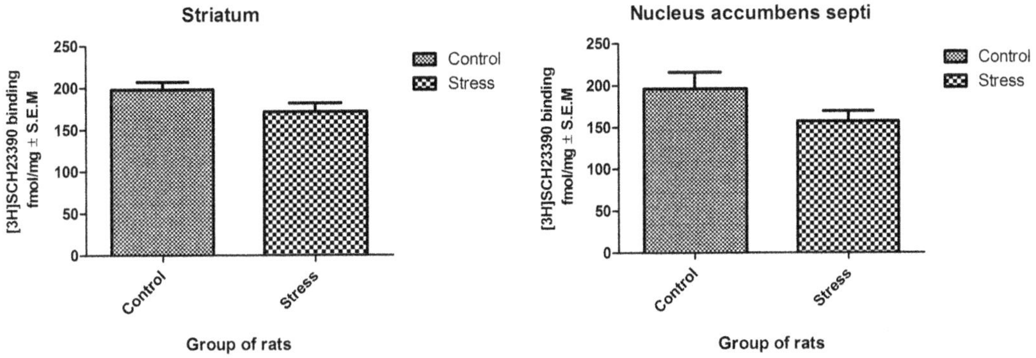

Fig. 6 Representative diagram showing [^{3}H]SCH23390 specific binding to dopamine D1 receptors in the striatum and nucleus accumbens septi in rats following the CMS procedure ($n = 10$ for each group of rats)

18. Calculate the specific radioligand binding to dopamine D1 receptors by subtracting the nonspecific binding values in adjacent brain slices from the total binding values. Representative autoradiograms showing [^{3}H]SCH23390 binding sites are shown in Fig. 5. Figure 6 shows a representative diagram of the analyzed and normalized results of [^{3}H]SCH23390 binding to dopamine D1 receptor protein.

3.3 In Situ Hybridization of Radiolabeled Probes Complementary to D1 and D2 Receptor mRNA Expressed in Rat Brain Tissue

Here, we describe an in situ hybridization technique suitable for the detection of dopamine D1 and D2 receptor mRNA levels in the brain. This is a universal technique that can be adapted to other mRNAs. Temporal changes in dopamine D1 or D2 receptor mRNA can be examined by collecting brain samples at different experimental time points. Here, we describe the in situ hybridization procedure which is similar for dopamine D1 or D2 mRNA expression. However, the appropriate, specific mixture of complementary oligonucleotides must be added. In situ hybridization using radiolabeled probes is a multistep technique taking 2–3 days.

3.3.1 Radiolabeling of Probes for Dopamine D1 or D2 Receptor mRNA

1. Dilute custom-made oligonucleotides in RNase-free water to a final concentration of 1 pmol/μl in an Eppendorf tube (or microfuge tube).
2. Vortex and spin briefly.
3. Add the following to a new PCR-clean Eppendorf tube: 21 μl of RNase-free H_2O, 10 μl of resuspended oligonucleotides (for the D1 or D2 dopamine receptor), 10 μl of 5× reaction buffer for Terminal Deoxynucleotidyl Transferase (Thermo Scientific, USA), 7.5 μl of [^{35}S]α-dATP (Hartman Analytic, Germany), and 1.5 μl of terminal transferase enzyme (Fermentas, Lithuania). The final reaction volume is 50 μl.
4. Vortex and centrifuge briefly at 12,000 × *g*.
5. Incubate the mixture in a water bath at 37 °C for 20–30 min.
6. After incubation, add the following reagents to the reaction mixture: 400 μl of TE buffer (pH 7.4), 2 μl of tRNA (25 mg/ml), 250 μl of chloroform and isoamyl alcohol mixture (50:1), and 250 μl phenol.
7. Shake vigorously for 20 s, and centrifuge for 15 min at 4 °C and 12,000 × *g*.
8. Gently collect the upper fraction (approximately 450 μl), and transfer it to a new Eppendorf tube.
9. Add 450 μl of chloroform and isoamyl alcohol mixture to the supernatant (50:1).
10. Shake vigorously for 20 s, and centrifuge at 12,000 × *g* for 15 min at 4 °C.
11. Collect the upper fraction, and transfer it to a new Eppendorf tube. Add 1 ml of ice-cold ethanol and 40 μl of 4 M NaCl.
12. Shake vigorously for 30 s, and leave the samples overnight at -20 °C.
13. The following day, precipitate the samples by centrifugation at 12,000 × *g* for 20 min at 4 °C (*see* **Note 9**).
14. Very gently, transfer the supernatant to a new Eppendorf tube (*see* **Note 10**).
15. Add 1 ml of 70 % ethanol to the probe pellet, and gently mix.
16. Centrifuge at 12,000 × *g* for 15 min.
17. After centrifugation, transfer the ethanol supernatant to a new Eppendorf tube.
18. Dry the pellet. Leave the Eppendorf tube (with pellet) opened and turned upside down until the ethanol has completely evaporated.
19. Resuspend the pellet in 100 μl of TE buffer.
20. Add 2 μl of 1 M DTT to remove potential disulphide bridges.

21. Check the probe radioactivity. Pipette 1 μl of resuspended probe into a polypropylene scintillation vial (put the pipette tip into the scintillation vial as well), and add 4 ml of high flash-point LSC-cocktail Ultima Gold (Perkin Elmer, USA).
22. Count the DPM values in a Beckman 6500 LS scintillation counter (*see* **Note 11**).

3.3.2 Prehybridization

All prehybridization steps should be performed in fume hood.

1. Prepare fresh TEA solution: resuspend 9 g of NaCl in 800 ml of DEPC-treated water, add 13.3 ml TEA and adjust the mixture to pH 8, and fill with water to a final volume of 1 l. Immediately prior to use, add 250 μl of acetic anhydride per 100 ml of TEA solution.
2. Fix the brain tissue sections in freshly prepared, cold 4 % formaldehyde solution in PBS on ice for 10 min.
3. Wash out the formaldehyde with cold PBS (on ice) for 5 min.
4. Incubate the tissue in cold TEA solution for 10 min on ice.
5. Dehydrate the tissue sections with short (20 s) incubations in increasing concentrations of ethanol as follows: 70, 95, and 99.8 % ethanol.
6. Incubate the tissue sections twice in chloroform for 10 min.
7. Incubate the tissue sections for short intervals (20 s) in decreasing concentrations of ethanol as follows: 99.8, 95, and 80 % ethanol.
8. Dry the tissue sections under a gentle stream of air for at least 4 h.

3.3.3 Hybridization of the Radiolabeled Probes Complementary to Dopamine D1 or D2 mRNA In Situ

1. Prior to the hybridization step, prepare the hybridization buffer as follows (per 50 ml):
   - Add 3.5 ml of DEPC-treated water, 10 ml of SSC buffer 20X solution, 25 ml of formamide, and 5 g of dextran sulfate sodium salt to a 50-ml plastic Falcon test tube. Close tightly and shake vigorously.
   - Mix in a hybridization oven at 60 °C until dextran is completely dissolved (a few hours).
   - Denature 2.5 ml of ssDNA (10 mg/ml) for 5 min at 95 °C, and place it on ice.
   - Once the dextran sulfate is completely dissolved, cool the hybridization buffer base to 37 °C. Then, add the following reagents: 1 ml of 5× Denhardt's solution (BioChemica, USA), 0.5 ml of yeast tRNA (25 mg/ml, Sigma Aldrich), 2.5 ml of denatured ssDNA (10 mg/ml), and 2.5 ml of 1 M DTT.
   - Mix the hybridization buffer in the hybridization oven at 37–40 °C for 4 h.

2. Add 50 μl of the radiolabeled probe mixture (*see* Sect. 3.3.1) to 50 ml of the hybridization buffer and mix very carefully. Do not shake the mixture, and avoid making air bubbles.
3. Apply 190 μl of the hybridization buffer enriched with radiolabeled probes onto the microscope slides with fixed, prehybridized tissue sections.
4. Cover each microscope slide with Parafilm (cut to fit the slide).
5. Place the prepared slides in a hybridization dish, and incubate the slides at 37–40 °C for 18 h in humid conditions.

3.3.4 Washing and Developing

1. Prepare 1× SSC.
2. Prepare 2× SSC with 50 % formamide: 250 ml of formamide, 50 ml of SSC buffer 20× solution, and 200 ml of DEPC-treated water.
3. Immerse the hybridized slides in glass containers with 1× SSC, and remove the Parafilm.
4. Wash out the hybridization buffer. Incubate the slides four times for 15 min in glass containers with 2× SSC with 50 % formamide at 42 °C with gentle shaking in a water bath.
5. Incubate the slides in glass containers with 1× SSC for 15 min at 25 °C.
6. Incubate the slides in deionized water for 5 min at 25 °C.
7. Dehydrate the slides by brief immersion (ca. 20 s) in increasing concentrations of ethanol: 70 and 99.8 %.
8. Dry the prepared slides overnight under a stream of air.
9. Load the dry slides into a cassette and expose them to an imaging plate for 10–15 days.
10. Develop the imaging plate using a FujiFilm BAS-5000 Phosphoimager (FujiFilm, Japan). Here, we recommend the FujiFilm system because it requires less time than traditional Kodak film exposure (approximately 20–28 days).
11. Analyze and normalize the results using the ImageGauge software (FujiFilm, Japan). Remember that in situ hybridization is a semiquantitative technique. Thus, if you have many slides, you must place slides from every experimental group into one cassette. Representative autoradiograms of in situ hybridized dopamine D1 and dopamine D2 receptor mRNAs are shown in Figs. 3 and 4, respectively.

4 Notes

1. Baseline test is the only last sucrose consumption test at the end of training period. Based on this test animals should be randomly matched into respective groups. All animals at the

end of training procedure should have similar sucrose consumption level. Earlier tests (especially the first ones) may not be very similar during adaptation test because rats are not trained yet and are not familiar with test conditions. Importantly, if rats drink during the adaptation procedure (e.g., 5 g of sucrose solution in each consecutive weekly test) and if in the final baseline test sucrose consumption rises to 15 g, it suggests that these rats are poorly adapted animals and consequently the researcher must exclude them from the study. Likewise, when consecutive tests during the adaptation period show that some rats drink variable amounts (e.g., 5 g in one test followed the next week by a consumption of 14 g and then 7 g in another test), the researcher should classify these rats as unstable drinking animals and exclude them from further experiments.

2. The bottles should be numbered and presented according to the matching numbers on the rat cages. Weigh every sucrose solution bottle twice: immediately before and after the sucrose test. Subtract the latter value from the former value. The same room should be used for adaptation tests and behavioral tests.
3. The mean values of sucrose intake in the baseline test should exceed 10 g for every matched group. To obtain mean group values exceeding 10 g, exclude animals with the lowest intake values (notably below 10 g) in the baseline test. Animals with highly variable intake values during the few last adaptation tests may also be excluded. In our experience, we excluded 15–20 % of rats after the baseline test and prior to randomization. Unstable drinking in each single rat (do not group the results from all rats per group) is established when there are more than 25 % changes in sucrose consumption as revealed by last few consecutive sucrose tests during training procedure. Be careful with interpretation of unstable drinking and always compare present test with the previous one in the same rat. It is important to note that during CMS experiment, the stressed group of rats should exhibit significant change in sucrose solution intake following stress regimen and behavioral experiment conditions. This change is not considered as unstable drinking. This criterion should be applied especially to control unstressed group of rats.
4. Mark the tail of every second rat on the rack and mark its cage before grouping. The marks will distinguish the rats for correct separation of every pair after the grouping procedure. Change the pattern of pairing on consecutive groupings so that every stressed rat has different social partners (i.e., each rat will be the host once and intruder once).

5. If the measured radioligand concentration is different from the expected value, adjust the concentration to the desired level.
6. (+) Butaclamol is an antagonist that binds with high specificity to the dopamine D2 receptor. It is used to obtain [^{3}H] domperidone nonspecific binding values. Prepare 1 mM stock solution, and then dilute with ion-enriched buffer containing the radioligand to a final concentration of 10 μM. If necessary, other specific dopamine D2 receptor compounds can be used.
7. Incubation duration depends on the desired strength of the signal. A longer incubation will result in a stronger signal. For [^{3}H]domperidone as well as [^{3}H]SCH23390 the optimal incubation time is 7–10 days (with an imaging plate). Avoid the over incubation of tissue sections with an imaging plate as it can increase the nonspecific signal.
8. *Cis*-(*Z*)-flupentixol is a dopamine D1 antagonist. It is used to obtain [^{3}H] SCH23390 nonspecific binding values. Prepare 1 mM stock solution and then dilute with an ion-enriched buffer containing the radioligand to a final concentration of 5 μM. If necessary, other specific dopamine D1 receptor compounds can be used.
9. Place all Eppendorf tubes in the same orientation within the centrifuge each time to track the position of the pellet. The pellet may be difficult to see.
10. Keep the discarded supernatant in a new Eppendorf tube in case the pellet is accidentally removed with the supernatant. If this occurs, centrifuge the supernatant again to reform the probe pellet.
11. Measured DPM values should range between 500,000 and 1,000,000 for one radiolabeled probe. Counts less than 500,000 dpm are indicative of a low efficiency labeling of the probe.

Acknowledgements

This work was partly supported by the following grants: (1) DeMeTer (project number POIG.01.01.02-12-004/09; 3.6), (2) statutory activity of Institute of Pharmacology Polish Academy of Sciences, and (3) Dariusz Zurawek is a holder of scholarship from the KNOW sponsored by Ministry of Science and Higher Education, Republic of Poland.

References

1. Lim BK, Huang KW, Grueter BA, Rothwell PE, Malenka RC (2012) Anhedonia requires MC4R-mediated synaptic adaptations in nucleus accumbens. Nature 487(7406):183–189
2. Duman RS, Voleti B (2012) Signaling pathways underlying the pathophysiology and treatment of depression: novel mechanisms for rapid-acting agents. Trends Neurosci 35(1):47–56
3. Manji HK, Drevets WC, Charney DS (2001) The cellular neurobiology of depression. Nat Med 7(5):541–547
4. Krishnan V, Han MH, Graham DL, Berton O, Renthal W, Russo SJ, Laplant Q, Graham A, Lutter M, Lagace DC, Ghose S, Reister R, Tannous P, Green TA, Neve RL, Chakravarty S, Kumar A, Eisch AJ, Self DW, Lee FS, Tamminga CA, Cooper DC, Gershenfeld HK, Nestler EJ (2007) Molecular adaptations underlying susceptibility and resistance to social defeat in brain reward regions. Cell 131(2):391–404
5. Faron-Górecka A, Kuśmider M, Solich J, Kolasa M, Szafran K, Żurawek D, Pabian P, Dziedzicka-Wasylewska M (2013) Involvement of prolactin and somatostatin in depression and the mechanism of action of antidepressant drugs. Pharmacol Rep 65(6):1640–1646
6. Kvetnansky R, Sabban EL, Palkovits M (2009) Catecholaminergic systems in stress: structural and molecular genetic approaches. Physiol Rev 89(2):535–606
7. Żurawek D, Faron-Górecka A, Kuśmider M, Kolasa M, Gruca P, Papp M, Dziedzicka-Wasylewska M (2013) Mesolimbic dopamine D_2 receptor plasticity contributes to stress resilience in rats subjected to chronic mild stress. Psychopharmacology (Berl) 227(4):583–593
8. Zangen A, Overstreet DH, Yadid G (1999) Increased catecholamine levels in specific brain regions of a rat model of depression: normalization by chronic antidepressant treatment. Brain Res 824(2):243–250
9. Willner P, Towell A, Sampson D, Sophokleous S, Muscat R (1987) Reduction of sucrose preference by chronic unpredictable mild stress, and its restoration by a tricyclic antidepressant. Psychopharmacology (Berl) 93(3):358–364
10. Feder A, Nestler EJ, Charney DS (2009) Psychobiology and molecular genetics of resilience. Nat Rev Neurosci 10(6):446–457
11. Haglund ME, Nestadt PS, Cooper NS, Southwick SM, Charney DS (2007) Psychobiological mechanisms of resilience: relevance to prevention and treatment of stress-related psychopathology. Dev Psychopathol 19(3):889–920
12. Willner P (2005) Chronic mild stress (CMS) revisited: consistency and behavioural-neurobiological concordance in the effects of CMS. Neuropsychobiology 52(2):90–110
13. Bergström A, Jayatissa MN, Mørk A, Wiborg O (2008) Stress sensitivity and resilience in the chronic mild stress rat model of depression; an in situ hybridization study. Brain Res 1196:41–52
14. Looi LM, Cheah PL (1992) In situ hybridisation: principles and applications. Malays J Pathol 14(2):69–76
15. Langdale JA (1993) In situ hybridization. In: Walbot V, Freeling M (eds) The maize handbook. Springer, New York, NY, pp 165–180
16. Halling KC, Wendel AJ (2008) In situ hybridization: principles and applications for pulmonary medicine. In: Zander DS et al. (eds) Molecular pathology of lung diseases, Vol. 1, pp 117–129
17. de Muro MA (2005) Probe design, production, and applications. In: Walker JM, Rapley R (eds) Medical biomethods handbook. Humana, Totowa, NJ, pp 13–23
18. Seeman P, Tallerico T, Ko F (2003) Dopamine displaces [^{3}H]domperidone from high-affinity sites of the dopamine D2 receptor, but not [^{3}H]raclopride or [^{3}H]spiperone in isotonic medium: Implications for human positron emission tomography. Synapse 49(4): 209–215
19. Czyrak A, Wedzony K, Michalska B, Fijał K, Dziedzicka-Wasylewska M, Maćkowiak M (1997) The corticosterone synthesis inhibitor metyrapone decreases dopamine D1 receptors in the rat brain. Neuroscience 79(2):489–495

Index

Mario Tiberi (ed.), *Dopamine Receptor Technologies*, Neuromethods, vol. 96, DOI 10.1007/978-1-4939-2196-6, © Springer Science+Business Media New York 2015

CPSIA information can be obtained
at www.ICGtesting.com
Printed in the USA
LVOW02*2133171215
467050LV00002B/94/P

9 781493 921959